Advances in Experimental Medicine and Biology

Volume 1024

Advances in Experimental Medicine and Biology presents multidisciplinary and dynamic findings in the broad fields of experimental medicine and biology. The wide variety in topics it presents offers readers multiple perspectives on a variety of disciplines including neuroscience, microbiology, immunology, biochemistry, biomedical engineering and cancer research. Advances in Experimental Medicine and Biology has been publishing exceptional works in the field for over 30 years and is indexed in Medline, Scopus, EMBASE, BIOSIS, Biological Abstracts, CSA, Biological Sciences and Living Resources (ASFA-1), and Biological Sciences. The series also provides scientists with up to date information on emerging topics and techniques.

More information about this series at http://www.springer.com/series/5584

Dakang Xu
Editor

Regulation of Inflammatory Signaling in Health and Disease

Springer

Editor
Dakang Xu
Faculty of Medical Laboratory Science, Ruijin Hospital
Shanghai Jiao Tong University School of Medicine
Shanghai, China

ISSN 0065-2598 ISSN 2214-8019 (electronic)
Advances in Experimental Medicine and Biology
ISBN 978-981-10-5986-5 ISBN 978-981-10-5987-2 (eBook)
DOI 10.1007/978-981-10-5987-2

Library of Congress Control Number: 2017952362

Printed on acid-free paper

This Springer imprint is published by Springer Nature
The registered company is Springer Nature Singapore Pte Ltd.
The registered company address is: 152 Beach Road, #21-01/04 Gateway East, Singapore 189721, Singapore

Preface

The innate immune system provides immediate defense against invading microbial pathogens and is found in all classes of living organisms, whereas the adaptive immune system is found only in vertebrates. This book covers recent research in innate immunity that has revealed a large number of receptors that sense the presence of microorganisms or cellular damage in tissues. In complex tissues, many of these sensing events occur simultaneously. Thus, downstream signaling pathways need to be integrated so that an appropriate inflammatory response can be initiated.

Current key concepts in innate immunity are discussed, including the formation of signalosomes, inflammasome formation and pyroptosis, methods of extrinsic cell communication, and examples of receptor cooperation. There is an overview of posttranslational modifications in different inflammatory signaling pathways and their essential roles in the regulation of inflammation. Emerging evidence suggests that epigenetic mechanisms play an important role in fine tuning the innate immune response. An understanding of epigenetic regulation of innate immune cell identity and function will enable elucidation of the relationship between gene-specific host defense and inflammatory diseases, as well as innate immune memory in health and disease. Inflammasomes determine the molecular and cellular processes of inflammation in response to microbial infection. Current data suggest that inflammasomes also have antimicrobial functions. Microbial factors are involved in regulating host inflammatory signaling pathways, the composition and load of the gut microbiota, the co-metabolism of the host and the microbiota, and the host immune system and physiology.

Novel aspects of functional genomics, epigenomics, transcriptomics, posttranslational modifications, the microbiome, and immunometabolism are reviewed in relation to inflammatory signaling and responses. Study of the inflammation that occurs in response to host-pathogen interactions is essential for the development of new therapies to improve human health. Here we review recent findings on the mechanisms underlying the regulation of inflammatory responses to pathogens, dysregulation of these responses in inflammatory disease, and the use of such mechanisms to boost or subdue the inflammatory response. Such as TLR agonists as adjuvants for cancer vaccines, small-molecule HDAC inhibitors and other epigenetic regulators as drug targets in inflammatory diseases, and the potential role of miR-155 as a diagnostic, prognostic, and therapeutic target in the treatment of multiple sclerosis.

The cGAS–STING pathway detects the presence of cytosolic DNA and triggers the expression of inflammatory genes in response to DNA damage. It is also linked to the tumor microenvironment, where it is paramount to immune clearance of tumors. To date, little attention has been paid to the aging-associated alteration in intercellular communication known as "inflammaging," a proinflammatory phenotype in organs that occurs with aging, suggesting an alternative senescence-associated secretory phenotype. Telomere attrition induces not only epithelial stem cell senescence, but also low-grade inflammation in lungs.

We also review crosstalk between the innate and adaptive immune systems, in particular in relation to the newly emerged innate lymphoid cells (ILCs). Innate lymphoid cells are involved in both innate and adaptive immunity, and contribute to tissue homeostasis, as well as defense against pathogens and inflammatory disorders.

Finally, I thank all the authors for their valuable contributions. It is my hope that this book will stimulate scientists and clinicians to continue to explore this fascinating field. Bridging the gaps in understanding between the fields of human and mouse immunology will provide new insights into inflammatory-mediated disease and immune defense. These innovative perspectives in basic and clinical research will aid in the translation of knowledge to the clinic.

Shanghai, China Dakang Xu

Contents

Contributors

Jesse J. Balic Centre for Innate Immunity and Infectious Diseases, Hudson Institute of Medical Research, Clayton, VIC, Australia

Department of Molecular Translational Science, Faculty of Medicine, Nursing and Health Sciences, Monash University, Clayton, VIC, Australia

Zijuan Chen Department of Immunology, School of Basic Medical Sciences, Fudan University, Shanghai, People's Republic of China

Key Laboratory of Medical, Molecular Virology of Ministries of Education and Health, School of Basic Medical Sciences, Fudan University, Shanghai, People's Republic of China

Dominic De Nardo Inflammation Division, The Walter and Eliza Hall Institute of Medical Research, Parkville, VIC, Australia

Department of Medical Biology, The University of Melbourne, Parkville, Australia

Michael P. Gantier Centre for Innate Immunity and Infectious Diseases, Hudson Institute of Medical Research, Clayton, VIC, Australia

Department of Molecular and Translational Science, Monash University, Clayton, VIC, Australia

Doudou Hao Department of Immunology, School of Basic Medical Sciences, Fudan University, Shanghai, People's Republic of China

Key Laboratory of Medical, Molecular Virology of Ministries of Education and Health, School of Basic Medical Sciences, Fudan University, Shanghai, People's Republic of China

Yiqun Hu Faculty of Medical Laboratory Science, Ruijin Hospital, Shanghai Jiao Tong University School of Medicine, Shanghai, China

Bing Huang Guangzhou Institute of Paediatrics, Guangzhou Women and Children's Medical Centre, Guangzhou Medical University, Guangzhou, Guangdong, China

Brendan Jenkins Centre for Innate Immunity and Infectious Diseases, Hudson Institute of Medical Research, Clayton, VIC, Australia

Department of Molecular Translational Science, Faculty of Medicine, Nursing and Health Sciences, Monash University, Clayton, VIC, Australia

Ji-Kun Li Department of General Surgery, Shanghai Jiao Tong University Affiliated First People's Hospital, Shanghai Jiao Tong University, Shanghai, China

Jun-Ping Liu Institute of Ageing Research, Hangzhou Normal University School of Medicine, Hangzhou, Zhejiang Province, China

Molecular Signaling Laboratory, Monash University Department of Immunology, Central Clinical School, Faculty of Medicine, Prahran, VIC, Australia

Hudson Institute of Medical Research, Clayton, VIC, Australia

Monash University Department of Molecular and Translational Science, Faculty of Medicine, Clayton, VIC, Australia

Claire E. McCoy Hudson Institute of Medical Research, Department of Molecular and Translational Science, Faculty of Medicine, Nursing and Health Sciences, Monash University, Clayton, VIC, Australia

Molecular and Cellular Therapeutics, Royal College of Surgeons in Ireland, Dublin 2, Ireland

Guangxun Meng CAS Key Laboratory of Molecular Virology & Immunology, Institut Pasteur of Shanghai, Chinese Academy of Sciences, Shanghai, China

Hong Sheng Ong Department of Immunology and Infectious Disease, John Curtin School of Medical Research, Australian National University, Canberra, ACT, Australia

Geneviève Pépin Centre for Innate Immunity and Infectious Diseases, Hudson Institute of Medical Research, Clayton, VIC, Australia

Department of Molecular and Translational Science, Monash University, Clayton, VIC, Australia

Yibo Si Department of Immunology, School of Basic Medical Sciences, Fudan University, Shanghai, People's Republic of China

Key Laboratory of Medical Molecular Virology of Ministries of Education and Health, School of Basic Medical Sciences, Fudan University, Shanghai, People's Republic of China

Bandar Ali Suliman College of Applied Medical Sciences, Taibah University, Madinah, Saudi Arabia

Molecular Biomedicine Program, Research Center, King Faisal Specialist Hospital & Research Center, Riyadh, Saudi Arabia

Lihui Wang Institute of Ageing Research, Hangzhou Normal University School of Medicine, Hangzhou, Zhejiang Province, China

Dakang Xu Faculty of Medical Laboratory Science, Ruijin Hospital, Shanghai Jiao Tong University School of Medicine, Shanghai, China

Dapeng Yan Department of Immunology, School of Basic Medical Sciences, Fudan University, Shanghai, People's Republic of China

Key Laboratory of Medical Molecular Virology of Ministries of Education and Health, School of Basic Medical Sciences, Fudan University, Shanghai, People's Republic of China

Feng Yan Faculty of Medical Laboratory Science, Ruijin Hospital, Shanghai Jiao Tong University School of Medicine, Shanghai, China

Institute of Ageing Research, Hangzhou Normal University School of Medicine, Hangzhou, China

Xiaomin Yao CAS Key Laboratory of Molecular Virology & Immunology, Institut Pasteur of Shanghai, Chinese Academy of Sciences, Shanghai, China

University of Chinese Academy of Sciences, Beijing, China

Howard Chi Ho Yim Department of Medicine, St George & Sutherland Clinical School, Faculty of Medicine, University of New South Wales, Australia, Kogarah, NSW, Australia

Le Ying Faculty of Medical Laboratory Science, Ruijin Hospital, Shanghai Jiao Tong University School of Medicine, Shanghai, China

Department of Tea Science, Zhejiang University, Hangzhou, China

Liang Yu Department of General Surgery, Shanghai Jiao Tong University Affiliated First People's Hospital, Shanghai Jiao Tong University, Shanghai, China

Centre for Innate Immunity and Infectious Diseases, Hudson Institute of Medical Research, Clayton, VIC, Australia

Department of Molecular Translational Science, Faculty of Medicine, Nursing and Health Sciences, Monash University, Clayton, VIC, Australia

Xianhua Yu Institute of Ageing Research, Hangzhou Normal University School of Medicine, Hangzhou, Zhejiang Province, China

Yihua Zhang Department of Immunology, School of Basic Medical Sciences, Fudan University, Shanghai, People's Republic of China

Key Laboratory of Medical Molecular Virology of Ministries of Education and Health, School of Basic Medical Sciences, Fudan University, Shanghai, People's Republic of China

Yunlu Zhang Department of Immunology, School of Basic Medical Sciences, Fudan University, Shanghai, People's Republic of China

Key Laboratory of Medical Molecular Virology of Ministries of Education and Health, School of Basic Medical Sciences, Fudan University, Shanghai, People's Republic of China

Yuxia Zhang Guangzhou Institute of Paediatrics, Guangzhou Women and Children's Medical Centre, Guangzhou Medical University, Guangzhou, Guangdong, China

Ruixue Zhou Department of Immunology, School of Basic Medical Sciences, Fudan University, Shanghai, People's Republic of China

Key Laboratory of Medical Molecular Virology of Ministries of Education and Health, School of Basic Medical Sciences, Fudan University, Shanghai, People's Republic of China

Chapter 1
Activation of the Innate Immune Receptors: Guardians of the Micro Galaxy

Activation and Functions of the Innate Immune Receptors

Dominic De Nardo

Abstract The families of innate immune receptors are the frontline responders to danger. These superheroes of the host immune systems populate innate immune cells, surveying the extracellular environment and the intracellular endolysosomal compartments and cytosol for exogenous and endogenous danger signals. As a collective the innate immune receptors recognise a wide array of stimuli, and in response they initiate specific signalling pathways leading to activation of transcriptional or proteolytic pathways and the production of inflammatory molecules to destroy foreign pathogens and/or resolve tissue injury. In this review, I will give an overview of the innate immune system and the activation and effector functions of the families of receptors it comprises. Current key concepts will be described throughout, including innate immune memory, formation of innate immune receptor signalosomes, inflammasome formation and pyroptosis, methods of extrinsic cell communication and examples of receptor cooperation. Finally, several open questions and future directions in the field of innate immunity will be presented and discussed.

Keywords Innate immunity • Innate immune receptors • Inflammasomes • Myddosome • PRRs • Innate immune memory • TLRs • NLRs • CLRs • RLRs

---

D. De Nardo (✉)
Inflammation Division, The Walter and Eliza Hall Institute of Medical Research, 1G Royal Parade, Parkville, VIC 3052, Australia

Department of Medical Biology, The University of Melbourne, Parkville 3010, Australia
e-mail: denardo.d@wehi.edu.au

D. Xu (ed.), *Regulation of Inflammatory Signaling in Health and Disease*, Advances in Experimental Medicine and Biology 1024,
DOI 10.1007/978-981-10-5987-2_1

1.1 Inflammation and the Innate Immune Response

Inflammation is a complex biological process critical for the host that mediates numerous pathological and physiological responses [1]. In response to microbial infection or the introduction of environmental irritants, the innate immune system acts to protect the host via the rapid detection and removal of foreign threats and the removal of dead/dying cells through inflammatory processes. Imbalances in homeostatic processes, such as during metabolic dysfunction or tissue injury, can also trigger sterile inflammatory responses. There are five classical signs used in the clinical description of inflammation: swelling, heat, pain, redness and loss of function [2]. On a cellular level, these symptoms reflect the actions of specialised tissue-resident innate immune cells (e.g. macrophages and dendritic cells) that send out signals (cytokines, chemokines and interferons) to mediate increased blood flow, allowing access to plasma proteins as well as infiltration of additional circulating white blood cells (e.g. neutrophils and monocytes/macrophages) to the tissue site of inflammation. Once the immune stimulus is removed, repair processes are initiated to restore tissue homeostasis. An effective inflammatory response is therefore acute and rapidly resolved following the removal of the stimuli. Low-level inflammation is also critical for physiological processes; for instance, the balance between the commensal microflora and the host immune system is necessary to maintain a healthy intestinal environment [3], as well as small systemic concentrations of type I IFNs under homeostatic conditions, which is important for priming immune cells for rapid responses to microbial insults [4, 5]. Controlled inflammation in the context of protection, repair and physiology is therefore beneficial for the host. In contrast, chronic inflammation can have detrimental outcomes for the host, often resulting from persistent infection or in response to sterile inflammation driven by metabolic abnormalities or the inability to remove environmental irritants [1]. Chronic inflammation is a hallmark of autoimmune and monogenic autoinflammatory diseases where, often, the proverbial innate immune receptor heroes are turned villains. Whilst the enduring inflammation in autoimmune disease results from continual activation of the immune system by endogenous stimuli, autoinflammatory diseases stem from specific gain-of-function genetic variations in single innate immune genes leading to unwarranted inflammation [6–8]. Some recent examples of autoinflammatory diseases include those stemming from genetic variations in genes encoding innate immune receptors, such as stimulator of interferon genes (STING)-associated vasculopathy with onset in infancy (SAVI) [9] and pyrin-associated autoinflammation with neutrophilic dermatosis (PAAND) [10]. Identification of such genetic variants has given molecular insights into the mechanisms of innate immune activation and signalling and may lead to better treatment options for affected patients.

The innate immune system is typically described as a non-specific form of host defence due to its rapid induction and apparent lack of any immunological memory. Recently, this conventional model was contested by findings that even in the absence of the adaptive immune response, myeloid cells (predominantly monocytes and macrophages) can adopt long-term inflammatory phenotypes via epigenetic reprogramming [11]. This so-called innate immunological memory likely evolved prior to the adaptive immune system to confer protection upon secondary infections. Indeed, even in *Drosophila* phagocytosis of apoptotic debris was found to reprogramme macrophages, priming them to rapidly respond to future infections or tissue injury [12]. This system is also active in both murine and human myeloid cells. An emerging body of work shows priming myeloid cells, such as monocytes, with the potent bacterial agent lipopolysaccharide (LPS), and results in mature macrophages displaying a sustained immunosuppressive phenotype (known as 'tolerance'). In contrast, a primary infection with *Candida albicans* increases subsequent immune responses against secondary exposures with the same or different pathogens (termed 'trained immunity') [13]. These findings highlight an intriguing new capability to the innate immune response.

In order to better understand the inflammatory response, the mechanisms mediating the activation and functions of the innate immune system receptors central to the inflammatory process must be understood.

1.2 Triggering the Innate Immune System

All living organisms possess some form of physiological response to danger. Innate immune receptors, also known as pattern recognition receptors (PRRs), are highly conserved germ-line proteins, present throughout evolution to mediate host defence via the recognition of foreign and host-derived danger signals. This evolutionary conservation is best exemplified by the expression of numerous families of PRRs in sponges [14, 15], the most primitive of multicellular organisms. Collectively, PRRs recognise a wide variety of danger signals arising from both exogenous and endogenous sources (see Table 1.1). Exogenous danger signals are represented predominantly by highly conserved components of microbes, termed pathogen-associated molecular patterns (PAMPs). This includes structural elements found within bacterial and fungal cell walls, such as LPS (Gram-negative bacteria), peptidoglycans (Gram-positive bacteria) and ß-glucans (fungi), as well as forms of viral and bacterial nucleic acids. In addition, some PRRs respond uniquely to viability-associated PAMPs of the microbes encountered (e.g. prokaryotic mRNA), termed vita-PAMPs [16, 17]. In addition to pathogen-derived molecules, some exogenous environmental irritants (e.g. silica crystals) can also activate PRRs. Host-derived molecules that accumulate or become modified following tissue injury, metabolic dysfunction and uncontrolled cell death, and those that mediate sterile inflammatory responses via PRRs, represent danger-associated molecular patterns (DAMPs). A classic example is uric acid, which is soluble and innocuous in blood under physiological conditions; however, following drastic increases in levels of circulating uric acid (hyperuricaemia) results in a chemical-phase transition of soluble uric acid into immunoreactive monosodium urate crystals that deposit within joints and periarticular tissues driving the inflammation seen in gout [18]. Another DAMP that is increasingly implicated in mediating sterile inflammation is DNA leaked from

Table 1.1 Localisation of innate immune receptors and their stimuli

PRR	Localisation	PAMPs	DAMPs	Synthetic ligands
Toll-like receptors (TLRs)				
TLR1	Plasma membrane	Triacyl lipoproteins	Unknown	Pam3CSK4
TLR2	Plasma membrane	Lipoproteins, zymosan, mannan, fungal ß-glucans, peptidoglycan, lipoteichoic acid, type 1 fimbria	Versican	Pam2CSK4, Pam3CSK4
TLR3	Endolysosomal membrane	Viral dsRNA	mRNA	Endosomal poly(I:C), poly(A:U)
TLR4	Plasma‡endolysosomal membrane	Extracellular LPS[••], type 1 fimbria	Oxidised low-density lipoprotein, amyloid-ß	Lipid A derivatives, purified LPS
TLR5	Plasma membrane	Flagellin	Unknown	Recombinant flagellin
TLR6	Plasma membrane	Diacyl lipoproteins, lipoteichoic acid, zymosan	Oxidised low-density lipoprotein, amyloid-ß, versican	Macrophage-activating lipopeptide 2, synthetic diacylated lipoproteins, Pam2CSK4
TLR7	Endolysosomal membrane	Viral and bacterial ssRNA	Immune complexes, self-RNA	Thiazoquinoline and imidazoquinoline compounds (e.g. R848, imiquimod)
TLR8	Endolysosomal membrane	Viral and bacterial ssRNA	Immune complexes, self-RNA	Thiazoquinoline and imidazoquinoline compounds (e.g. R848, imiquimod)
TLR9	Endolysosomal membrane	Viral and bacterial CpG DNA, DNA:RNA hybrids	Chromatin IgG immune complexes	Class A, B and C CpG oligodeoxynucleotides
TLR10[^]	Plasma membrane	Unknown	Unknown	Unknown
TLR11[*]	Endolysosomal membrane	Profilin and flagellin	Unknown	Unknown
TLR12[*]	Endolysosomal membrane	Profilin	Unknown	Unknown
TLR13[*]	Endolysosomal membrane	Bacterial 23S ribosomal RNA (rRNA)	Unknown	23S rRNA-derived oligoribonucleotide

NOD-like receptors (NLRs)				
NOD1	Cytosolic	Peptidoglycan-containing iE-DAP motifs		Synthetic iE-DAP peptides
NOD2	Cytosolic	Muramyl dipeptide (MDP)		Mifamurtide
NLRP1◆	Cytosolic	*B. anthracis* lethal toxin*, *T. gondii* effector proteins*	Unknown	Purified *B. anthracis* lethal toxin*
NLRP3◆†	Cytosolic	*S. hygroscopicus* pore-forming toxin (nigericin), influenza A virus PB1-FB peptide, influenza M2 protein	Cholesterol crystals, monosodium urate (MSU) crystals, calcium pyrophosphate dihydrate (CPPD) crystals, extracellular ATP, islet amyloid polypeptide (IAPP), amyloid-ß, extracellular ASC specks	Alum crystals, TiO_2 and SiO_2 nanoparticles, silica crystals
NLRP6	Cytosolic	Unknown	Unknown	Unknown
NLRP7	Cytosolic	Unknown	Unknown	Unknown
NLRP12	Cytosolic	Unknown	Unknown	Unknown
NAIP^ NAIP1,2,5,6*	Cytosolic	Flagellin, components of the type 3 secretion system*	Unknown	Recombinant flagellin
NLRC4◆†	Cytosolic	Flagellin, components of the type 3 secretion system	Unknown	Recombinant flagellin

(continued)

Table 1.1 (continued)

PRR	Localisation	PAMPs	DAMPs	Synthetic ligands
PYRIN				
PYRIN◆[†]	Cytosolic	Rho GTPase modification: *C. difficile* toxin B (TcdB), *C. botulinum* C3 toxin, *V. parahaemolyticus* VopS, *H. somni* IbpA, *B. cenocepacia* type 6 secretion system components	Unknown	Purified TcdB
RIG-I-like receptors (RLRs)				
RIG-I	Cytosolic	Short viral dsRNA	Unknown	5′ppp-dsRNA, cytosolic poly(I:C)
MDA5	Cytosolic	Long viral dsRNA	Unknown	Cytosolic poly(I:C)
LGP2	Cytosolic	dsRNA?	Unknown	Unknown
C-type lectin-like receptors (CLRs)				
Dectin-1	Plasma membrane	Fungal ß-glucans, zymosan	Unknown	Unknown
Dectin-2	Plasma membrane	Fungal α-mannose, *M. furfur* Furfurman	Unknown	Unknown
Mincle (Clec4e)	Plasma membrane	Mycobacterial cord factor/trehalose-6,6-dimycolate (TDM)	SAP130	Synthetic trehalose-6,6-dibehenate (TDB)
Cytosolic DNA receptors (CDRs)				
AIM2◆	Cytosolic/nucleus	Viral dsDNA, bacterial dsDNA	mtDNA Self-DNA Damaged nuclear DNA	Poly(dA:dT)

cGAS	Cytosolic	Viral dsDNA	mtDNA	Transfected dsDNA, poly(dA:dT), DNA-intercalating agents
		Bacterial dsDNA	Nuclear DNA	
		dsDNA provirus	Circulating tumour DNA	
		DNA:RNA hybrids	Damaged DNA	
		Y-DNA	Oxidised DNA	
			Self-DNA	
STING†	Cytosolic: ER‡ Golgi membrane	Bacterial cyclic dinucleotides (CDNs): 3′3-cGAMP, cyclic di-GMP, cyclic di-AMP	Endogenous second messenger CDN: 2′3′-cGAMP (produced via cGAS in response to dsDNA)	Synthetic CDNs, CMA*, DMXXA*

^human specific, *mouse specific, ‡ traffics to, ◆ forms an inflammasome, † receptor can be activated indirectly, ¨ intracellular LPS directly binds caspase-4, caspase-5 and caspase-11 membrane-compromised mitochondria, i.e. mitochondrial-derived DNA (mtDNA), which is linked to the activation of several PRRs [19]. Rather than responding to specific signals, some PRRs are activated more generally in response to perturbations in cellular homeostasis, recently coined homeostasis-altering molecular processes (HAMPs) [20]. Alterations to the normal cellular environment can be sensed following pathogen infection, for instance, modification to the actin cytoskeleton by bacterial toxins, as well as via induction of endogenous mechanisms, such as the alteration of intracellular ion levels.

1.3 Detecting Danger

The vertebrate innate immune system comprises families of PRRs expressed primarily on highly specialised immune cell such as monocytes, macrophages, dendritic cells (DCs) and natural killer (NK) cells. PRRs are also expressed in other cell types that commonly encounter potential danger signals, including epithelial cells. The PRR families can be broadly classified according to their localisation on either the plasma and endolysosomal membranes or within the cytosolic compartment. Whilst Toll-like receptors (TLRs) and C-type lectin receptors (CLRs) represent the membrane-bound receptors, retinoic acid-inducible gene I (RIG-I)-like receptors (RLRs), nucleotide-binding oligomerisation domain (NOD)-like receptors (NLRs) and several cytosolic DNA receptors (CDRs) represent those expressed in the cytosol. An overview of the major PRR families is presented in Fig. 1.1. Conceptually, the PRRs act as sensor proteins, which engage an adaptor(s) molecule, mediating the activity of downstream effectors, which results in production of inflammatory signals. Activation of most PRRs leads to precise intracellular signalling cascades culminating in specific transcriptional responses driven primarily through nuclear factor-kappa B (NF-κB) and the interferon regulatory factor (IRF) families of transcription factors, ultimately resulting in the production of inflammatory cytokines (e.g. TNF, IL-6), chemokines and type I interferons (IFNs). In contrast, rather than initiating transcription, some PRRs are able to form oligomeric protein structures, termed inflammasomes, that instigate proteolytic maturation of members of the IL-1 family of cytokines (i.e. IL-1ß and IL-18), as well as mediate the inflammatory form of programmed cell death termed pyroptosis. The eventual secretion of inflammatory mediators downstream of innate immune pathways mobilises recruitment of an army of host immune cells and facilitates acute inflammatory processes.

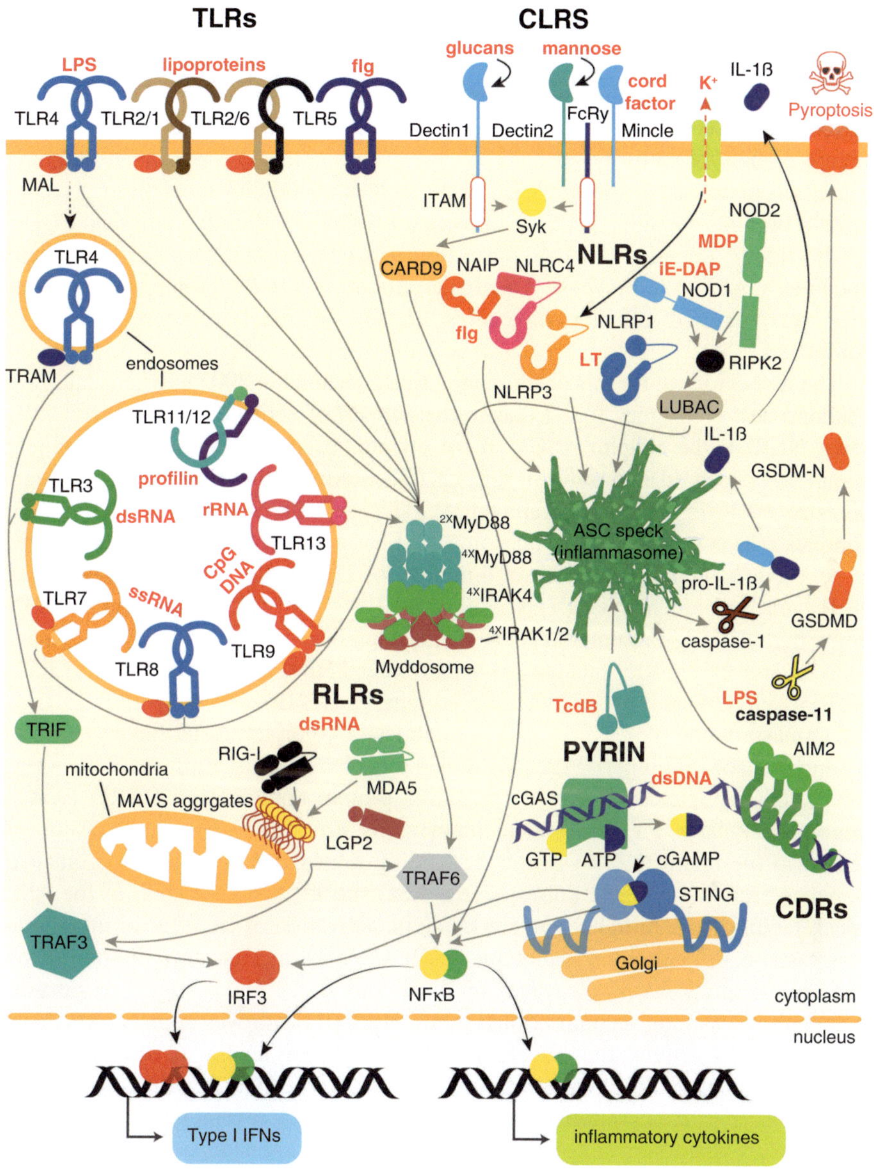

Fig. 1.1 Activation of the innate immune receptors and their signalling pathways (see text for details)

1.4 Innate Immune Signalosomes

An interesting concept emerging in innate immunity is the prerequisite for some PRRs to assemble into higher-order signalosomes in order to transduce their downstream signalling pathways [21, 22]. A feature common to these innate immune signalling platforms is the presence of molecules containing specialised protein–protein binding domains of the death domain (DD) superfamily comprising DDs, death effector domains (DEDs), pyrin domains (PYDs) and caspase recruitment domains (CARDs) [23]. These domains are found within the protein structures of PRR adaptor and effector molecules of several innate immune signalling complexes.

The first evidence for this phenomenon was presented in 2002 when the late Jürg Tschopp and colleagues found that in the cytosol of activated immune cells, the senor NLRP1, the adaptor ASC and the effector caspase-1 formed large protein complexes observed as significant shifts in the elution profiles of these proteins during size exclusion chromatography [24]. This sizeable inflammatory signalling complex was coined the 'inflammasome' and has subsequently been demonstrated to be a requirement downstream of a number of PRRs, which to date includes the NLR family members, NLRP1, NLRP3 and NLRC4, as well as the pyrin and HIN domain-containing protein (PYHIN) family member, the absence in melanoma 2 (AIM2) and the PYRIN receptor. Whether other PRRs, namely, NLRP6, NLRP7, NLRP12, RIG-I and IFN-γ-inducible protein 16 (IFI16), can also form genuine inflammasome complexes remains to be established. Upon activation, inflammasome-forming sensors oligomerise creating a seed for subsequent recruitment of the adaptor protein, apoptosis-associated speck-like protein containing a CARD (ASC), which forms large PYD-dependent filamentous structures [25]. This recruitment results in the majority of cytosolic ASC being relocalised to a single perinuclear polymer protein aggregate known as the ASC speck [26]. Formation of the ASC speck facilitates recruitment and autocatalytic activation of the effector inflammatory caspase, caspase-1 via homotypic CARD–CARD interactions to initiate cleavage and maturation (via cleavage) of IL-1ß and IL-18 and instigate pyroptosis. Although it was long recognised that pyroptotic cell death was entirely dependent upon caspase-1 activity, the exact molecular mechanisms leading to cellular rupture remained a mystery. The major breakthrough came when Gasdermin-D (GSDMD) was identified as the executioner protein in pyroptosis, by several groups in parallel, using independent screening approaches [27, 28]. These studies revealed that in addition to the pro-forms of IL-1ß and IL-18, GSDMD is also a caspase-1 substrate following inflammasome activation. Caspase-1 cleavage of GSDMD results in the release of an active amino (N)-terminal fragment termed Gasdermin-N that specifically binds to lipids (phosphoinositide and cardiolipin) of the inner leaflet of the plasma membrane. The binding of the Gasdermin-N fragment generates cytotoxic oligomeric pores that perforate the cellular membrane leading to cell swelling and lysis [29]. The specific lipid-binding preferences of Gasdermin-N to the inner leaflet of the plasma membrane ensure the intrinsic destruction of infected cells whilst limiting damage to surrounding tissue and other host cells. Interestingly, Gasdermin-N reportedly displays some bactericidal activity, which is likely explained by the presence of cardiolipin as a component of bacterial outer membranes. It is conceivable that Gasdermin-N released following cellular lysis of infected cells could subsequently insert into extracellular bacterial membranes to cause cell lysis.

Several members of the RLR family also form higher-order signalosome platforms upon activation [30]. Following binding of ssRNA in the cytosol, monomers of RIG-I or melanoma differentiation-associated gene 5 (MDA-5) interact via tandem N-terminal CARD (2CARD) interactions forming tetrameric structures. This oligomerisation seeds the recruitment of the adaptor protein mitochondrial antiviral signalling (MAVS) [31], which subsequently forms CARD-dependent helical filament structures analogous to those of ASC formed upon inflammasome activation. Filamentous MAVS drives downstream signalling events leading to NF-κB and IRF3 transcriptional responses.

With the exception of TLR3, all the TLRs emit signals via engaging the adaptor molecule, myeloid differentiation primary response gene 88 (MyD88) upon activation. MyD88 contains two critical domains that fulfil its role as an adaptor: (1) a carboxyl (C)-terminal Toll/interleukin-1 (IL-1) receptor (TIR) domain that enables coupling to the intracellular region of TLRs or the IL-1R and (2) a DD at the N-terminus that facilitates protein–protein interactions with the four members of the IL-1 receptor-associated kinase (IRAK) family of serine/threonine effector kinases [32]. Although the key players and kinetics of TLR/IL-1R proximal events were identified some time ago, our understanding of the mechanisms controlling this process was greatly enhanced following the revelation that in solution the DDs of MyD88 form specific oligomeric complexes upon addition of IRAK4 DDs [33, 34]. This stable oligomeric complex was coined the Myddosome. The crystal structure of the complex revealed a tower-shaped oligomer of four layers: two upper layers comprising two and four MyD88 DDs, respectively, followed by a third layer comprising four IRAK4 DDs and a third and final layer of four IRAK2 DDs assembling as a left-handed helical structure [33]. Architecturally, the Myddosome is stabilised via specific type III DD–DD interactions, whilst the specificity of its formation is enhanced via complementary surface electrostatic charges and contour between adjacent layers. The finding that IRAK2 DDs only binds to stable MyD88–IRAK4 DD complexes rather than with the individual DDs of either MyD88 or IRAK4 suggests a sequential ordered assembly process of the Myddosome. Discovery of the Myddosome revealed new insights into mechanisms of IRAK activation, and more mechanistic intricacies of TLR signalling are sure to emerge [35–37].

1.5 The Families of Innate Immune Receptors

1.5.1 Toll-Like Receptors (TLRs)

TLRs were the first family of innate immune receptors described. To date ten TLRs have been identified in humans (TLRs1–10) with twelve expressed in mice (TLRs1–9 and TLRs11–13, with TLR10 a pseudogene) [32]. The TLRs are type I transmembrane proteins anchored into either the plasma membrane or endolysosomal membranes. Their cellular localisation is indicative of the ligands they bind, with cell surface TLRs recognising typically bacterial outer membrane components, whilst microbial nucleic acids are sensed from TLRs localised within acidified endolysosomal compartments (see Table 1.1). Most of the TLRs are well characterised, however the ligand and precise function of TLR10 remains in question with some reports suggesting that it signals as a heterodimer with TLR2 and upstream of MyD88 to induce signals in response to influenza infection, whilst others have shown that it plays an inhibitory role [38–42]. Structurally, the TLRs consist of three major domains: (1) a ligand recognition domain consisting of folded leucine-rich repeats (LRRs) at the amino (N)-terminus, (2) a central transmembrane region and (3) a cytoplasmic TIR domain at the carboxyl (C)-terminus [43]. Variations in the LRRs of the different TLRs are thought to give specificity to the ligands they sense. Ligand binding induces the dimerisation of most TLRs, whilst TLR7, TLR8 and TLR9 have been shown to exist as preformed homodimers. As well as forming homodimers, some TLRs can recognise additional ligands via the formation of heterodimers or in cooperation with co-receptors or accessory proteins. Whilst TLR2/TLR1 heterodimers specifically recognise triacylated lipoproteins, TLR2/TLR6 dimers sense diacylated lipoproteins [44–46]. TLR2 is also thought to act synergistically with Dectin-1 (see below) in response to fungal ß-glucans leading to robust production of pro-inflammatory cytokines [47, 48]. In mice TLR11/TLR12 dimers were shown to be responsible for engaging profilin from the outer membrane of *Toxoplasma gondii* [49]. Additionally, TLR4/TLR6 heterodimers in concert with the scavenger receptor, CD36 can form a trimeric complex able to recognise oxidised low-density lipoproteins and amyloid-ß to elicit sterile inflammatory responses [50].

In the presence of a ligand, TLR dimers undergo conformational changes that enable the intracellular TIR domains to engage specific adaptor molecules for initiating downstream signal transduction pathways. Four adaptors have been identified that induce positive signalling: MyD88, MAL (also known as TIRAP), TRIF and TRAM. Broadly, MyD88-depedent pathways predominantly trigger production of inflammatory cytokines, whilst TRIF-mediated signalling elicits secretion of type I IFNs. Whilst all TLRs, excluding TLR3, utilise MyD88 for signalling [51, 52], TLRs 2, 4, 7 and 9 have been shown to require the addition of MAL as a bridging adaptor [53–57]. MAL is thought to facilitate stronger binding between MyD88 and TIR domains that have incompatible electrostatic surface charges [58]. Formation of the Myddosome (see above) allows transient recruitment of the E3 ubiquitin (Ub) ligase and TNFR-associated factor 6 (TRAF6) to the receptor complex through interactions with IRAK1/IRAK2 via C-terminal TRAF6-binding motifs [59]. TRAF6 is subsequently activated and released into the cytosol where it mediates activation of the NF-kappa B inhibitor (IκB) kinase (IKK) and mitogen-activated protein kinase (MAPK) signalling complexes leading to translocation of the major transcription factors, NF-κB, cyclic AMP-responsive element-binding protein (CREB), activator protein 1 (AP1) and IRF5 for induction of pro-inflammatory cytokines. In plasmacytoid dendritic cells (pDCs) specifically, MyD88 signalling can also lead to type I IFN production via IRAK1-dependent IRF7 activation [60–62].

A role for the TIR adaptor TRIF has been demonstrated downstream of both TLR3 and TLR4 [63–65]. Whilst TLR3 signals exclusively via TRIF, TLR4 requires the bridging adaptor TRAM to facilitate TRIF-dependent signalling [66]. Interestingly, activation of TLR4 induces signalling via MAL/MyD88 at the plasma membrane; however, the TRAM/TRIF pathway is only initiated following CD14-dependent endocytosis of the receptor complex [67, 68]. TRIF signals primarily through TRAF3 enabling TANK-binding kinase 1 (TBK-1) activity and successive IRF3-medaited transcription of type I IFNs [69]. In addition, TRAF6 is recruited, mediating a modest NF-κB response via receptor-interacting serine/threonine-protein kinase 1 (RIPK1) [70].

1.5.2 NOD-Like Receptors (NLRs)

The NLRs comprise the largest family of cytosolic receptors with a common central nucleotide oligomerisation domain (NOD), C-terminal LRRs of variable lengths and variable N-terminal protein-interacting domains that further classify the NLRs into four subfamilies [71]: (1) NLRAs have an acidic transactivating domain and contains class II major histocompatibility complex transactivator (CITTA); (2) NLRBs express a baculovirus inhibitor of apoptosis protein repeat (BIR) domain and is composed of NLR family apoptosis inhibitory proteins (NAIPs); (3) NLRCs harbour a CARD and include NOD1, NOD2, NLRC3, NLRC4 and NLRC5; and (iv) NLRPs with a PYD represent the biggest subgroup with NLRP1–14. Major structural differences within the NLRP family include NLRP10, which lacks the C-terminal LRRs and NLRP1 that contains an extended C-terminus with additional 'function to find' domain (FIIND) and CARD. Finally, NLRX1 is an orphan receptor that unlike the other cytosolic NLRs is anchored into the mitochondrial outer membrane via a unique N-terminal mitochondrial targeting sequence. The NLRs play various immune functions including response to infection, formation of inflammasomes (NLRP1, NLRP3, NLRC4) [72], regulation of antigen presentation (NLRC5, CIITA) [73, 74], regulation of microbiota homeostasis (NLRP6) [75–77] as well as modulatory roles in NF-κB (NLRP6, NLRP12, NLRC3), MAVS (NLRX1) [78] and STING (NLRC3, NLRX1) [79, 80] responses. Below, I will summarise the known roles for NLRs in activation of innate immune pathways.

NOD1 and NOD2 receptors sense specific motifs within the peptidoglycan (PGN) carbohydrate components of bacterial outer membranes within the cytosol. NOD1 recognises PGN containing the D-glutamyl-meso-diaminopimelic acid (iE-DAP) moiety most commonly present in Gram-negative bacteria [81, 82], whilst NOD2 detects the PGN muramyl dipeptide (MDP) structure found in almost all Gram-negative and Gram-positive bacteria [83]. Ligand binding to either NOD1 or NOD2 causes auto-oligomerisation of receptor molecules leading to CARD–CARD interactions with the kinase, RIPK2 that recruits the ubiquitin ligases, X-linked inhibitor of apoptosis protein (XIAP), baculoviral IAP repeat containing 2 and 3 (BIRC2, BIRC3) and the linear ubiquitin chain assembly complex (LUBAC) [84]. This induces subsequent activation of AP1- and NF-κB-dependent transcription downstream of the MAPK and IKK complexes, respectively, for the production of pro-inflammatory cytokines [84, 85].

NLRP1 is expressed as a single protein in humans, whilst mice express three paralogs due to multiple gene duplication events: NLRP1a, NLRP1b and NLRP1c [86]. Physiological activation has only been demonstrated for murine NLRP1b in response to anthrax lethal toxin from *Bacillus anthracis* and effector proteins from *Toxoplasma gondii* [87]. However, in mice deficiency of NLRP1 is associated with spontaneous obesity [88], whilst SNPs in human NLRP1 have led to implications in several autoimmune and autoinflammatory diseases, including vitiligo as well as inflammatory skin disorders relating to carcinoma [89–92]. A unique feature of the NLRP1 structure is the presence of both a FIIND and CARD at the C-terminus. In humans, the PYD is thought to structurally auto-inhibit NLRP1 via physically blocking accessibility to the CARD. However, following activation of NLRP1, the FIIND undergoes spontaneous proteolysis liberating a smaller C-terminal fragment containing the partial FIIND and the exposed CARD [92]. This cleavage fragment is therefore free to bind ASC via atypical CARD–CARD interactions to form a functional inflammasome.

NLRP3 is activated by a diverse array of stimuli (see Table 1.1), including endogenous and environmental crystals (e.g. cholesterol crystals), particulates (nanoparticles) and protein aggregates (amyloid-ß), as well as extracellular ATP, specific influenza viral components and the bacterial pore-forming toxin nigericin from *Streptomyces hygroscopicus* [93]. Hence, it is generally accepted that NLRP3 is activated downstream of a more generalised form of cellular stress. The suggested models to date include mitochondria dysfunction, intracellular ROS production, lysosomal rupture and changes to intracellular ion levels [94]. The most reproducible and consistent finding suggests that efflux of potassium leading to reduced intracellular levels is upstream of NLRP3 activation. This unified view is strengthened by the observations that addition of culture medium with reduced potassium alone is sufficient to activate immune cells and, conversely, high-level potassium medium can inhibit NLRP3 activation [95–97]. Additionally, potassium efflux appears to be associated with most NLRP3 activators [96]. Activation of the NLRP3 inflammasome is dependent upon a prerequisite priming step (e.g. via TLR activation) that enables both a transcriptional induction of NLRP3 expression and post-translational modifications that license its inflammasome activation [93].

NLRC4 is an indirect sensor of cytosolic bacterial flagellin, as well as the needle and rod subunits of the type 3 secretion system (T3SS). Direct binding of these DAMPs is to NAIPs that subsequently recruit and activate the NLRC4 inflammasome. In humans, only one NAIP exists that recognises *Salmonella* flagellin and the bacterial T3SS needle protein, whilst in mice multiple NAIPs are able to recognise flagellin (NAIP4 or NAIP5) plus the T3SS needle (NAIP1) and rod (NAIP2) components [98–100]. Upon recruitment to a single NAIP, NLRC4 undergoes significant conformational change leading to oligomerisation of 10–12 monomers of NLRC4 into a circular structure [101, 102]. NLRC4 is thought to require a specific serine phosphorylation event at residue 533 for its activation [103, 104]. Although the precise mechanistic details remain unclear, the NAIP/NLRC4 oligomeric complex then recruits ASC via CARD–CARD interactions and subsequently activates caspase-1 and its downstream effector functions.

Activation of NLRP7 and NLRP12 inflammasomes is implicated in bacterial infection. NLRP7 appears to be required for cytoplasmic detection of bacterial lipopeptides [105, 106], whilst NLRP12 may recognise T3SS components during *Yersinia pestis* infection [107]. However, more work is required to determine if these receptors play other roles in bacterial detection and indeed form a physiological inflammasomes upon activation.

1.5.3 PYRIN

Structurally, PYRIN is related to the tripartite motif (TRIM)-containing family of proteins due to the presence of two central B-box zinc finger domains and a coiled-coil domain, as well as a C-terminal B30.2 PRY-SPRY domain (absent from murine PYRIN). However, unlike other TRIM proteins, PYRIN expresses an N-terminal PYD domain enabling assembly of a functional ASC-dependent inflammasome [108]. The activation of the PYRIN inflammasome is triggered in response to specific bacterial toxins and effectors (see Table 1.1), such as toxin B from *Clostridium difficile* (TcdB) and *Vibrio parahaemolyticus* VopS [109]. Interestingly, rather than direct sensing of these bacterial proteins, PYRIN is triggered downstream of their ability to inactivate Rho GTPase, thereby inducing actin depolymerisation [109]. Mechanistically, PYRIN is held in the cytoplasm in an inactive state bound by 14-3-3 molecules to a specific motif containing a constitutively phosphorylated serine residue at position 242 [10, 110]. Serine phosphorylation is mediated via RhoA-dependent activation of the protein kinase C-family members, PKN1 and PKN2 [110]. Hence, RhoA inactivation results in loss of PKN kinase activity, dephosphorylation of serine 242, removal of 14-3-3 and activation of the PYRIN inflammasome.

1.5.4 RIG-I-Like Receptors (RLRs)

The family of RLRs consists of RIG-I, MDA5 and laboratory of genetics and physiology 2 (LGP2) that survey the cytosolic compartment for viral dsRNA [111–113]. RIG-I and MDA5 recognise specific RNA from different viral infections [114], and whilst RIG-I senses short dsRNA (up to ~1 kb), MDA5 binds larger dsRNA fragments (>2 kb) [115]. RIG-I sensing of dsRNA is enhanced by the presence of a 5′ triphosphate (PPP) moiety leading to increased type I IFN-inducing activity [116]. Both RIG-I and MDA5 are structurally similar, expressing two N-terminal CARD domains, a central helicase/ATP domain and a regulatory domain (RD) at the C-terminus. LGP2 also contains the helicase domain and the RD but lacks the 2CARD that is required to trigger downstream signalling. Hence, LGP2 was initially thought to act as a negative regulator by sequestering dsRNA and/or binding to RIG-I [117]; however, later studies suggested that LGP2 may act in concert with RIG-I and MDA5 to induce positive signalling [112]. RIG-I and MDA5 signal via the adaptor molecule MAVS (see above) [118–121], which recruits the E3 ubiquitin ligases TRAF3 and TRAF6, which elicit activation of TBK-1/IRF3 and NF-κB, respectively, culminating in a strong type I IFN antiviral response [122].

1.5.5 C-Type Lectin Receptors (CLRs)

CLRs comprise a large family of calcium-dependent carbohydrate-binding receptors. To date, only activation of the type II CLRs, namely, Dectin-1, Dectin-2 and Mincle (also known as Clec4e), is implicated in antifungal immune responses. These CLRs are transmembrane proteins expressed on the surface of immune cells containing a single N-terminal extracellular carbohydrate-recognition domain allowing detection of ß-glucans and α-mannose components of fungi by Dectin-1 and Dectin-2, respectively [123, 124] and, in the case of Mincle, an immunostimulatory factor from *Mycobacterium tuberculosis* [125] (see Table 1.1). Upon activation, these receptors engage spleen tyrosine kinase (Syk) via an immunoreceptor tyrosine-based activation motif (ITAM). This occurs directly in the case of Dectin-1 which contains this motif within its intracellular region, whilst Dectin-2 and Mincle must associate with the ITAM-containing adaptor FcRγ [126]. The recruitment and activation of Syk lead to signalling via the CARD9/Bcl-10/MALT-1 complex and downstream activity of NF-κB, resulting in production of pro-inflammatory cytokines [127, 128].

1.5.6 Cytosolic DNA Receptors (CDRs)

Two major cytoplasmic signalling pathways exist within the innate immune system downstream of dsDNA sensing: the cyclic GMP–AMP synthase (cGAS)/STING pathway and the AIM2 inflammasome pathway. Several other receptors have also been implicated in sensing intracellular DNA, such as IFI16- and DNA-dependent activator of IRFs (DAI); however, their precise functions remain unclear.

AIM2 is a member of the PYHIN family containing two major structural domains, a C-terminal HIN-200 DNA-binding domain and an N-terminal PYD to mediate protein–protein interactions with the adaptor molecule ASC. Several groups concurrently identified AIM2 as an inflammasome-forming sensor in response to cytosolic dsDNA [129–132], which occurs most prominently following bacterial infection [133]. Interestingly, AIM2 was recently shown to recognise radiation-induced DNA damage from within the nucleus [134]. Unlike the inflammasome-forming NLRs that can undergo self-oligomerisation by virtue of their NODs, AIM2 forms filaments along dsDNA via regular HIN domain binding, leading to assembly of PYDs from multiple AIM2 molecules and the polymerisation of ASC into filaments that eventually aggregate into ASC specks [135, 136]. This process facilitates the effector functions of caspase-1, maturation of IL-1β and IL-18 and pyroptotic cell death.

cGAS is a cytosolic sensor of dsDNA most commonly in the context of viral infection [137]. However, rather than directly inducing a downstream signalling cascade in response to DNA, the enzymatic activity of cGAS produces an endogenous second messenger that engages STING to elicit a strong antiviral immune response [138, 139]. Upon binding cytosolic DNA, cGAS forms dimers and undergoes structural changes facilitating the recruitment and catalysis of GTP and ATP into the cyclic dinucleotide (CDN) [140, 141], cyclic guanosine monophosphate–adenosine monophosphate (cyclic GMP–AMP or 2′3′-cGAMP) [142–146]. Interestingly, unlike other PRRs cGAS is not able to discriminate between the origins of dsDNA it encounters. Hence, cGAS can also produce cGAMP in response to host-derived DNA, such as from the mitochondria, nucleus or oxidised DNA [147, 148]. Not surprisingly, the cGAS/STING pathway is implicated in a number of autoimmune disorders including Aicardi–Goutières syndrome and systemic lupus erythematosus [149–152]. In addition to microbial and host dsDNA, cGAS activity has been reported in response to other forms of DNA, such as RNA/DNA hybrids and specific HIV-1-derived Y-DNA [153, 154] (see Table 1.1).

Preformed STING dimers reside within the endoplasmic reticulum (ER) membrane by virtue of four transmembrane domains at the N-terminus of each STING monomer [138]. STING dimers contain a V-shaped pocket that enables the binding of CDNs [155, 156]. In addition to endogenous 2′3′-cGAMP, STING was originally shown to bind CDNs of bacterial origin, such as 3′5′-cGAMP produced upon *Listeria monocytogenes* infection [157]. The binding of CDNs to STING leads to recruitment of TANK-binding kinase-1 (TBK-1) and trafficking of the dimeric STING/TBK-1 complex to perinuclear Golgi compartments. TBK-1 then phosphorylates the transcription factor IRF3, leading to nuclear translocation and strong induction of type I IFN genes [158]. In addition, STING activation also results in downstream NF-κB activation to further promote maximal transcription of IFNβ and for the production of pro-inflammatory cytokines.

1.6 Spreading Inflammatory Signals

An effective innate immune response is dependent upon the transfer of inflammatory signals between an activated host cell and neighbouring immune cells. This occurs most notably via production of cytokines, chemokines and type I IFNs upon their transcription and subsequent secretion into the extracellular environment. These inflammatory signals can act on the host cell themselves (autocrine), on proximal cells (paracrine) and occasionally even on distal cells (endocrine). As their name suggests, chemokines act to induce directed chemotaxis of neighbouring immune cells to sites of inflammation. Whilst pro- and anti-inflammatory cytokines orchestrate the magnitude and specificity of general responses to microbial infections, type I IFNs, such as IFNα and IFNβ, are particularly important for mounting an immediate and effective antiviral response. Type I IFNs mediate their effects by binding to the IFNα/β receptor (IFNAR), which is comprised of an IFNAR1/IFNAR2 heterodimer and results in the transcription of numerous IFN-stimulated genes (ISGs) encoding proteins with diverse antiviral functions [159, 160]. Although effective, this form of cell-to-cell communication is often dependent upon transcription, translation and secretion of new proteins, which takes time. In addition, initiation of innate immune pathways in infected cells is commonly targeted by the invading pathogens that have evolved immune evasion strategies to escape detection. Hence, more rapid methods of host cell-to-cell communication have evolved to initiate and amplify inflammatory signals in uninfected cells. For instance, the transfer of messages packaged within exosomes can be an effective means of communication between cells. However, perhaps the most rapid form of intercellular communication is the transfer of ions and signalling molecules through gap junctions, which are small channels directly connecting the cytoplasm of two adjacent cells. Such a mechanism has been noted to propagate both NF-κB- and IFN-dependent responses [161, 162]. Recently, it was demonstrated that the endogenous cGAS-derived second messenger, 2′3′-cGAMP, is efficiently trafficked through gap junctions from activated immune cells to neighbouring cells to directly engage and activate STING, thereby propagating the antiviral response [163]. A similar mechanism exists whereby transfer of cGAMP from HIV-infected host cells to recipient cells via HIV envelope membrane fusion sites results in an antiviral response in target cells and protection from HIV infection [164]. Interestingly, in other studies, infected immune cells were also shown to transmit an antiviral state to bystander cells through packaging of cGAMP into progeny virus particles that subsequently go on to infect surrounding cells and elicit STING-dependent responses [165, 166]. Mechanisms involving the transfer of cGAMP may be of particular relevance for cells that express no or low amount of cGAS but have sufficient STING levels, such as in certain epithelial cell populations.

Following inflammasome activation, plasma membrane integrity is severely compromised via the pore-forming activity of GSDMD, which ultimately results in cell lysis and leakage of its contents into the peripheral environment. This includes the release of preformed ASC specks, which were recently demonstrated to retain their activity in the extracellular space leading to further activation of caspase-1 and IL-1ß, thus promoting an inflammatory environment [167, 168]. Furthermore, in a similar manner to other large protein aggregates, extracellular ASC specks were shown to be taken up by macrophages acting as DAMPs to trigger subsequent activation of the NLRP3 inflammasome via lysosomal rupture [168]. Of note, ASC specks were observed in the bronchial lavage fluid of chronic obstructive pulmonary disease patients and the serum of autoinflammatory disease patients with cryopyrin-associated periodic syndromes (CAPS) [167, 168]. These findings suggest that whilst release of ASC specks represents a rapid way for cells to signal danger to their local microenvironment, in cases of chronic inflammation, ASC specks may contribute to a sustained inflammatory response.

1.7 Receptor Cooperation

Innate immune responses to pathogenic infection must be finely orchestrated in order to provoke a suitable reaction. This can include the cooperation between multiple PRRs to distinct stimuli and multiple PRRs to the same stimuli from distinct locations (Fig. 1.2). Through in vitro cell stimulation assays with synthetic or purified PRR ligands, we have learnt a great deal about the mechanisms and function of

Fig. 1.2 Localisation of DAMP sensing

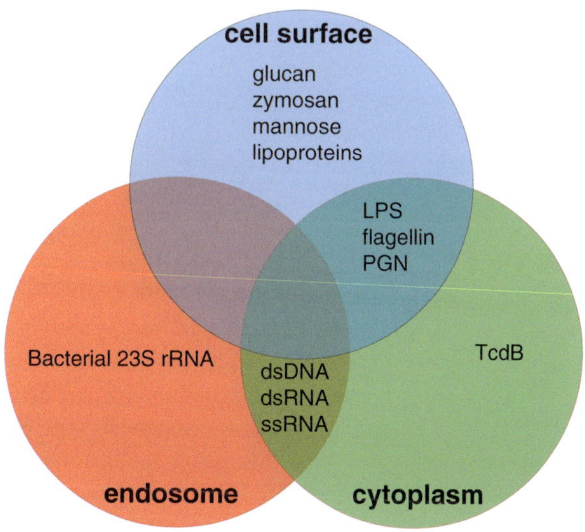

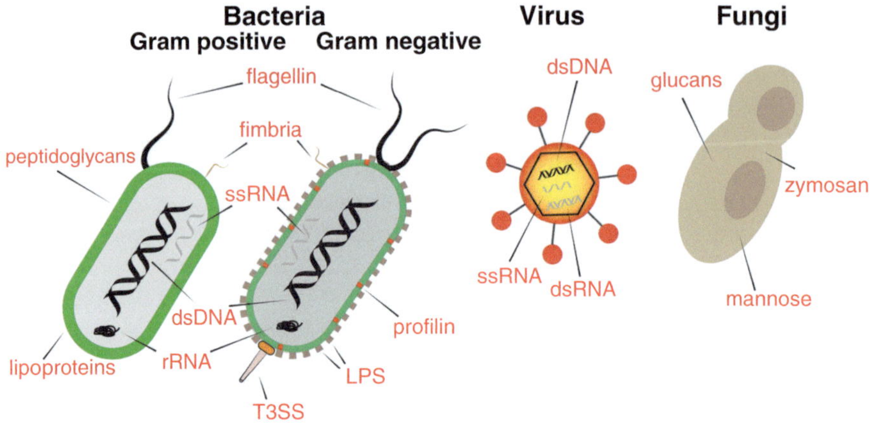

Fig. 1.3 Multiple DAMPs expressed by pathogens

the innate immune system. These assays are often performed using singular ligands in order to reduce the influence of other factors and reveal specific mechanisms of regulation and activation. However, this often leads us to think of the activation of PRRs in isolation, when in fact pathogens invariably carry numerous ligands that can potentially activate multiple PRRs simultaneously and/or sequentially during infection (Fig. 1.3 and Table 1.1). For instance, Gram-negative bacteria can harbour multiple ligands, including LPS on the outer cell membrane detected by TLR4, flagella recognised by TLR5, dsDNA sensed in endolysosomes via TLR9 or in the cytosol via cGAS as well as prokaryotic CDNs (e.g. cyclic di-GMP), directly activating STING. Hence, coordinating multiple PRR responses must be tightly regulated in order to induce an effective and specific immune response. For instance, it is well documented that a prerequisite step for maturation of IL-1ß by the inflammasome is an NF-κB transcriptional response to induce the expression of the pro-form of IL-1ß. This is likely mediated via initial detection of pathogens by TLRs at the cell surface or in endosomal compartments prior to secondary activation of inflammasome-forming receptors in the cytosol. One example is the Gram-positive bacterium, *Listeria monocytogenes*, which triggers TLR2 activation at the cell surface and subsequently NLRP3, AIM2 and NLRC4 responses within the cytosol [169–172].

In addition to multiple danger signals provoking activation of several PRRs upon infection, some singular ligands can also engage multiple PRRs from differing cellular locations in order to effectively destroy the foreign threat. An excellent example of this is during the detection of the potent immunogenic agent, LPS (also known as endotoxin) from multiple locations. LPS accounts for the majority of the structure of outer cell walls of Gram-negative bacteria and as such is detected in the extracellular environment via TLR4 in concert with the co-receptor CD14 and MD2 during infection. This recognition triggers a strong inflammatory response via activation of both NF-κB and type I IFN responses. Interestingly, LPS leakage into the host cytosol also elicits a more drastic all or nothing inflammasome response via activation of the so-called 'noncanonical' inflammasome pathway following direct binding of LPS to caspase-11 (caspase-4 and caspase-5 in humans) [173–180]. This process was shown to be dependent upon lysis of intracellular vacuoles by the IFN-inducible GTPases, guanylate-binding proteins (GBPs), thus exposing bacteria to the cytosol for activation of caspase-11 [181]. Activation of caspase-11 or caspase-4/caspase-5 mediates GSDMD cleavage and pore formation driving two separate intracellular pathways within the host cell: (1) pyroptosis and (2) activation of the NLRP3 inflammasome driving caspase-1-depedent IL-1ß release. The activation of NLRP3 downstream of caspase-11 or caspase-4/caspase-5 appears to be mediated via the efflux of intracellular potassium through GSDMD-dependent pore formation as was recently demonstrated for NLRP3 activation downstream of the activity of plasma membrane pore-forming MLKL [95]. Another fantastic example of effective PRR crosstalk is in response to infection by the Gram-positive bacterium, *Francisella novicida*. Upon entry into the cytosolic compartment, *F. novicida* activates a type I IFN response via cGAS/STING (potentially via low levels of dsDNA release), inducing IRF1-dependent expression of the IFN-inducible GBPs and IRGB10, which target the bacterial cell membrane to allow significant levels of dsDNA access into the cytosol [182–184]. Cytosolic *F. novicida* DNA then binds and activates the AIM2 inflammasome. Interestingly, the bacteriolysis activity of GBPs and IRGB10 also facilitates LPS-induced caspase-11 activation and subsequent NLRP3 inflammasome formation [183].

1.8 Open Questions and Future Directions

In this review, I have presented an overview of the activation of the PRRs in the context of innate immune function. The innate immune system is infinitely complex, and this review merely scratches the surface of the precise mechanisms controlling these processes and the exciting and emerging concepts in this fundamental and rapidly developing field of research. Below, I will discuss some of the pertinent open questions relating to this review.

Innate immunological memory is a recent and interesting concept shown to provide important protection against bacterial and fungal infection. Importantly, innate immune memory has also been linked to atherosclerosis progression [185]. Hence, understanding the exact mechanisms of innate immune memory in specific contexts is of therapeutic interest and begs the question: can innate immune memory be exploited for clinical benefits? Recently, innate immune memory was found to form the molecular basis for protection provided by various vaccines, suggesting that manipulating this process could provide benefit in other therapeutic settings [186]. In atherosclerosis, oxidised low-density lipoprotein particles, which drive damaging local inflammation, have been shown to induce a long-term pro-inflammatory phenotype in human monocyte-derived macrophages in vitro [185]. Meanwhile, high-density lipoproteins (HDLs) are well characterised to be anti-inflammatory in human and murine myeloid cells via instigating epigenetic down-modulation of the pro-inflammatory transcriptional landscape [187, 188]. Hence, understanding if HDL can instal counteractive anti-inflammatory innate immune memory in monocyte-derived macrophages of the heart would give great therapeutic power to HDL-based therapies and potential modulators of this process.

Another concept that has recently regained significance is the relationship between inflammasome-mediated cytokine release and cell death: is the release of IL-1 cytokines downstream of inflammasome activation merely cellular cargo passively lost from activated cells undergoing GSDMD-mediated pyroptosis? Or is this cytokine release an active cellular process? IL-1ß lacks a conventional secretory signal sequence, and, indeed, in the majority of studies on inflammasome activation in innate immune cells, it appears these events are intrinsically linked. However, it was recently shown that neutrophils are able to secrete IL-1ß upon activation of NLRP3 or NLRC4 without undergoing pyroptosis [189]. Additionally, in response to LPS stimulation, human monocytes can secrete IL-1ß in the absence of pyroptosis via an alternative caspase-8-induced NLRP3 inflammasome [190]. These findings are supported by recent work in which a direct caspase-1 dimerisation system (inflammasome independent), in combination with single-cell cytokine and viability measurements, was used to demonstrate that IL-1ß is released from live cells following caspase-1 activation, in the absence of cell death [191]. Further work is required to understand if these processes are indeed independent events or if this uncoupling is merely cell type and/or context specific.

The requirement for some PRRs to form large oligomeric signalosomes to induce downstream signalling is an interesting theme. ASC- and MAV-dependent protein aggregation is well established and readily reported upon physiological activation of either the inflammasome or RIG-I/MDA5, respectively. However, formation of the Myddosome structure has only been resolved during crystallisation studies of the MyD88, IRAK4 and IRAK2 DDs in vitro and as such raises the question of whether or not the Myddosome holds any physiological relevance for TLR activation. To date, the only evidence to suggest the biological existence of the Myddosome is the occurrence of natural MyD88 variants expressing amino acid changes within the DD that would disrupt the specific interaction required for Myddosome assembly. Endogenous interactions between Myddosome components are commonly reported using standard biochemical techniques [57, 192]; however, direct visual evidence of Myddosome formation in living cells remains to be achieved. This may be due to the potentially small size of the Myddosome complex, whilst ASC and MAVS aggregates form to sizes of up to ~0.4–1 μm and are readily visible by current confocal imaging techniques. Visual confirmation of physiological Myddosome assembly may be more feasible in the near future with the rapid advances being made in high-resolution imaging techniques, such as lattice light sheet technology.

An open question in the field remains: what is the precise mechanism and interactor(s) that mediate activation of the NLRP3 inflammasome? This seemly simple question has been a persistent thorn in the side of inflammasome researchers, despite a plethora of published examinations on the topic. Although activators of NLRP3 seem to converge upstream of potassium efflux, what directly binds or controls activation of NLRP3 in response to potassium efflux remains a mystery. Several post-translational modifications have been proposed in NLRP3 activation, suggesting that this could be indirectly mediated by potassium efflux. To date, fully understanding what the genuine mechanism of NLRP3 activation entails is still being significantly pursued by researchers.

Acknowledgements I am grateful to C.M. De Nardo for proofreading of the manuscript. I acknowledge the exceptional work by many investigators in the field who were not cited here due to space limitations.

References

1. Medzhitov R (2008) Origin and physiological roles of inflammation. Nature 454(7203):428–435. doi:10.1038/nature07201
2. Takeuchi O, Akira S (2010) Pattern recognition receptors and inflammation. Cell 140(6):805–820. doi:10.1016/j.cell.2010.01.022
3. Thaiss CA, Levy M, Suez J, Elinav E (2014) The interplay between the innate immune system and the microbiota. Curr Opin Immunol 26:41–48. doi:10.1016/j.coi.2013.10.016
4. Abt MC, Osborne LC, Monticelli LA, Doering TA, Alenghat T, Sonnenberg GF, Paley MA, Antenus M, Williams KL, Erikson J, Wherry EJ, Artis D (2012) Commensal bacteria calibrate the activation threshold of innate antiviral immunity. Immunity 37(1):158–170. doi:10.1016/j.immuni.2012.04.011
5. Kawashima T, Kosaka A, Yan H, Guo Z, Uchiyama R, Fukui R, Kaneko D, Kumagai Y, You DJ, Carreras J, Uematsu S, Jang MH, Takeuchi O, Kaisho T, Akira S, Miyake K, Tsutsui H, Saito T, Nishimura I, Tsuji NM (2013) Double-stranded RNA of intestinal commensal but not pathogenic bacteria triggers production of protective interferon-beta. Immunity 38(6):1187–1197. doi:10.1016/j.immuni.2013.02.024
6. Broderick L, De Nardo D, Franklin BS, Hoffman HM, Latz E (2015) The inflammasomes and autoinflammatory syndromes. Annu Rev Pathol 10:395–424. doi:10.1146/annurev-pathol-012414-040431
7. de Jesus AA, Canna SW, Liu Y, Goldbach-Mansky R (2015) Molecular mechanisms in genetically defined autoinflammatory diseases: disorders of amplified danger signaling. Annu Rev Immunol 33:823–874. doi:10.1146/annurev-immunol-032414-112227
8. Grateau G, Hentgen V, Stojanovic KS, Jeru I, Amselem S, Steichen O (2013) How should we approach classification of autoinflammatory diseases? Nat Rev Rheumatol 9(10):624–629. doi:10.1038/nrrheum.2013.101
9. Liu Y, Jesus AA, Marrero B, Yang D, Ramsey SE, Montealegre Sanchez GA, Tenbrock K, Wittkowski H, Jones OY, Kuehn HS, Lee CC, DiMattia MA, Cowen EW, Gonzalez B, Palmer I, DiGiovanna JJ, Biancotto A, Kim H, Tsai WL, Trier AM, Huang Y, Stone DL, Hill S, Kim HJ, St Hilaire C, Gurprasad S, Plass N, Chapelle D, Horkayne-Szakaly I, Foell D, Barysenka A, Candotti F, Holland SM, Hughes JD, Mehmet H, Issekutz AC, Raffeld M, McElwee J, Fontana JR, Minniti CP, Moir S, Kastner DL, Gadina M, Steven AC, Wingfield PT, Brooks SR, Rosenzweig SD, Fleisher TA, Deng Z, Boehm M, Paller AS, Goldbach-Mansky R (2014) Activated STING in a vascular and pulmonary syndrome. N Engl J Med 371(6):507–518. doi:10.1056/NEJMoa1312625
10. Masters SL, Lagou V, Jeru I, Baker PJ, Van Eyck L, Parry DA, Lawless D, De Nardo D, Garcia-Perez JE, Dagley LF, Holley CL, Dooley J, Moghaddas F, Pasciuto E, Jeandel PY, Sciot R, Lyras D, Webb AI, Nicholson SE, De Somer L, van Nieuwenhove E, Ruuth-Praz J, Copin B, Cochet E, Medlej-Hashim M, Megarbane A, Schroder K, Savic S, Goris A,

Amselem S, Wouters C, Liston A (2016) Familial autoinflammation with neutrophilic dermatosis reveals a regulatory mechanism of pyrin activation. Sci Transl Med 8(332):332ra345. doi:10.1126/scitranslmed.aaf1471
11. Netea MG, Joosten LA, Latz E, Mills KH, Natoli G, Stunnenberg HG, O'Neill LA, Xavier RJ (2016) Trained immunity: a program of innate immune memory in health and disease. Science 352(6284):aaf1098. doi:10.1126/science.aaf1098
12. Weavers H, Evans IR, Martin P, Wood W (2016) Corpse engulfment generates a molecular memory that primes the macrophage inflammatory response. Cell 165(7):1658–1671. doi:10.1016/j.cell.2016.04.049
13. Saeed S, Quintin J, Kerstens HH, Rao NA, Aghajanirefah A, Matarese F, Cheng SC, Ratter J, Berentsen K, van der Ent MA, Sharifi N, Janssen-Megens EM, Ter Huurne M, Mandoli A, van Schaik T, Ng A, Burden F, Downes K, Frontini M, Kumar V, Giamarellos-Bourboulis EJ, Ouwehand WH, van der Meer JW, Joosten LA, Wijmenga C, Martens JH, Xavier RJ, Logie C, Netea MG, Stunnenberg HG (2014) Epigenetic programming of monocyte-to-macrophage differentiation and trained innate immunity. Science 345(6204):1251086. doi:10.1126/science.1251086
14. Gauthier ME, Du Pasquier L, Degnan BM (2010) The genome of the sponge Amphimedon Queenslandica provides new perspectives into the origin of toll-like and interleukin 1 receptor pathways. Evol Dev 12(5):519–533. doi:10.1111/j.1525-142X.2010.00436.x
15. Yuen B, Bayes JM, Degnan SM (2014) The characterization of sponge NLRs provides insight into the origin and evolution of this innate immune gene family in animals. Mol Biol Evol 31(1):106–120. doi:10.1093/molbev/mst174
16. Mourao-Sa D, Roy S, Blander JM (2013) Vita-PAMPs: signatures of microbial viability. Adv Exp Med Biol 785:1–8. doi:10.1007/978-1-4614-6217-0_1
17. Sander LE, Davis MJ, Boekschoten MV, Amsen D, Dascher CC, Ryffel B, Swanson JA, Muller M, Blander JM (2011) Detection of prokaryotic mRNA signifies microbial viability and promotes immunity. Nature 474(7351):385–389. doi:10.1038/nature10072
18. Martinon F (2010) Mechanisms of uric acid crystal-mediated autoinflammation. Immunol Rev 233(1):218–232. doi:10.1111/j.0105-2896.2009.00860.x
19. West AP, Shadel GS (2017) Mitochondrial DNA in innate immune responses and inflammatory pathology. Nat Rev Immunol 17(6):363–375. doi:10.1038/nri.2017.21
20. Liston A, Masters SL (2017) Homeostasis-altering molecular processes as mechanisms of inflammasome activation. Nat Rev Immunol 17(3):208–214. doi:10.1038/nri.2016.151
21. Kagan JC, Magupalli VG, Wu H (2014) SMOCs: supramolecular organizing centres that control innate immunity. Nat Rev Immunol 14(12):821–826. doi:10.1038/nri3757
22. Qiao Q, Wu H (2015) Supramolecular organizing centers (SMOCs) as signaling machines in innate immune activation. Sci China Life Sci 58(11):1067–1072. doi:10.1007/s11427-015-4951-z
23. Ferrao R, Wu H (2012) Helical assembly in the death domain (DD) superfamily. Curr Opin Struct Biol 22(2):241–247. doi:10.1016/j.sbi.2012.02.006
24. Martinon F, Burns K, Tschopp J (2002) The inflammasome: a molecular platform triggering activation of inflammatory caspases and processing of proIL-beta. Mol Cell 10(2):417–426
25. Dick MS, Sborgi L, Ruhl S, Hiller S, Broz P (2016) ASC filament formation serves as a signal amplification mechanism for inflammasomes. Nat Commun 7:11929. doi:10.1038/ncomms11929
26. Fernandes-Alnemri T, Wu J, Yu JW, Datta P, Miller B, Jankowski W, Rosenberg S, Zhang J, Alnemri ES (2007) The pyroptosome: a supramolecular assembly of ASC dimers mediating inflammatory cell death via caspase-1 activation. Cell Death Differ 14(9):1590–1604. doi:10.1038/sj.cdd.4402194
27. Kayagaki N, Stowe IB, Lee BL, O'Rourke K, Anderson K, Warming S, Cuellar T, Haley B, Roose-Girma M, Phung QT, Liu PS, Lill JR, Li H, Wu J, Kummerfeld S, Zhang J, Lee WP, Snipas SJ, Salvesen GS, Morris LX, Fitzgerald L, Zhang Y, Bertram EM, Goodnow CC, Dixit

VM (2015) Caspase-11 cleaves gasdermin D for non-canonical inflammasome signalling. Nature 526(7575):666–671. doi:10.1038/nature15541
28. Shi J, Zhao Y, Wang K, Shi X, Wang Y, Huang H, Zhuang Y, Cai T, Wang F, Shao F (2015) Cleavage of GSDMD by inflammatory caspases determines pyroptotic cell death. Nature 526(7575):660–665. doi:10.1038/nature15514
29. Ding J, Wang K, Liu W, She Y, Sun Q, Shi J, Sun H, Wang DC, Shao F (2016) Pore-forming activity and structural autoinhibition of the gasdermin family. Nature 535(7610):111–116. doi:10.1038/nature18590
30. Xu H, He X, Zheng H, Huang LJ, Hou F, Yu Z, de la Cruz MJ, Borkowski B, Zhang X, Chen ZJ, Jiang QX (2014a) Structural basis for the prion-like MAVS filaments in antiviral innate immunity. elife 3:e01489. doi:10.7554/eLife.01489
31. Hou F, Sun L, Zheng H, Skaug B, Jiang QX, Chen ZJ (2011) MAVS forms functional prion-like aggregates to activate and propagate antiviral innate immune response. Cell 146(3):448–461. doi:10.1016/j.cell.2011.06.041
32. De Nardo D (2015) Toll-like receptors: activation, signalling and transcriptional modulation. Cytokine 74(2):181–189. doi:10.1016/j.cyto.2015.02.025
33. Lin SC, Lo YC, Wu H (2010) Helical assembly in the MyD88-IRAK4-IRAK2 complex in TLR/IL-1R signalling. Nature 465(7300):885–890. doi:10.1038/nature09121
34. Motshwene PG, Moncrieffe MC, Grossmann JG, Kao C, Ayaluru M, Sandercock AM, Robinson CV, Latz E, Gay NJ (2009) An oligomeric signaling platform formed by the toll-like receptor signal transducers MyD88 and IRAK-4. J Biol Chem 284(37):25404–25411. doi:10.1074/jbc.M109.022392
35. Ferrao R, Zhou H, Shan Y, Liu Q, Li Q, Shaw DE, Li X, Wu H (2014) IRAK4 dimerization and trans-autophosphorylation are induced by Myddosome assembly. Mol Cell 55(6):891–903. doi:10.1016/j.molcel.2014.08.006
36. Gay NJ, Gangloff M, O'Neill LA (2011) What the Myddosome structure tells us about the initiation of innate immunity. Trends Immunol 32(3):104–109. doi:10.1016/j.it.2010.12.005
37. Gay NJ, Symmons MF, Gangloff M, Bryant CE (2014) Assembly and localization of toll-like receptor signalling complexes. Nat Rev Immunol 14(8):546–558. doi:10.1038/nri3713
38. Hasan U, Chaffois C, Gaillard C, Saulnier V, Merck E, Tancredi S, Guiet C, Briere F, Vlach J, Lebecque S, Trinchieri G, Bates EE (2005) Human TLR10 is a functional receptor, expressed by B cells and plasmacytoid dendritic cells, which activates gene transcription through MyD88. J Immunol 174(5):2942–2950
39. Hess NJ, Felicelli C, Grage J, Tapping RI (2017) TLR10 suppresses the activation and differentiation of monocytes with effects on DC-mediated adaptive immune responses. J Leukoc Biol 101(5):1245–1252. doi:10.1189/jlb.3A1116-492R
40. Jiang S, Li X, Hess NJ, Guan Y, Tapping RI (2016) TLR10 is a negative regulator of both MyD88-dependent and -independent TLR signaling. J Immunol 196(9):3834–3841. doi:10.4049/jimmunol.1502599
41. Oosting M, Cheng SC, Bolscher JM, Vestering-Stenger R, Plantinga TS, Verschueren IC, Arts P, Garritsen A, van Eenennaam H, Sturm P, Kullberg BJ, Hoischen A, Adema GJ, van der Meer JW, Netea MG, Joosten LA (2014) Human TLR10 is an anti-inflammatory pattern-recognition receptor. Proc Natl Acad Sci U S A 111(42):E4478–E4484. doi:10.1073/pnas.1410293111
42. Regan T, Nally K, Carmody R, Houston A, Shanahan F, Macsharry J, Brint E (2013) Identification of TLR10 as a key mediator of the inflammatory response to Listeria monocytogenes in intestinal epithelial cells and macrophages. J Immunol 191(12):6084–6092. doi:10.4049/jimmunol.1203245
43. Gay NJ, Gangloff M (2007) Structure and function of toll receptors and their ligands. Annu Rev Biochem 76:141–165. doi:10.1146/annurev.biochem.76.060305.151318
44. Morr M, Takeuchi O, Akira S, Simon MM, Muhlradt PF (2002) Differential recognition of structural details of bacterial lipopeptides by toll-like receptors. Eur J Immunol 32(12):3337–3347. doi:10.1002/1521-4141(200212)32:12<3337::AID-IMMU3337>3.0.CO;2-#

45. Takeuchi O, Kawai T, Muhlradt PF, Morr M, Radolf JD, Zychlinsky A, Takeda K, Akira S (2001) Discrimination of bacterial lipoproteins by toll-like receptor 6. Int Immunol 13(7):933–940
46. Takeuchi O, Sato S, Horiuchi T, Hoshino K, Takeda K, Dong Z, Modlin RL, Akira S (2002) Cutting edge: role of toll-like receptor 1 in mediating immune response to microbial lipoproteins. J Immunol 169(1):10–14
47. Dillon S, Agrawal S, Banerjee K, Letterio J, Denning TL, Oswald-Richter K, Kasprowicz DJ, Kellar K, Pare J, van Dyke T, Ziegler S, Unutmaz D, Pulendran B (2006) Yeast zymosan, a stimulus for TLR2 and Dectin-1, induces regulatory antigen-presenting cells and immunological tolerance. J Clin Invest 116(4):916–928. doi:10.1172/JCI27203
48. Yadav M, Schorey JS (2006) The beta-glucan receptor Dectin-1 functions together with TLR2 to mediate macrophage activation by mycobacteria. Blood 108(9):3168–3175. doi:10.1182/blood-2006-05-024406
49. Andrade WA, Souza Mdo C, Ramos-Martinez E, Nagpal K, Dutra MS, Melo MB, Bartholomeu DC, Ghosh S, Golenbock DT, Gazzinelli RT (2013) Combined action of nucleic acid-sensing toll-like receptors and TLR11/TLR12 heterodimers imparts resistance to Toxoplasma gondii in mice. Cell Host Microbe 13(1):42–53. doi:10.1016/j.chom.2012.12.003
50. Stewart CR, Stuart LM, Wilkinson K, van Gils JM, Deng J, Halle A, Rayner KJ, Boyer L, Zhong R, Frazier WA, Lacy-Hulbert A, El Khoury J, Golenbock DT, Moore KJ (2010) CD36 ligands promote sterile inflammation through assembly of a toll-like receptor 4 and 6 heterodimer. Nat Immunol 11(2):155–161. doi:10.1038/ni.1836
51. Burns K, Martinon F, Esslinger C, Pahl H, Schneider P, Bodmer JL, Di Marco F, French L, Tschopp J (1998) MyD88, an adapter protein involved in interleukin-1 signaling. J Biol Chem 273(20):12203–12209
52. Kawai T, Adachi O, Ogawa T, Takeda K, Akira S (1999) Unresponsiveness of MyD88-deficient mice to endotoxin. Immunity 11(1):115–122
53. Bonham KS, Orzalli MH, Hayashi K, Wolf AI, Glanemann C, Weninger W, Iwasaki A, Knipe DM, Kagan JC (2014) A promiscuous lipid-binding protein diversifies the subcellular sites of toll-like receptor signal transduction. Cell 156(4):705–716. doi:10.1016/j.cell.2014.01.019
54. Fitzgerald KA, Palsson-McDermott EM, Bowie AG, Jefferies CA, Mansell AS, Brady G, Brint E, Dunne A, Gray P, Harte MT, McMurray D, Smith DE, Sims JE, Bird TA, O'Neill LA (2001) Mal (MyD88-adapter-like) is required for toll-like receptor-4 signal transduction. Nature 413(6851):78–83. doi:10.1038/35092578
55. Horng T, Barton GM, Flavell RA, Medzhitov R (2002) The adaptor molecule TIRAP provides signalling specificity for toll-like receptors. Nature 420(6913):329–333. doi:10.1038/nature01180
56. Horng T, Barton GM, Medzhitov R (2001) TIRAP: an adapter molecule in the toll signaling pathway. Nat Immunol 2(9):835–841. doi:10.1038/ni0901-835
57. Kagan JC, Medzhitov R (2006) Phosphoinositide-mediated adaptor recruitment controls toll-like receptor signaling. Cell 125(5):943–955. doi:10.1016/j.cell.2006.03.047
58. Dunne A, Ejdeback M, Ludidi PL, O'Neill LA, Gay NJ (2003) Structural complementarity of toll/interleukin-1 receptor domains in toll-like receptors and the adaptors mal and MyD88. J Biol Chem 278(42):41443–41451. doi:10.1074/jbc.M301742200
59. Qian Y, Commane M, Ninomiya-Tsuji J, Matsumoto K, Li X (2001) IRAK-mediated translocation of TRAF6 and TAB2 in the interleukin-1-induced activation of NFkappa B. J Biol Chem 276(45):41661–41667. doi:10.1074/jbc.M102262200
60. Honda K, Ohba Y, Yanai H, Negishi H, Mizutani T, Takaoka A, Taya C, Taniguchi T (2005) Spatiotemporal regulation of MyD88-IRF-7 signalling for robust type-I interferon induction. Nature 434(7036):1035–1040. doi:10.1038/nature03547
61. Honda K, Yanai H, Mizutani T, Negishi H, Shimada N, Suzuki N, Ohba Y, Takaoka A, Yeh WC, Taniguchi T (2004) Role of a transductional-transcriptional processor complex involving MyD88 and IRF-7 in toll-like receptor signaling. Proc Natl Acad Sci U S A 101(43):15416–15421. doi:10.1073/pnas.0406933101

62. Uematsu S, Sato S, Yamamoto M, Hirotani T, Kato H, Takeshita F, Matsuda M, Coban C, Ishii KJ, Kawai T, Takeuchi O, Akira S (2005) Interleukin-1 receptor-associated kinase-1 plays an essential role for toll-like receptor (TLR)7- and TLR9-mediated interferon-{alpha} induction. J Exp Med 201(6):915–923. doi:10.1084/jem.20042372
63. Oshiumi H, Matsumoto M, Funami K, Akazawa T, Seya T (2003) TICAM-1, an adaptor molecule that participates in toll-like receptor 3-mediated interferon-beta induction. Nat Immunol 4(2):161–167. doi:10.1038/ni886
64. Yamamoto M, Sato S, Hemmi H, Hoshino K, Kaisho T, Sanjo H, Takeuchi O, Sugiyama M, Okabe M, Takeda K, Akira S (2003a) Role of adaptor TRIF in the MyD88-independent toll-like receptor signaling pathway. Science 301(5633):640–643. doi:10.1126/science.1087262
65. Yamamoto M, Sato S, Mori K, Hoshino K, Takeuchi O, Takeda K, Akira S (2002) Cutting edge: a novel toll/IL-1 receptor domain-containing adapter that preferentially activates the IFN-beta promoter in the toll-like receptor signaling. J Immunol 169(12):6668–6672
66. Yamamoto M, Sato S, Hemmi H, Uematsu S, Hoshino K, Kaisho T, Takeuchi O, Takeda K, Akira S (2003b) TRAM is specifically involved in the toll-like receptor 4-mediated MyD88-independent signaling pathway. Nat Immunol 4(11):1144–1150. doi:10.1038/ni986
67. Kagan JC, Su T, Horng T, Chow A, Akira S, Medzhitov R (2008) TRAM couples endocytosis of toll-like receptor 4 to the induction of interferon-beta. Nat Immunol 9(4):361–368. doi:10.1038/ni1569
68. Zanoni I, Ostuni R, Marek LR, Barresi S, Barbalat R, Barton GM, Granucci F, Kagan JC (2011) CD14 controls the LPS-induced endocytosis of toll-like receptor 4. Cell 147(4):868–880. doi:10.1016/j.cell.2011.09.051
69. Sato S, Sugiyama M, Yamamoto M, Watanabe Y, Kawai T, Takeda K, Akira S (2003) Toll/IL-1 receptor domain-containing adaptor inducing IFN-beta (TRIF) associates with TNF receptor-associated factor 6 and TANK-binding kinase 1, and activates two distinct transcription factors, NF-kappa B and IFN-regulatory factor-3, in the toll-like receptor signaling. J Immunol 171(8):4304–4310
70. Gohda J, Matsumura T, Inoue J (2004) Cutting edge: TNFR-associated factor (TRAF) 6 is essential for MyD88-dependent pathway but not toll/IL-1 receptor domain-containing adaptor-inducing IFN-beta (TRIF)-dependent pathway in TLR signaling. J Immunol 173(5):2913–2917
71. Ting JP, Lovering RC, Alnemri ES, Bertin J, Boss JM, Davis BK, Flavell RA, Girardin SE, Godzik A, Harton JA, Hoffman HM, Hugot JP, Inohara N, Mackenzie A, Maltais LJ, Nunez G, Ogura Y, Otten LA, Philpott D, Reed JC, Reith W, Schreiber S, Steimle V, Ward PA (2008) The NLR gene family: a standard nomenclature. Immunity 28(3):285–287. doi:10.1016/j.immuni.2008.02.005
72. Latz E, Xiao TS, Stutz A (2013) Activation and regulation of the inflammasomes. Nat Rev Immunol 13(6):397–411. doi:10.1038/nri3452
73. Broxmeyer HE, Cooper S, Hangoc G, Chang CH (2006) Class II transactivator-mediated regulation of major histocompatibility complex class II antigen expression is important for hematopoietic progenitor cell suppression by chemokines and iron-binding proteins. Exp Hematol 34(8):1078–1084. doi:10.1016/j.exphem.2006.04.008
74. Yao Y, Wang Y, Chen F, Huang Y, Zhu S, Leng Q, Wang H, Shi Y, Qian Y (2012) NLRC5 regulates MHC class I antigen presentation in host defense against intracellular pathogens. Cell Res 22(5):836–847. doi:10.1038/cr.2012.56
75. Elinav E, Strowig T, Kau AL, Henao-Mejia J, Thaiss CA, Booth CJ, Peaper DR, Bertin J, Eisenbarth SC, Gordon JI, Flavell RA (2011) NLRP6 inflammasome regulates colonic microbial ecology and risk for colitis. Cell 145(5):745–757. doi:10.1016/j.cell.2011.04.022
76. Levy M, Thaiss CA, Zeevi D, Dohnalova L, Zilberman-Schapira G, Mahdi JA, David E, Savidor A, Korem T, Herzig Y, Pevsner-Fischer M, Shapiro H, Christ A, Harmelin A, Halpern Z, Latz E, Flavell RA, Amit I, Segal E, Elinav E (2015) Microbiota-modulated metabolites shape the intestinal microenvironment by regulating NLRP6 Inflammasome signaling. Cell 163(6):1428–1443. doi:10.1016/j.cell.2015.10.048

77. Wlodarska M, Thaiss CA, Nowarski R, Henao-Mejia J, Zhang JP, Brown EM, Frankel G, Levy M, Katz MN, Philbrick WM, Elinav E, Finlay BB, Flavell RA (2014) NLRP6 inflammasome orchestrates the colonic host-microbial interface by regulating goblet cell mucus secretion. Cell 156(5):1045–1059. doi:10.1016/j.cell.2014.01.026
78. Allen IC, Moore CB, Schneider M, Lei Y, Davis BK, Scull MA, Gris D, Roney KE, Zimmermann AG, Bowzard JB, Ranjan P, Monroe KM, Pickles RJ, Sambhara S, Ting JP (2011) NLRX1 protein attenuates inflammatory responses to infection by interfering with the RIG-I-MAVS and TRAF6-NF-kappaB signaling pathways. Immunity 34(6):854–865. doi:10.1016/j.immuni.2011.03.026
79. Guo H, Konig R, Deng M, Riess M, Mo J, Zhang L, Petrucelli A, Yoh SM, Barefoot B, Samo M, Sempowski GD, Zhang A, Colberg-Poley AM, Feng H, Lemon SM, Liu Y, Zhang Y, Wen H, Zhang Z, Damania B, Tsao LC, Wang Q, Su L, Duncan JA, Chanda SK, Ting JP (2016) NLRX1 sequesters STING to negatively regulate the interferon response, thereby facilitating the replication of HIV-1 and DNA viruses. Cell Host Microbe 19(4):515–528. doi:10.1016/j.chom.2016.03.001
80. Zhang L, Mo J, Swanson KV, Wen H, Petrucelli A, Gregory SM, Zhang Z, Schneider M, Jiang Y, Fitzgerald KA, Ouyang S, Liu ZJ, Damania B, Shu HB, Duncan JA, Ting JP (2014) NLRC3, a member of the NLR family of proteins, is a negative regulator of innate immune signaling induced by the DNA sensor STING. Immunity 40(3):329–341. doi:10.1016/j.immuni.2014.01.010
81. Chamaillard M, Hashimoto M, Horie Y, Masumoto J, Qiu S, Saab L, Ogura Y, Kawasaki A, Fukase K, Kusumoto S, Valvano MA, Foster SJ, Mak TW, Nunez G, Inohara N (2003) An essential role for NOD1 in host recognition of bacterial peptidoglycan containing diaminopimelic acid. Nat Immunol 4(7):702–707. doi:10.1038/ni945
82. Girardin SE, Boneca IG, Carneiro LA, Antignac A, Jehanno M, Viala J, Tedin K, Taha MK, Labigne A, Zahringer U, Coyle AJ, DiStefano PS, Bertin J, Sansonetti PJ, Philpott DJ (2003a) Nod1 detects a unique muropeptide from gram-negative bacterial peptidoglycan. Science 300(5625):1584–1587. doi:10.1126/science.1084677
83. Girardin SE, Boneca IG, Viala J, Chamaillard M, Labigne A, Thomas G, Philpott DJ, Sansonetti PJ (2003b) Nod2 is a general sensor of peptidoglycan through muramyl dipeptide (MDP) detection. J Biol Chem 278(11):8869–8872. doi:10.1074/jbc.C200651200
84. Philpott DJ, Sorbara MT, Robertson SJ, Croitoru K, Girardin SE (2014) NOD proteins: regulators of inflammation in health and disease. Nat Rev Immunol 14(1):9–23. doi:10.1038/nri3565
85. Strober W, Murray PJ, Kitani A, Watanabe T (2006) Signalling pathways and molecular interactions of NOD1 and NOD2. Nat Rev Immunol 6(1):9–20. doi:10.1038/nri1747
86. Sastalla I, Crown D, Masters SL, McKenzie A, Leppla SH, Moayeri M (2013) Transcriptional analysis of the three Nlrp1 paralogs in mice. BMC Genomics 14:188. doi:10.1186/1471-2164-14-188
87. Gorfu G, Cirelli KM, Melo MB, Mayer-Barber K, Crown D, Koller BH, Masters S, Sher A, Leppla SH, Moayeri M, Saeij JP, Grigg ME (2014) Dual role for inflammasome sensors NLRP1 and NLRP3 in murine resistance to Toxoplasma gondii. MBio 5(1):e01117–e01113. doi:10.1128/mBio.01117-13
88. Murphy AJ, Kraakman MJ, Kammoun HL, Dragoljevic D, Lee MK, Lawlor KE, Wentworth JM, Vasanthakumar A, Gerlic M, Whitehead LW, DiRago L, Cengia L, Lane RM, Metcalf D, Vince JE, Harrison LC, Kallies A, Kile BT, Croker BA, Febbraio MA, Masters SL (2016) IL-18 production from the NLRP1 Inflammasome prevents obesity and metabolic syndrome. Cell Metab 23(1):155–164. doi:10.1016/j.cmet.2015.09.024
89. D'Osualdo A, Reed JC (2012) NLRP1, a regulator of innate immunity associated with vitiligo. Pigment Cell Melanoma Res 25(1):5–8. doi:10.1111/j.1755-148X.2011.00942.x
90. Jin Y, Riccardi SL, Gowan K, Fain PR, Spritz RA (2010) Fine-mapping of vitiligo susceptibility loci on chromosomes 7 and 9 and interactions with NLRP1 (NALP1). J Invest Dermatol 130(3):774–783. doi:10.1038/jid.2009.273

91. Levandowski CB, Mailloux CM, Ferrara TM, Gowan K, Ben S, Jin Y, McFann KK, Holland PJ, Fain PR, Dinarello CA, Spritz RA (2013) NLRP1 haplotypes associated with vitiligo and autoimmunity increase interleukin-1beta processing via the NLRP1 inflammasome. Proc Natl Acad Sci U S A 110(8):2952–2956. doi:10.1073/pnas.1222808110
92. Zhong FL, Mamai O, Sborgi L, Boussofara L, Hopkins R, Robinson K, Szeverenyi I, Takeichi T, Balaji R, Lau A, Tye H, Roy K, Bonnard C, Ahl PJ, Jones LA, Baker P, Lacina L, Otsuka A, Fournie PR, Malecaze F, Lane EB, Akiyama M, Kabashima K, Connolly JE, Masters SL, Soler VJ, Omar SS, McGrath JA, Nedelcu R, Gribaa M, Denguezli M, Saad A, Hiller S, Reversade B (2016) Germline NLRP1 mutations cause skin inflammatory and cancer susceptibility syndromes via inflammasome activation. Cell 167(1):187–202. e117. doi:10.1016/j.cell.2016.09.001
93. Broz P, Dixit VM (2016) Inflammasomes: mechanism of assembly, regulation and signalling. Nat Rev Immunol 16(7):407–420. doi:10.1038/nri.2016.58
94. De Nardo D, De Nardo CM, Latz E (2014a) New insights into mechanisms controlling the NLRP3 inflammasome and its role in lung disease. Am J Pathol 184(1):42–54. doi:10.1016/j.ajpath.2013.09.007
95. Conos SA, Chen KW, De Nardo D, Hara H, Whitehead L, Nunez G, Masters SL, Murphy JM, Schroder K, Vaux DL, Lawlor KE, Lindqvist LM, Vince JE (2017) Active MLKL triggers the NLRP3 inflammasome in a cell-intrinsic manner. Proc Natl Acad Sci U S A 114(6):E961–E969. doi:10.1073/pnas.1613305114
96. Munoz-Planillo R, Kuffa P, Martinez-Colon G, Smith BL, Rajendiran TM, Nunez G (2013) K(+) efflux is the common trigger of NLRP3 inflammasome activation by bacterial toxins and particulate matter. Immunity 38(6):1142–1153. doi:10.1016/j.immuni.2013.05.016
97. Petrilli V, Papin S, Dostert C, Mayor A, Martinon F, Tschopp J (2007) Activation of the NALP3 inflammasome is triggered by low intracellular potassium concentration. Cell Death Differ 14(9):1583–1589. doi:10.1038/sj.cdd.4402195
98. Halff EF, Diebolder CA, Versteeg M, Schouten A, Brondijk TH, Huizinga EG (2012) Formation and structure of a NAIP5-NLRC4 inflammasome induced by direct interactions with conserved N- and C-terminal regions of flagellin. J Biol Chem 287(46):38460–38472. doi:10.1074/jbc.M112.393512
99. Vance RE (2015) The NAIP/NLRC4 inflammasomes. Curr Opin Immunol 32:84–89. doi:10.1016/j.coi.2015.01.010
100. Zhao Y, Yang J, Shi J, Gong YN, Lu Q, Xu H, Liu L, Shao F (2011) The NLRC4 inflammasome receptors for bacterial flagellin and type III secretion apparatus. Nature 477(7366):596–600. doi:10.1038/nature10510
101. Hu Z, Zhou Q, Zhang C, Fan S, Cheng W, Zhao Y, Shao F, Wang HW, Sui SF, Chai J (2015) Structural and biochemical basis for induced self-propagation of NLRC4. Science 350(6259):399–404. doi:10.1126/science.aac5489
102. Zhang L, Chen S, Ruan J, Wu J, Tong AB, Yin Q, Li Y, David L, Lu A, Wang WL, Marks C, Ouyang Q, Zhang X, Mao Y, Wu H (2015) Cryo-EM structure of the activated NAIP2-NLRC4 inflammasome reveals nucleated polymerization. Science 350(6259):404–409. doi:10.1126/science.aac5789
103. Matusiak M, Van Opdenbosch N, Vande Walle L, Sirard JC, Kanneganti TD, Lamkanfi M (2015) Flagellin-induced NLRC4 phosphorylation primes the inflammasome for activation by NAIP5. Proc Natl Acad Sci U S A 112(5):1541–1546. doi:10.1073/pnas.1417945112
104. Qu Y, Misaghi S, Izrael-Tomasevic A, Newton K, Gilmour LL, Lamkanfi M, Louie S, Kayagaki N, Liu J, Komuves L, Cupp JE, Arnott D, Monack D, Dixit VM (2012) Phosphorylation of NLRC4 is critical for inflammasome activation. Nature 490(7421):539–542. doi:10.1038/nature11429
105. Khare S, Dorfleutner A, Bryan NB, Yun C, Radian AD, de Almeida L, Rojanasakul Y, Stehlik C (2012) An NLRP7-containing inflammasome mediates recognition of microbial lipopeptides in human macrophages. Immunity 36(3):464–476. doi:10.1016/j.immuni.2012.02.001

106. Radian AD, Khare S, Chu LH, Dorfleutner A, Stehlik C (2015) ATP binding by NLRP7 is required for inflammasome activation in response to bacterial lipopeptides. Mol Immunol 67(2 Pt B):294–302. doi:10.1016/j.molimm.2015.06.013
107. Vladimer GI, Weng D, Paquette SW, Vanaja SK, Rathinam VA, Aune MH, Conlon JE, Burbage JJ, Proulx MK, Liu Q, Reed G, Mecsas JC, Iwakura Y, Bertin J, Goguen JD, Fitzgerald KA, Lien E (2012) The NLRP12 inflammasome recognizes Yersinia pestis. Immunity 37(1):96–107. doi:10.1016/j.immuni.2012.07.006
108. Yu JW, Wu J, Zhang Z, Datta P, Ibrahimi I, Taniguchi S, Sagara J, Fernandes-Alnemri T, Alnemri ES (2006) Cryopyrin and pyrin activate caspase-1, but not NF-kappaB, via ASC oligomerization. Cell Death Differ 13(2):236–249. doi:10.1038/sj.cdd.4401734
109. Xu H, Yang J, Gao W, Li L, Li P, Zhang L, Gong YN, Peng X, Xi JJ, Chen S, Wang F, Shao F (2014b) Innate immune sensing of bacterial modifications of Rho GTPases by the Pyrin inflammasome. Nature 513(7517):237–241. doi:10.1038/nature13449
110. Park YH, Wood G, Kastner DL, Chae JJ (2016) Pyrin inflammasome activation and RhoA signaling in the autoinflammatory diseases FMF and HIDS. Nat Immunol 17(8):914–921. doi:10.1038/ni.3457
111. Kang DC, Gopalkrishnan RV, Wu Q, Jankowsky E, Pyle AM, Fisher PB (2002) mda-5: an interferon-inducible putative RNA helicase with double-stranded RNA-dependent ATPase activity and melanoma growth-suppressive properties. Proc Natl Acad Sci U S A 99(2):637–642. doi:10.1073/pnas.022637199
112. Satoh T, Kato H, Kumagai Y, Yoneyama M, Sato S, Matsushita K, Tsujimura T, Fujita T, Akira S, Takeuchi O (2010) LGP2 is a positive regulator of RIG-I- and MDA5-mediated antiviral responses. Proc Natl Acad Sci U S A 107(4):1512–1517. doi:10.1073/pnas.0912986107
113. Yoneyama M, Kikuchi M, Natsukawa T, Shinobu N, Imaizumi T, Miyagishi M, Taira K, Akira S, Fujita T (2004) The RNA helicase RIG-I has an essential function in double-stranded RNA-induced innate antiviral responses. Nat Immunol 5(7):730–737. doi:10.1038/ni1087
114. Kato H, Takeuchi O, Sato S, Yoneyama M, Yamamoto M, Matsui K, Uematsu S, Jung A, Kawai T, Ishii KJ, Yamaguchi O, Otsu K, Tsujimura T, Koh CS, Reis e Sousa C, Matsuura Y, Fujita T, Akira S (2006) Differential roles of MDA5 and RIG-I helicases in the recognition of RNA viruses. Nature 441(7089):101–105. doi:10.1038/nature04734
115. Kato H, Takeuchi O, Mikamo-Satoh E, Hirai R, Kawai T, Matsushita K, Hiiragi A, Dermody TS, Fujita T, Akira S (2008) Length-dependent recognition of double-stranded ribonucleic acids by retinoic acid-inducible gene-I and melanoma differentiation-associated gene 5. J Exp Med 205(7):1601–1610. doi:10.1084/jem.20080091
116. Hornung V, Ellegast J, Kim S, Brzozka K, Jung A, Kato H, Poeck H, Akira S, Conzelmann KK, Schlee M, Endres S, Hartmann G (2006) 5'-Triphosphate RNA is the ligand for RIG-I. Science 314(5801):994–997. doi:10.1126/science.1132505
117. Rothenfusser S, Goutagny N, DiPerna G, Gong M, Monks BG, Schoenemeyer A, Yamamoto M, Akira S, Fitzgerald KA (2005) The RNA helicase Lgp2 inhibits TLR-independent sensing of viral replication by retinoic acid-inducible gene-I. J Immunol 175(8):5260–5268
118. Kawai T, Takahashi K, Sato S, Coban C, Kumar H, Kato H, Ishii KJ, Takeuchi O, Akira S (2005) IPS-1, an adaptor triggering RIG-I- and Mda5-mediated type I interferon induction. Nat Immunol 6(10):981–988. doi:10.1038/ni1243
119. Meylan E, Curran J, Hofmann K, Moradpour D, Binder M, Bartenschlager R, Tschopp J (2005) Cardif is an adaptor protein in the RIG-I antiviral pathway and is targeted by hepatitis C virus. Nature 437(7062):1167–1172. doi:10.1038/nature04193
120. Seth RB, Sun L, Ea CK, Chen ZJ (2005) Identification and characterization of MAVS, a mitochondrial antiviral signaling protein that activates NF-kappaB and IRF 3. Cell 122(5):669–682. doi:10.1016/j.cell.2005.08.012
121. Xu LG, Wang YY, Han KJ, Li LY, Zhai Z, Shu HB (2005) VISA is an adapter protein required for virus-triggered IFN-beta signaling. Mol Cell 19(6):727–740. doi:10.1016/j.molcel.2005.08.014

122. Kawai T, Akira S (2008) Toll-like receptor and RIG-I-like receptor signaling. Ann N Y Acad Sci 1143:1–20. doi:10.1196/annals.1443.020
123. Brown GD, Herre J, Williams DL, Willment JA, Marshall AS, Gordon S (2003) Dectin-1 mediates the biological effects of beta-glucans. J Exp Med 197(9):1119–1124. doi:10.1084/jem.20021890
124. Gazi U, Rosas M, Singh S, Heinsbroek S, Haq I, Johnson S, Brown GD, Williams DL, Taylor PR, Martinez-Pomares L (2011) Fungal recognition enhances mannose receptor shedding through Dectin-1 engagement. J Biol Chem 286(10):7822–7829. doi:10.1074/jbc.M110.185025
125. Ishikawa E, Ishikawa T, Morita YS, Toyonaga K, Yamada H, Takeuchi O, Kinoshita T, Akira S, Yoshikai Y, Yamasaki S (2009a) Direct recognition of the mycobacterial glycolipid, trehalose dimycolate, by C-type lectin Mincle. J Exp Med 206(13):2879–2888. doi:10.1084/jem.20091750
126. Kerrigan AM, Brown GD (2010) Syk-coupled C-type lectin receptors that mediate cellular activation via single tyrosine based activation motifs. Immunol Rev 234(1):335–352. doi:10.1111/j.0105-2896.2009.00882.x
127. Geijtenbeek TB, Gringhuis SI (2009) Signalling through C-type lectin receptors: shaping immune responses. Nat Rev Immunol 9(7):465–479. doi:10.1038/nri2569
128. Gross O, Gewies A, Finger K, Schafer M, Sparwasser T, Peschel C, Forster I, Ruland J (2006) Card9 controls a non-TLR signalling pathway for innate anti-fungal immunity. Nature 442(7103):651–656. doi:10.1038/nature04926
129. Burckstummer T, Baumann C, Bluml S, Dixit E, Durnberger G, Jahn H, Planyavsky M, Bilban M, Colinge J, Bennett KL, Superti-Furga G (2009) An orthogonal proteomic-genomic screen identifies AIM2 as a cytoplasmic DNA sensor for the inflammasome. Nat Immunol 10(3):266–272. doi:10.1038/ni.1702
130. Fernandes-Alnemri T, Yu JW, Datta P, Wu J, Alnemri ES (2009) AIM2 activates the inflammasome and cell death in response to cytoplasmic DNA. Nature 458(7237):509–513. doi:10.1038/nature07710
131. Hornung V, Ablasser A, Charrel-Dennis M, Bauernfeind F, Horvath G, Caffrey DR, Latz E, Fitzgerald KA (2009) AIM2 recognizes cytosolic dsDNA and forms a caspase-1-activating inflammasome with ASC. Nature 458(7237):514–518. doi:10.1038/nature07725
132. Roberts TL, Idris A, Dunn JA, Kelly GM, Burnton CM, Hodgson S, Hardy LL, Garceau V, Sweet MJ, Ross IL, Hume DA, Stacey KJ (2009) HIN-200 proteins regulate caspase activation in response to foreign cytoplasmic DNA. Science 323(5917):1057–1060. doi:10.1126/science.1169841
133. Rathinam VA, Jiang Z, Waggoner SN, Sharma S, Cole LE, Waggoner L, Vanaja SK, Monks BG, Ganesan S, Latz E, Hornung V, Vogel SN, Szomolanyi-Tsuda E, Fitzgerald KA (2010) The AIM2 inflammasome is essential for host defense against cytosolic bacteria and DNA viruses. Nat Immunol 11(5):395–402. doi:10.1038/ni.1864
134. Hu B, Jin C, Li HB, Tong J, Ouyang X, Cetinbas NM, Zhu S, Strowig T, Lam FC, Zhao C, Henao-Mejia J, Yilmaz O, Fitzgerald KA, Eisenbarth SC, Elinav E, Flavell RA (2016) The DNA-sensing AIM2 inflammasome controls radiation-induced cell death and tissue injury. Science 354(6313):765–768. doi:10.1126/science.aaf7532
135. Jin T, Perry A, Jiang J, Smith P, Curry JA, Unterholzner L, Jiang Z, Horvath G, Rathinam VA, Johnstone RW, Hornung V, Latz E, Bowie AG, Fitzgerald KA, Xiao TS (2012) Structures of the HIN domain: DNA complexes reveal ligand binding and activation mechanisms of the AIM2 inflammasome and IFI16 receptor. Immunity 36(4):561–571. doi:10.1016/j.immuni.2012.02.014
136. Lu A, Li Y, Yin Q, Ruan J, Yu X, Egelman E, Wu H (2015) Plasticity in PYD assembly revealed by cryo-EM structure of the PYD filament of AIM2. Cell Discov 1. doi:10.1038/celldisc.2015.13

137. Li XD, Wu J, Gao D, Wang H, Sun L, Chen ZJ (2013) Pivotal roles of cGAS-cGAMP signaling in antiviral defense and immune adjuvant effects. Science 341(6152):1390–1394. doi:10.1126/science.1244040
138. Ishikawa H, Barber GN (2008) STING is an endoplasmic reticulum adaptor that facilitates innate immune signalling. Nature 455(7213):674–678. doi:10.1038/nature07317
139. Ishikawa H, Ma Z, Barber GN (2009b) STING regulates intracellular DNA-mediated, type I interferon-dependent innate immunity. Nature 461(7265):788–792. doi:10.1038/nature08476
140. Sun L, Wu J, Du F, Chen X, Chen ZJ (2013) Cyclic GMP-AMP synthase is a cytosolic DNA sensor that activates the type I interferon pathway. Science 339(6121):786–791. doi:10.1126/science.1232458
141. Wu J, Sun L, Chen X, Du F, Shi H, Chen C, Chen ZJ (2013) Cyclic GMP-AMP is an endogenous second messenger in innate immune signaling by cytosolic DNA. Science 339(6121):826–830. doi:10.1126/science.1229963
142. Ablasser A, Goldeck M, Cavlar T, Deimling T, Witte G, Rohl I, Hopfner KP, Ludwig J, Hornung V (2013a) cGAS produces a 2'-5'-linked cyclic dinucleotide second messenger that activates STING. Nature 498(7454):380–384. doi:10.1038/nature12306
143. Civril F, Deimling T, de Oliveira Mann CC, Ablasser A, Moldt M, Witte G, Hornung V, Hopfner KP (2013) Structural mechanism of cytosolic DNA sensing by cGAS. Nature 498(7454):332–337. doi:10.1038/nature12305
144. Diner EJ, Burdette DL, Wilson SC, Monroe KM, Kellenberger CA, Hyodo M, Hayakawa Y, Hammond MC, Vance RE (2013) The innate immune DNA sensor cGAS produces a noncanonical cyclic dinucleotide that activates human STING. Cell Rep 3(5):1355–1361. doi:10.1016/j.celrep.2013.05.009
145. Gao P, Ascano M, Wu Y, Barchet W, Gaffney BL, Zillinger T, Serganov AA, Liu Y, Jones RA, Hartmann G, Tuschl T, Patel DJ (2013a) Cyclic [G(2',5')pA(3',5')p] is the metazoan second messenger produced by DNA-activated cyclic GMP-AMP synthase. Cell 153(5):1094–1107. doi:10.1016/j.cell.2013.04.046
146. Zhang X, Shi H, Wu J, Zhang X, Sun L, Chen C, Chen ZJ (2013) Cyclic GMP-AMP containing mixed phosphodiester linkages is an endogenous high-affinity ligand for STING. Mol Cell 51(2):226–235. doi:10.1016/j.molcel.2013.05.022
147. Gehrke N, Mertens C, Zillinger T, Wenzel J, Bald T, Zahn S, Tuting T, Hartmann G, Barchet W (2013) Oxidative damage of DNA confers resistance to cytosolic nuclease TREX1 degradation and potentiates STING-dependent immune sensing. Immunity 39(3):482–495. doi:10.1016/j.immuni.2013.08.004
148. White MJ, McArthur K, Metcalf D, Lane RM, Cambier JC, Herold MJ, van Delft MF, Bedoui S, Lessene G, Ritchie ME, Huang DC, Kile BT (2014) Apoptotic caspases suppress mtDNA-induced STING-mediated type I IFN production. Cell 159(7):1549–1562. doi:10.1016/j.cell.2014.11.036
149. Ablasser A, Hemmerling I, Schmid-Burgk JL, Behrendt R, Roers A, Hornung V (2014) TREX1 deficiency triggers cell-autonomous immunity in a cGAS-dependent manner. J Immunol 192(12):5993–5997. doi:10.4049/jimmunol.1400737
150. Ahn J, Gutman D, Saijo S, Barber GN (2012) STING manifests self DNA-dependent inflammatory disease. Proc Natl Acad Sci U S A 109(47):19386–19391. doi:10.1073/pnas.1215006109
151. Gray EE, Treuting PM, Woodward JJ, Stetson DB (2015) Cutting edge: cGAS is required for lethal autoimmune disease in the Trex1-deficient mouse model of Aicardi-Goutieres syndrome. J Immunol 195(5):1939–1943. doi:10.4049/jimmunol.1500969
152. Pokatayev V, Hasin N, Chon H, Cerritelli SM, Sakhuja K, Ward JM, Morris HD, Yan N, Crouch RJ (2016) RNase H2 catalytic core Aicardi-Goutieres syndrome-related mutant invokes cGAS-STING innate immune-sensing pathway in mice. J Exp Med 213(3):329–336. doi:10.1084/jem.20151464
153. Herzner AM, Hagmann CA, Goldeck M, Wolter S, Kubler K, Wittmann S, Gramberg T, Andreeva L, Hopfner KP, Mertens C, Zillinger T, Jin T, Xiao TS, Bartok E, Coch C,

Ackermann D, Hornung V, Ludwig J, Barchet W, Hartmann G, Schlee M (2015) Sequence-specific activation of the DNA sensor cGAS by Y-form DNA structures as found in primary HIV-1 cDNA. Nat Immunol 16(10):1025–1033. doi:10.1038/ni.3267

154. Mankan AK, Schmidt T, Chauhan D, Goldeck M, Honing K, Gaidt M, Kubarenko AV, Andreeva L, Hopfner KP, Hornung V (2014) Cytosolic RNA:DNA hybrids activate the cGAS-STING axis. EMBO J 33(24):2937–2946. doi:10.15252/embj.201488726
155. Gao P, Ascano M, Zillinger T, Wang W, Dai P, Serganov AA, Gaffney BL, Shuman S, Jones RA, Deng L, Hartmann G, Barchet W, Tuschl T, Patel DJ (2013b) Structure-function analysis of STING activation by c[G(2',5')pA(3',5')p] and targeting by antiviral DMXAA. Cell 154(4):748–762. doi:10.1016/j.cell.2013.07.023
156. Shu C, Yi G, Watts T, Kao CC, Li P (2012) Structure of STING bound to cyclic di-GMP reveals the mechanism of cyclic dinucleotide recognition by the immune system. Nat Struct Mol Biol 19(7):722–724. doi:10.1038/nsmb.2331
157. Burdette DL, Monroe KM, Sotelo-Troha K, Iwig JS, Eckert B, Hyodo M, Hayakawa Y, Vance RE (2011) STING is a direct innate immune sensor of cyclic di-GMP. Nature 478(7370):515–518. doi:10.1038/nature10429
158. Barber GN (2014) STING-dependent cytosolic DNA sensing pathways. Trends Immunol 35(2):88–93. doi:10.1016/j.it.2013.10.010
159. de Weerd NA, Samarajiwa SA, Hertzog PJ (2007) Type I interferon receptors: biochemistry and biological functions. J Biol Chem 282(28):20053–20057. doi:10.1074/jbc.R700006200
160. Schoggins JW (2014) Interferon-stimulated genes: roles in viral pathogenesis. Curr Opin Virol 6:40–46. doi:10.1016/j.coviro.2014.03.006
161. Kasper CA, Sorg I, Schmutz C, Tschon T, Wischnewski H, Kim ML, Arrieumerlou C (2010) Cell-cell propagation of NF-kappaB transcription factor and MAP kinase activation amplifies innate immunity against bacterial infection. Immunity 33(5):804–816. doi:10.1016/j.immuni.2010.10.015
162. Patel SJ, King KR, Casali M, Yarmush ML (2009) DNA-triggered innate immune responses are propagated by gap junction communication. Proc Natl Acad Sci U S A 106(31):12867–12872. doi:10.1073/pnas.0809292106
163. Ablasser A, Schmid-Burgk JL, Hemmerling I, Horvath GL, Schmidt T, Latz E, Hornung V (2013b) Cell intrinsic immunity spreads to bystander cells via the intercellular transfer of cGAMP. Nature 503(7477):530–534. doi:10.1038/nature12640
164. Xu S, Ducroux A, Ponnurangam A, Vieyres G, Franz S, Musken M, Zillinger T, Malassa A, Ewald E, Hornung V, Barchet W, Haussler S, Pietschmann T, Goffinet C (2016) cGAS-mediated innate immunity spreads intercellularly through HIV-1 Env-induced membrane fusion sites. Cell Host Microbe 20(4):443–457. doi:10.1016/j.chom.2016.09.003
165. Bridgeman A, Maelfait J, Davenne T, Partridge T, Peng Y, Mayer A, Dong T, Kaever V, Borrow P, Rehwinkel J (2015) Viruses transfer the antiviral second messenger cGAMP between cells. Science 349(6253):1228–1232. doi:10.1126/science.aab3632
166. Gentili M, Kowal J, Tkach M, Satoh T, Lahaye X, Conrad C, Boyron M, Lombard B, Durand S, Kroemer G, Loew D, Dalod M, Thery C, Manel N (2015) Transmission of innate immune signaling by packaging of cGAMP in viral particles. Science 349(6253):1232–1236. doi:10.1126/science.aab3628
167. Baroja-Mazo A, Martin-Sanchez F, Gomez AI, Martinez CM, Amores-Iniesta J, Compan V, Barbera-Cremades M, Yague J, Ruiz-Ortiz E, Anton J, Bujan S, Couillin I, Brough D, Arostegui JI, Pelegrin P (2014) The NLRP3 inflammasome is released as a particulate danger signal that amplifies the inflammatory response. Nat Immunol 15(8):738–748. doi:10.1038/ni.2919
168. Franklin BS, Bossaller L, De Nardo D, Ratter JM, Stutz A, Engels G, Brenker C, Nordhoff M, Mirandola SR, Al-Amoudi A, Mangan MS, Zimmer S, Monks BG, Fricke M, Schmidt RE, Espevik T, Jones B, Jarnicki AG, Hansbro PM, Busto P, Marshak-Rothstein A, Hornemann S, Aguzzi A, Kastenmuller W, Latz E (2014) The adaptor ASC has extracellular and 'prionoid' activities that propagate inflammation. Nat Immunol 15(8):727–737. doi:10.1038/ni.2913

169. Kim S, Bauernfeind F, Ablasser A, Hartmann G, Fitzgerald KA, Latz E, Hornung V (2010) Listeria monocytogenes is sensed by the NLRP3 and AIM2 inflammasome. Eur J Immunol 40(6):1545–1551. doi:10.1002/eji.201040425
170. Ozoren N, Masumoto J, Franchi L, Kanneganti TD, Body-Malapel M, Erturk I, Jagirdar R, Zhu L, Inohara N, Bertin J, Coyle A, Grant EP, Nunez G (2006) Distinct roles of TLR2 and the adaptor ASC in IL-1beta/IL-18 secretion in response to Listeria monocytogenes. J Immunol 176(7):4337–4342
171. Sauer JD, Witte CE, Zemansky J, Hanson B, Lauer P, Portnoy DA (2010) Listeria monocytogenes triggers AIM2-mediated pyroptosis upon infrequent bacteriolysis in the macrophage cytosol. Cell Host Microbe 7(5):412–419. doi:10.1016/j.chom.2010.04.004
172. Wu J, Fernandes-Alnemri T, Alnemri ES (2010) Involvement of the AIM2, NLRC4, and NLRP3 inflammasomes in caspase-1 activation by Listeria monocytogenes. J Clin Immunol 30(5):693–702. doi:10.1007/s10875-010-9425-2
173. Baker PJ, Boucher D, Bierschenk D, Tebartz C, Whitney PG, D'Silva DB, Tanzer MC, Monteleone M, Robertson AA, Cooper MA, Alvarez-Diaz S, Herold MJ, Bedoui S, Schroder K, Masters SL (2015) NLRP3 inflammasome activation downstream of cytoplasmic LPS recognition by both caspase-4 and caspase-5. Eur J Immunol 45(10):2918–2926. doi:10.1002/eji.201545655
174. Casson CN, Yu J, Reyes VM, Taschuk FO, Yadav A, Copenhaver AM, Nguyen HT, Collman RG, Shin S (2015) Human caspase-4 mediates noncanonical inflammasome activation against gram-negative bacterial pathogens. Proc Natl Acad Sci U S A 112(21):6688–6693. doi:10.1073/pnas.1421699112
175. Hagar JA, Powell DA, Aachoui Y, Ernst RK, Miao EA (2013) Cytoplasmic LPS activates caspase-11: implications in TLR4-independent endotoxic shock. Science 341(6151):1250–1253. doi:10.1126/science.1240988
176. Kayagaki N, Wong MT, Stowe IB, Ramani SR, Gonzalez LC, Akashi-Takamura S, Miyake K, Zhang J, Lee WP, Muszynski A, Forsberg LS, Carlson RW, Dixit VM (2013) Noncanonical inflammasome activation by intracellular LPS independent of TLR4. Science 341(6151):1246–1249. doi:10.1126/science.1240248
177. Ruhl S, Broz P (2015) Caspase-11 activates a canonical NLRP3 inflammasome by promoting K(+) efflux. Eur J Immunol 45(10):2927–2936. doi:10.1002/eji.201545772
178. Schmid-Burgk JL, Gaidt MM, Schmidt T, Ebert TS, Bartok E, Hornung V (2015) Caspase-4 mediates non-canonical activation of the NLRP3 inflammasome in human myeloid cells. Eur J Immunol 45(10):2911–2917. doi:10.1002/eji.201545523
179. Shi J, Zhao Y, Wang Y, Gao W, Ding J, Li P, Hu L, Shao F (2014) Inflammatory caspases are innate immune receptors for intracellular LPS. Nature 514(7521):187–192. doi:10.1038/nature13683
180. Vigano E, Diamond CE, Spreafico R, Balachander A, Sobota RM, Mortellaro A (2015) Human caspase-4 and caspase-5 regulate the one-step non-canonical inflammasome activation in monocytes. Nat Commun 6:8761. doi:10.1038/ncomms9761
181. Meunier E, Dick MS, Dreier RF, Schurmann N, Kenzelmann Broz D, Warming S, Roose-Girma M, Bumann D, Kayagaki N, Takeda K, Yamamoto M, Broz P (2014) Caspase-11 activation requires lysis of pathogen-containing vacuoles by IFN-induced GTPases. Nature 509(7500):366–370. doi:10.1038/nature13157
182. Man SM, Karki R, Malireddi RK, Neale G, Vogel P, Yamamoto M, Lamkanfi M, Kanneganti TD (2015) The transcription factor IRF1 and guanylate-binding proteins target activation of the AIM2 inflammasome by Francisella infection. Nat Immunol 16(5):467–475. doi:10.1038/ni.3118
183. Man SM, Karki R, Sasai M, Place DE, Kesavardhana S, Temirov J, Frase S, Zhu Q, Malireddi RK, Kuriakose T, Peters JL, Neale G, Brown SA, Yamamoto M, Kanneganti TD (2016) IRGB10 liberates bacterial Ligands for sensing by the AIM2 and Caspase-11-NLRP3 inflammasomes. Cell 167(2):382–396. e317. doi:10.1016/j.cell.2016.09.012

184. Meunier E, Wallet P, Dreier RF, Costanzo S, Anton L, Ruhl S, Dussurgey S, Dick MS, Kistner A, Rigard M, Degrandi D, Pfeffer K, Yamamoto M, Henry T, Broz P (2015) Guanylate-binding proteins promote activation of the AIM2 inflammasome during infection with Francisella novicida. Nat Immunol 16(5):476–484. doi:10.1038/ni.3119
185. Bekkering S, Joosten LA, van der Meer JW, Netea MG, Riksen NP (2015) The epigenetic memory of monocytes and macrophages as a novel drug target in atherosclerosis. Clin Ther 37(4):914–923. doi:10.1016/j.clinthera.2015.01.008
186. Blok BA, Arts RJ, van Crevel R, Benn CS, Netea MG (2015) Trained innate immunity as underlying mechanism for the long-term, nonspecific effects of vaccines. J Leukoc Biol 98(3):347–356. doi:10.1189/jlb.5RI0315-096R
187. De Nardo D, Labzin LI, Kono H, Seki R, Schmidt SV, Beyer M, Xu D, Zimmer S, Lahrmann C, Schildberg FA, Vogelhuber J, Kraut M, Ulas T, Kerksiek A, Krebs W, Bode N, Grebe A, Fitzgerald ML, Hernandez NJ, Williams BR, Knolle P, Kneilling M, Rocken M, Lutjohann D, Wright SD, Schultze JL, Latz E (2014b) High-density lipoprotein mediates anti-inflammatory reprogramming of macrophages via the transcriptional regulator ATF3. Nat Immunol 15(2):152–160. doi:10.1038/ni.2784
188. Xue J, Schmidt SV, Sander J, Draffehn A, Krebs W, Quester I, De Nardo D, Gohel TD, Emde M, Schmidleithner L, Ganesan H, Nino-Castro A, Mallmann MR, Labzin L, Theis H, Kraut M, Beyer M, Latz E, Freeman TC, Ulas T, Schultze JL (2014) Transcriptome-based network analysis reveals a spectrum model of human macrophage activation. Immunity 40(2):274–288. doi:10.1016/j.immuni.2014.01.006
189. Chen KW, Gross CJ, Sotomayor FV, Stacey KJ, Tschopp J, Sweet MJ, Schroder K (2014) The neutrophil NLRC4 inflammasome selectively promotes IL-1beta maturation without pyroptosis during acute Salmonella challenge. Cell Rep 8(2):570–582. doi:10.1016/j.celrep.2014.06.028
190. Gaidt MM, Ebert TS, Chauhan D, Schmidt T, Schmid-Burgk JL, Rapino F, Robertson AA, Cooper MA, Graf T, Hornung V (2016) Human monocytes engage an alternative inflammasome pathway. Immunity 44(4):833–846. doi:10.1016/j.immuni.2016.01.012
191. Conos SA, Lawlor KE, Vaux DL, Vince JE, Lindqvist LM (2016) Cell death is not essential for caspase-1-mediated interleukin-1beta activation and secretion. Cell Death Differ 23(11):1827–1838. doi:10.1038/cdd.2016.69
192. Sun J, Li N, Oh KS, Dutta B, Vayttaden SJ, Lin B, Ebert TS, De Nardo D, Davis J, Bagirzadeh R, Lounsbury NW, Pasare C, Latz E, Hornung V, Fraser ID (2016) Comprehensive RNAi-based screening of human and mouse TLR pathways identifies species-specific preferences in signaling protein use. Sci Signal 9(409):ra3. doi:10.1126/scisignal.aab2191

Chapter 2
Posttranslational Modification Control of Inflammatory Signaling

Yibo Si, Yunlu Zhang, Zijuan Chen, Ruixue Zhou, Yihua Zhang, Doudou Hao, and Dapeng Yan

Abstract Inflammation is usually the defensive reaction of the immune system to the invasion of pathogen and the exogenous objects. The activation of inflammation helps our body to eliminate pathogenic microbe, virus, and parasite harming our health, while under many circumstances inflammation is the direct cause of the pathological damage in tissues and dysfunction of organs. The posttranslational modification (PTM) of the inflammatory pathways, such as TLR pathways, RLR pathways, NLR pathway, intracellular DNA sensors, intracellular RNA sensors, and inflammasomes, is crucial in the regulation of these signaling trails. Ubiquitination, phosphorylation, polyubiquitination, methylation, and acetylation are the main forms of the PTM, and they respectively play different roles in signaling regulation. The effects of the PTM range from the production of pro-inflammatory factors and the interaction between adaptors and receptors to cell translocation in response to the infectious or other dangerous factors. In this chapter, we will have an overview of the different ways of the posttranslational modifications in different inflammatory signaling pathways and their essential roles in regulation of inflammation.

Keywords Posttranslational modification • TLR pathways • RLR pathways • NLR pathway • Intracellular DNA sensors • Intracellular RNA sensors • Inflammasomes

---

Yibo Si and Yunlu Zhang contributed equally to this work.

Y. Si • Y. Zhang • Z. Chen • R. Zhou • Y. Zhang • D. Hao • D. Yan (✉)
Department of Immunology, School of Basic Medical Sciences, Fudan University, Shanghai 200032, People's Republic of China

Key Laboratory of Medical Molecular Virology of Ministries of Education and Health, School of Basic Medical Sciences, Fudan University, Shanghai 200032, People's Republic of China
e-mail: dapengyan@fudan.edu.cn

D. Xu (ed.), *Regulation of Inflammatory Signaling in Health and Disease*, Advances in Experimental Medicine and Biology 1024,
DOI 10.1007/978-981-10-5987-2_2

2.1 Introduction

The definition of inflammatory signaling pathways is the mechanism of how inflammatory signals transduce and activate relative cells and thus induce inflammation. The innate immune system contributes a broad range of inflammatory signaling pathways with its pattern recognition receptors (PRRs). The PRRs represent many receptors that are able to recognize pathogen-associated molecular patterns (PAMPs) encoded by the pathogenic microbe, virus, and parasites living in the nature. The formation of the pairing PAMPs and PRRs is the result of the fight between our bodies and pathogens through the evolution. Besides PAMPs, the PRRs can also recognize the danger-associated molecular patterns (DAMPs) released from damaged cells in our bodies, the most important mechanism of pathological injury in tissues and organs when diseases happen. When PAMPs or DAMPs are derived and detected by PRRs, downstream molecular interaction occurs and eventually activates the inflammatory pathways. The best-known PRRs are Toll-like receptors, which are well studied these years. The classic TLR pathways include the interaction between myeloid differentiation primary response protein 88 (MyD88) adaptor-like (Mal) and MyD88, thus activating TNF receptor-associated factor 6 (TRAF6) and eventually activating the translocation of nuclear factor-κB (NF-κB), while there is an exception when it comes to TLR3. Other PRRs, such as retinoic acid-inducible gene I-like receptors (RIG-I-like receptors, RLRs) and nucleotide-binding oligomerization domain-containing protein I-like receptors (NLRs), also play essential roles in the recognition and elimination of pathogens attacking our body. RLR pathways are likely to recruit melanoma differentiation-associated proten-5 (MAV5) and caspase activation and recruitment domain-containing protein 9 (CARD9) that eventually activates the IRF3 and NF-κB, respectively; NLR pathways function as inflammatory signaling pathways using NOD-1/NOD-2 and receptor-interacting protein 2 (RIP2) to initiate NF-κB and cause inflammation. The receptors mentioned above are the initiators of their own inflammatory signaling pathways, and once they have detected the PAMPs or DAMPs that can be recognized and bind to them, they can recruit and activate downstream molecules (peptide, protein, or enzyme) which are responsible for the inflammation formation. Recently, more attentions are paid to other PRRs. Cytoplasmic sensors including cyclic GMP-AMP synthase (cGAS), DNA-dependent activator of IRFs(DAI), and IFN-inducible protein-16 (IFI16) are a series of molecules that are able to detect ds-DNA produced by pathogens, and all downstream molecular of them is stimulator of interferon genes (STING) which eventually trigger interferon production; cytoplasmic RNA sensors including RIG-I and MAVS are molecules monitoring abnormal cytoplasmic RNA and consequently activating IRF3. Also, inflammasome signaling via NLR family pyrin domain-containing protein 3 (NLRP3), NLR family CARD domain-containing protein 4 (NLRC4), AIM2, and so on can activate the assembly of inflammasomes, which eventually activate caspase-1 and accordingly produce inflammatory factors.

Posttranslational modification (PTM) is a process that is operating (adding or cutting) proteins, enzymes, or peptides with different functional chemical groups with the help of specific enzymes, which happens after the translation of these proteins. There are diverse PTMs, among which phosphorylation and ubiquitination are the two most prevalent modification forms involved in many signaling pathway regulations. Besides phosphorylation and ubiquitination, polyubiquitination, methylation, acetylation, sumoylation, and succinylation also participate in some protein modifications. PTM exerts profound influence on the fate of proteins. Different PTM types have different roles to different proteins, such as translocation, secretion, function, and elimination. These effects control a group of cells whose proteins are modified by PTM mechanism, and the response of these cells ranges from division, differentiation, and migration to apoptosis.

2.2 The PTM in TLR Pathways

The Toll-like receptor (TLR) pathways constitute an important mechanism in the activation of innate immune cells including monocytes, macrophages, and dendritic cells, and the activation of these cells can result in the formation of inflammation. PTMs are broadly and essentially involved in signal regulating of TLR pathways.

2.2.1 The Overview of TLR Pathways

Toll-like receptors (TLRs) are type I transmembrane glycoproteins which play an important role in recognizing infectious factors as well as the DAMPs released by the apoptotic cells and damaged cells. Ten TLRs have been found up to date termed as TLR1–10. TLRs 1, 2, 4, 5, 6, and 10 are located in the cell surface, while TLRs 3, 7, 8, and 9 have an endosomal localization [1]. The signaling pathways initialized by Mal (MyD88 adaptor-like) are called MyD88-dependent pathways which trigger TLR2/4 receptor-related pathways. While TLR3 and endosomal TLR4 receptor-related TLR pathways are resulted from TRAM/TRIF complex, so-called TRIF pathways. In MyD88-dependent pathways, after TLR2/4 activated by PAMP/DAMP, Mal is recruited to the endosomal parts of TLRs and sequentially recruits the TNF-receptor-associated factor 6 (TRAF6). TRAF6 then dissociates from the receptor and forms a complex with TAK1 (transforming growth factor β-activating kinase), TAB1 (TAK1-binding protein 1), and TAB2 (TAK1-binding protein) at the plasma membrane, which induces the phosphorylation of TAB2 and TAK1 [2]. After that, the complex TRAF6/TAK1/TAB1/TAB2 translocates to the cytosol from cell membrane leading to the ubiquitination of TRAF6 and activation of TAK1. Activated TAK1 mediates the phosphorylation of IκB kinase (IKK) composed of IKKα, IKKβ, and the modulating part IKKγ/NEMO. Phosphorylated IKK complex is activated and thus phosphorylates the IκBs, which triggers the polyubiquitination of IκBs

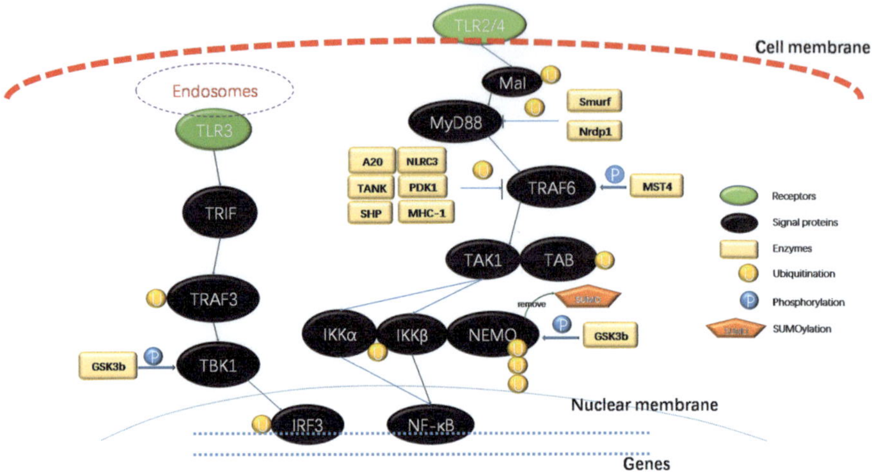

Fig. 2.1 Inflammatory signaling transduction and PTMs of TLR pathways

causing its degradation. Eventually, degraded IκBs result in the activation of NF-κB which activates the pro-inflammatory genes and causes the production of inflammatory factors such as IL-6. TAK1 also activates the MAP kinase kinase 3 MKK3/6-p38 signaling cascade, leading to the activation of cAMP response element binding (CREB) nuclear transcription factor and the MKK4/7-Jun N-terminal kinase (JNK) mediating the activation of the transcription factor activator protein-1 (AP-1) [2] which triggers the production of pro-inflammatory factors cooperatively with NF-κB. In TRIF pathways lacking the existence of MyD88, activated TLR3/endosomal TLR4 recruits TRAM and TRIF to their TIR domain to form a complex. Then the complex continuously recruits TRAF3 and TBK-1 (TANK-binding kinase 1). Activated TBK-1 can phosphorylate IRF3 to initiate its translocation to the nucleus and thus triggers the production of interferon, which plays a crucial role in antiviral response and local inflammation.

2.2.2 *The PTMs in TLR Pathways*

In the two main signaling pathways of TLRs, PTMs take the responsibility as the molecular mechanism in cascading signal transduction in the cells (Fig. 2.1). The phosphorylation of NF-κB inhibitor (IκB) kinase a (IKKa), IKKb, IκBa, and IRF3 is indispensable to their normal function. Additionally, phosphorylation occurring at a conserved pLxIS motif of TRIF is necessary for the translocation of IRF3 as well as IFN production [3]. Autophosphorylation of TBK1 modified by glycogen synthase kinase 3b (GSK3b) is essential of IRF3 activation [4]. Numbers of proteins are demonstrated that they act as the enzymes involved in phosphorylation-modification, such as Src homology-containing protein tyrosine phosphatase-1(SHP1) [5] and mammalian STE20-like protein kinase 4(MST4) [6]. For example, the kinase MST4 can phosphorylate the TRAF6 molecules that plays an important role in TLR-MyD88-dependent pathways causing its dysfunction in oligomerization and autoubiquitination, thus causing the repression of the TLR pathway. MST4 knockdown mice show severe inflammation and death in septic shock, which can be restored by heterozygous deletion of *Traf6* [6]. Moreover, in the NF-κB signaling pathway, NEMO, an adaptor protein involved in the activation of the IKK, is phosphorylated by GSK-3β (glycogen synthase kinase-3β) at S (8, 17, 31, 43), which facilitates the NF-κB activity [7]. The ubiquitination also participates in the regulation of TLR pathways. The form of lysine 63(K63)-linked ubiquitination/polyubiquitination is related to the signal transduction, while it is known that K48-linked ubiquitination is responsible for proteasomal degradation [8]. In the TLR pathways, K63-linked polyubiquitination of TRAF6, TAB2/3, NEMO, and TRAF3 is required for NF-κB and IRF3 activation [9, 10]. K48-linked ubiquitination of IκB leads to the proteasomal degradation of IκB and thus activates the NF-κB. Linear ubiquitination of NEMO (NF-κB essential modulator) is involved in the activation of NF-κB. The LUBAC ligase composed of the two RING finger proteins HOIL-1 L and HOIP activates the canonical NF-κB pathway by binding to NEMO and conjugates linear polyubiquitin chains onto specific Lys residues in the CC2-LZ domain of NEMO. In *HOIL-1* knockout mice and cells derived from these mice, NF-κB signaling induced by pro-inflammatory cytokines such as TNF-α and IL-1β was suppressed [11]. NEMO is able to bind many types of ubiquitin chains including K63, K48, K27, and linear-linked ubiquitin chains [8]. Additionally, TLR pathways can be inhibited via ubiquitination modification: Mal can be inhibited by suppressor of cytokine signaling 1 (SOCS1) [12]. MyD88 can be inhibited by neuroregulin receptor degradation protein 1 (Nrdp1) [13] and smad ubiquitin regulator factor proteins (Smurf). Another example for the suppression effect of ubiquitination in TLR pathways is that PDK1 inhibits tumor necrosis factor (TNF) receptor-associated factor 6 (TRAF6) ubiquitination by interrupting the complex between transforming growth factor beta-activated kinase 1 (TAK1) and TAK1-binding protein 2 (TAB2), which negatively regulates TAK1 activity [14]. TRAF6 can be inhibited by A20 [15], TRAF family member-associated NF-κB activator (TANK) [16], β-arrestin [17], SHP [18], MHC-1, and NLRC3 [19]. To some other special PTM forms, SUMO-specific protease 6 (SENP6) inhibits TLR inflammatory responses by catalyzing the de-SUMOylation of IKKγ/NEMO [20].

2.3 The PTMs in RLR Pathways

The RLR pathways (retinoic acid-inducible gene I-like receptors) are PRRs that detect RNA and DNA produced by intruding virus locating in cytoplasm. Basically, the triggered cellular RLRs include two members termed as RIG-I (retinoic acid-inducible gene I) and MDA5 (melanoma differentiation-associated protein 5).

While the regulation of RLR pathways is dominated by a load of factors, PTM is indispensable.

2.3.1 *The Overview of RLR Pathways*

The sensors belonging to the RLR family are intracellular localized. Within the RLR pathways, RIG-I recognizes RNA produced by virus with short double-stranded RNA (dsRNA) stretches and a 5′-triphosphate or 5′-diphosphate moiety, while MDA5 detects longer dsRNA or viral RNA. RIG-I and MDA5 are DExD/H--box-containing RNA helicases with similar structures: both possess two N-terminal caspase activation and recruitment domains (CARDs) for initiating downstream signaling and a central helicase domain and carboxyl-terminal domain (CTD), which are both required for RNA binding [21]. Binding with RNA, RIG-I/MDA5 is activated and thus binds to the mitochondrial antiviral-signaling protein (MAVS). MAVS then recruits the TBK1 (TANK-binding kinase 1) and IKKε which respectively phosphorylates the IRF3/7 and aggregates IKKα-IKKβ-IKKγ complex. Sequentially, phosphorylated IRF and NF-κB are translocated into the nucleus, triggering the transcription of antiviral genes and inflammatory genes. Besides RLR pathways, proteins of RLRs play their roles in many other signal transduction trails, such as RNA sensor-mediated inflammation formation. Among them, reversible Ser/Thr phosphorylation and K63- and K48-linked polyubiquitination are critical signals to control pro-inflammatory cytokine induction triggered by RLRs [21].

2.3.2 *The PTM Involved in RLR Pathways*

2.3.2.1 Phosphorylation

Phosphorylation of the relevant proteins in RIG-1 signaling pathway plays a critical role not only in recruiting and activating downstream proteins but also in activation of other PTM like ubiquitination. Siglec-G, a member of the immunoglobulin-like lectin family, binds to activated RIG-I and recruits tyrosine phosphatase SHP2 (phosphatase Src homology 2) which in turn phosphorylates Siglec-G leading to the promotion of interaction between Siglec-G and RIG-1. Then Siglec-G/SHP2/RIG-I complex recruits and phosphorylates c-Cbl (Casitas B-lineage lymphoma, a E3 ligase), which induces the K48-linked ubiquitination and degradation of RIG-1 at K813 and IRF3 in a dose-dependent manner [22] (Fig. 2.2). Lyn, a member of STKs (Src-family-tyrosine kinases) family, interacts with the helicase domain of RIG-I by its SH2/SH3 domain forming Lyn-RIG-I-MAVS trimolecular complex, which activates the signaling pathway by phosphorylation. Lyn itself is also activated by phosphorylation at T396 [23]. In addition, the Thr-170 phosphorylation, a phosphorylation closely located in proximity to Lys-172 which regulates the RLRs pathways by

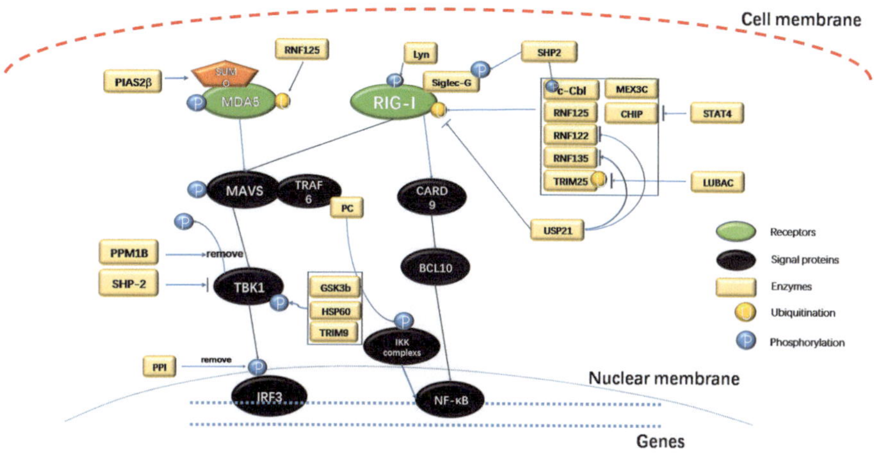

Fig. 2.2 Inflammatory signaling transduction and PTMs of RLR pathways

promoting the ubiquitination of RIG-I, antagonizes the Lys-172 ubiquitination and leads to the suppression of RLR activation [24]. Phosphorylation of Ser-8 of RIG-I is displayed in uninfected cells and decreases in virus-infected cells, serving as an inhibitory modification of RLRs to prevent cells from unwanted inflammatory response via RLRs [25].

In expect of RIG-I itself, the phosphorylation of other molecules in the RIG-I signaling pathway also effectively regulates the on-off of the pathway.

MAVS MAVS is located in the outer mitochondrial membrane. The interaction with it is indispensable for the phosphorylation and activation of IRF3 and TBK1, while both MAVS polymerization and phosphorylation are required for MAVS-IRF3 interaction. Phosphorylation of MAVS at S442 depending on its polymerization and TRAF-mediated polyubiquitin synthesis is critical for IRF3 activation but dispensable for TBK1 and IKK activation [3]. PC, a member of biotin-containing enzyme family, interacts with MAVS/TRAF6, which promotes phosphorylation of IKK (IκB kinase complex) and IκBα and induces the redistribution in nucleus of NF-κB [26].

TBK1 As a key kinase in both RNA and DNA sensing pathway, TBK1 is attenuated by diverse molecules with different mechanisms. Negative regulation of the activity of TBK1 is critical to avoid excessive generation of IFH-β and other inflammatory cytokines. PPM1B (protein phosphatase Mg^{2+}-/Mn^{2+}-dependent 1B) dephosphorylates TBK1 at S172 and decreases the IRF3 activation [27]. Moreover, SHP-2 interacts with TBK1 kinase domain to block its activity as a kinase [28]. In zebra fish, STAT6 also negatively regulates the phosphorylation of TBK1, which was not found in mammalian cells [29]. TRIM9s (TRIM9 short) interacts with TBK1 to facilitate phosphorylation of TBK1 via the RING domain of TRIM9s. In addition, TRIM9s, bridging GSK3β and TBK1, also initiates TBK1 oligomerization which is critical to its activation [30].

IRF3 MAVS also recruits TBK1 to phosphorylate IRF3 typically at 5T/S cluster (396–405) and S385–386. Then phosphorylated mono-IRF3s dissociate from MAVS to form dimers with a C-terminus-C-terminus interaction [22]and translocate to the nucleus. PPI, one of the most abundant phosphatases in eukaryotic, regulates the RIG-I signaling pathway by dephosphorylating IRF3 [31]. HSP60 (heat shock protein 60, also known as HSPD1) interacting with IRF3 by directly phosphorylating IRF3 that triggers its dimerization [32].

2.3.2.2 Ubiquitination

Modification of RIG-I 2CARD with K63-linked ubiquitination is important for signaling activity. The tetrameric architecture of 2CARD mentioned above binds to six Ub molecules in the asymmetric unit. Both the activation and degradation of RIG-I are attenuated by several host molecules.

RNF Family RNF125, a RING-type E3 ubiquitin ligase, polyubiquitylates RIG-I at K181 and leads to K48-linked proteasome-dependent degradation, which impedes the RIG-I signaling pathway as a negative feedback regulation [33]. Similarly, the TM domain of RNF122 directly binds to the 2CARD domain of RIG-I which mediates K48-linked ubiquitination at K115/K146 of 2CARD domain of RIG-I. Located on ER membrane, RNF122 can also mediate self-ubiquitin degradation [34]. Conversely, RNF135, also called REUL, positively regulates the RIG-I pathway via facilitating the K63-linked polyubiquitination of RIG-I [35]. In addition, RNf121 facilitates the proteasomal degradation of IκBα, whose degradation is critical for NF-κB to translocate into the nucleus and trigger IFN-β expression [36]. Moreover, RNF5 promotes the K48-linked polyubiquitination that degrades MAVS and STING on L150 [37].

TRIM Family RLR-MAVS-dependent signal pathways are under strict control of PTM including polyubiquitination and phosphorylation. The ubiquitin E3 ligase TRIM25 modifies RIG-I with K63-linked polyubiquitination [38]. TRIM25 (tripartite motif 25) composed of a RING domain at N-terminal, one or two B-box domains, a coiled-coil dimerization domain, and a C-terminal SPRY domain binding to 2CARD domain, catalyzes RIG-I at K172 via K63-linked polyubiquitination. Furthermore, the coiled-coil domain of TRIM25 is a stable, antiparallel dimer possessing two catalytic RING domains on opposite ends of an elongated rod. RING dimerization is required for catalysis, TRIM25-mediated RIG-I ubiquitination, interferon induction, and antiviral activity [39]. The interaction between TRIM25 and RIG-I delivering the Lys 63-linked ubiquitin moiety to the N-terminal CARDs of RIG-I results in an increase in RIG-I downstream signaling activity. The ubiquitination of RIG-I enables its oligomerization and thus induces RIG-I interacting with MAVS. Mutations of conserved residues of RIG-I that disrupt its ubiquitin binding also abrogate its ability to activate IRF3 demonstrating that CARD domain of RIG-I is the structural basis for the IRF-inducing ability of RLRs [40]. NS1 (non-structural protein 1), a virulence factor of *influenza A virus*, not only binds to

TRIM25 directly impeding the K63-linked ubiquitination of RIG-I but also prevents RIG-I from sensing the viral dsRNAs [41]. In addition, TRIM56 and TRIM32, both positive modulators in STING-dependent IRF3 activation, facilitate the K63-linked ubiquitination of STING required for recruitment and activation of TBK 1 [42, 43]. Conversely, TRIM23 interacts with NEMO and catalyzes the conjugation of K27-linked polyubiquitin chain onto NEMO, leading to its proteasome-dependent degradation [44]. TRIM21, bound to K63 E2 conjugating enzyme UBC13, facilitates TAK1-TAB1-TAB2, IKKα-IKKβ-NEMO, and IRF3/5/7 complex activation via K63-linked ubiquitination [45]. Additionally, TRIM4 is another E3 ligases identified as enzymes modifying RIG-I with K63-polyubiquitin chains [46].

Other Protein Families MEX3C interacts with RIG-I directly via its 382–599 domain, while its RING domain ubiquitylates RIG-I at K99/K169 causing the activation of RIG-I [47]. STAT4, in spite of a classical kinase, blocks the degradation of RIG-I via interacting with CHIP and disrupts the association between RIG-I and CHIP, which induces K48-linked ubiquitination and degradation of RIG-I [48]. LUBAC (linear ubiquitin chain assembly complex) natively regulates RIG-I signaling pathway in two ways. One is that LUBAC, directly bound to TRIM25 which facilitates the K63 ubiquitination of RIG-I (mentioned above), catalyzes the K48-linked polyubiquitination of TRIM25 resulting in the degradation of TRIM25 [49]. The other is that LUBAC catalyzes the linear ubiquitination of NEMO which plays a critical role in the activation of NF-κB signaling pathway, while the linear-ubiquitinated NEMO impedes the interaction between MAVS and TRAF3 [50]. C-Cbl, an E3 ligase, promotes the K48-linked polyubiquitination and proteasomal degradation of RIG-I [51], which is phosphorylated by Siglec-G. USP21 functions in few ways. It blocks the K63-linked ubiquitination of RIG-I, promotes the deubiquitination of RIG-I and MAVS, and also inhibits TRIM25- and RNF135-mediated RIG-I polyubiquitination and activation [52]. USP15 interacts with the CARD domain and C-terminal of RIG-I by its UCH domain while deubiquitinates the USP15 by its active sites, His862 [53]. P97, forming Np14-Ufd1-p97 complex, promotes the K48-linked ubiquitination of RIG-I by facilitating RNF125. Though Np14 binds to RIG-I at CARD of non-ubiquitination RIG-I to promote the K63-linked ubiquitination of RIG-I at K172/K170, then the activation of K63-linked progress triggers the interaction among Np14, Ufd1, and p97, which functions as a negative feedback regulation [54]. Smurf 2(Smad ubiquitin regulatory factor 2), another E3 ubiquitin ligase, physically interacts with MAVS and triggers the K48-linked ubiquitination and degradation of MAVS [55]. Similarly, PCBP1 or PCBP2 recruits AIP4, a HECT domain-containing E3 ligase, which induces the K48-linked polyubiquitylation and degradation of MAVS [56]. RAUL, a HECT domain E3 ubiquitin ligase, induces the K48-linked polyubiquitination of IRF3/7 that blocks the RIG-I signaling pathway [57].

2.3.2.3 Others

Besides phosphorylation and ubiquitination, there are other special PTMs required for the regulation of IRG-I signaling pathway. FAT10 negatively regulates the pathway by preventing RIG-I from ubiquitin-dependent activation and interaction with IRF3. Also excessive FAT10, bound to the CARD domain of RIG-I directly, sequesters the active form of RIG-I into insoluble precipitate [58]. What's more, several DUBs removing K63-linked polyubiquitin from RIG-I to balance pro-inflammatory responses are identified, including USP3 (ubiquitin-specific protease 3) [59], CYLD [60], and USP21 [52]. In addition, it is reported that ISG15-modified RIG-I displays lower activity to trigger the downstream response [61]. Similar to ubiquitylation, SUMOylation is a multistep reaction that covalently conjugates a 12-kDa small ubiquitin-like modifier (SUMO) to target proteins by a single E1-activating enzyme (Aos1/Uba2), a unique E2-conjugating enzyme (Ubc9), and an array of different E3 ligases [62]. The direct interaction between UBC9 and target proteins is required for SUMOylation to transfer SUMO from E1. Therefore, when UBC9 binds to RIG-I directly, activated SUMOylation of RIG-I alters the protein folding to assist RIG-I ubiquitylation and enhances the interaction with MAVS [63]. IRTKS (insulin receptor tyrosine kinase substrate) can also recruit Ubc9 to SUMOylate PCBP2 (a negative regulator of MAVS) by SUMO2 at K37 of PCBP2, which is required for cytoplasmic translocation of PCBP2 then leading to MAVS degradation [64]. Similarly, Ndfip1 promotes the degradation of MAVS by triggering the autoubiquitination of Smurf1, an E3 ubiquitin ligase with multiple HECT domains, which then interacts with and finally degrades MAVS [65]. When it comes to IRF family, α-LA (LA, lipoic acid, 1,2-dithiolano-3-pentanoic acid; $C_8H_{14}O_2S_2$) promotes SUMOylation of IRF-1 by SUMO1 which negatively regulates its transcription activity [66]. Both of IRF-3 and IRF-7 can be SUMOylated by SUMO 1/2/3 at K152 of IRF3 and K406 of IRF7, which induces a lower level of IFN mRNA [67]. In addition, amidation of RIG-I inhibits IFN induction, so the deamidation catalyzed by phosphoribosylformyglycinamide synthase (PFAS) of RIG-I at Q10, N245, and N445 residues converts the suppression. Upon viral RNA stimuli, the CARD domain of RIG-I interacts with the GAT domain of vGAT (glutamine aminotransferase); the latter was a pseudo enzyme recruiting PFAS to RIG-I which triggers the downstream pathway [68]. Moreover, as in uninfected cells, RIG-I is acetylated which restricts not only RIG-I translocation to intracellular membrane for interaction with MAVS but also the oligomerization of RIG-I, in infected cells, HDAC6 (histone deacetylase 6) deacetylates RIG-I to activate RIG-I depending IRF3 activation [69, 70].

2.4 The PTMs of NLR Pathways

NLR pathways refer to nucleotide-binding oligomerization domain containing protein 1-like receptor-related signaling pathways. NLR proteins are cytosolically located detecting PAMPs invading into the host cells. Up to now, more than 20 genes are identified in human genome [71], and certain roles of several NLR proteins including NOD1, NOD2, NLRP3, and NLRC4 are studied. Among different signaling transduction ways, PTM displays its roles in them.

2.4.1 The Overview of NLR

NLR proteins include NOD1/2, Nalps, Ipaf, and Naip. The essential N-terminal domains of NLR can be identified as characters of different proteins, NOD1/2 and Ipaf with CARD, Nalps with PYD (pyrin domain), and Naip with BIR (baculovirus inhibitor of apoptosis protein repeat domain) [72]. NOD1/2 are responsible for detecting microbe-derived peptidoglycan [73], and once activated, they can recruit a serine-threonine kinase called RIP2 (receptor-interacting protein 2) via CRAD-CARD interactions [74].Then, RIP2 directly interacts with the IKKγ (NEMO) and promotes its ubiquitylation; thus IκBα is degraded, and NF-κB is able to translocate into nuclear to promote pro-inflammatory chemokine production.

2.4.2 The PTMs Involved in NLR Pathways

Since RIP2 is the key protein linking NOD and NF-κB activation, RIP2 itself arises a lot of attentions for its capacity in regulating NLR pathways (Fig. 2.3). It has been proved that RIP2 undergoes autophosphorylation at Tyr 474 via tyrosine kinase activity in RIP2 itself, and this phosphorylation is necessary for effective NOD-related NLR signaling transduction [75]. Other phosphorylation sites identified include S168/176 in RIP2 kinase domain [76, 77], S363/393 in interdomain region of RIP2 [77], and S527/529/531/539 in the flexible C-terminal region of RIP2 [78, 79]. Additionally, RIP2 is also related to the ubiquitination of NEMO (IKKγ), important for sequent IκBα degradation and NF-κB activation [80]. Pellino3-mediated K63-linked polyubiquitination of RIP2 plays a positive role in NOD2-RIP2-NF-κB pathway [81]. Ubiquitination of RIP2 mediated by cIAPs (cellular inhibitors of apoptosis) is a promotive factor of NLR-related NF-κB and JNK pathways [82]. Similarly, XIAP mediating the ubiquitination of RIP2 also increases the activation of RIP2-mediated NF-κB [83].

While some PTMs are responsible for enabling and increasing NLR pathway activation, some others negatively regulate them. RIP2 is polyubiquitinated by ITCH, an E3 ubiquitin ligase, which inhibits the NF-κB activation and JNK

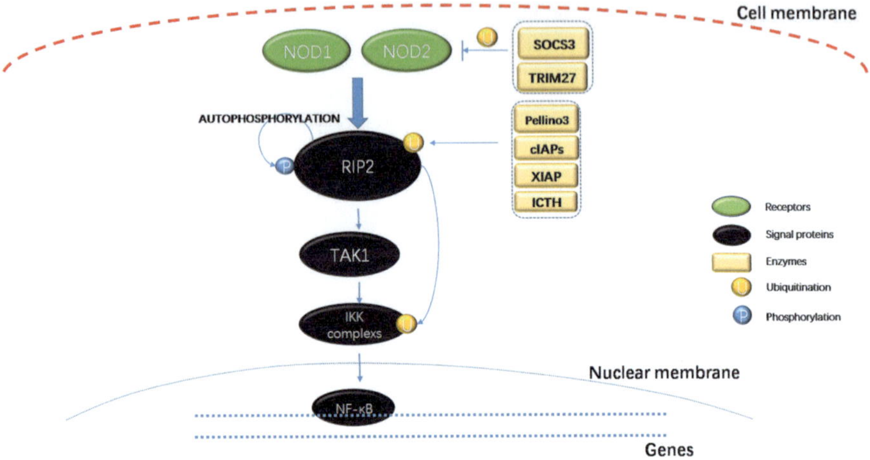

Fig. 2.3 Inflammatory signaling transduction and PTMs of NLR pathways

activation via phosphorylation of JNK and MAPK [84]. For NODs, SOCS3 and TRIM27 induce polyubiquitination and proteasomal degradation of NOD2 [85, 86] and thus attenuate the whole signal transduction mediated by NOD2.

2.5 Intracellular DNA Sensors

2.5.1 Overview of Cytosol DNA Sensor

Intracellular DNA sensors recognize the dsDNA and ssDNA of virus that invade into the cytoplasm, which then mediate a type I IFN response and other inflammatory cytokines such as TNF-α and IL-6 through recruitment and activation of their adaptor proteins, like cGAMP, STING, TBK1, and so on. One of critical DNA sensors that we have investigated most is cGAS (cyclic GMP-AMP synthase). In addition, many other DNA sensors in the cytoplasm have been identified, including IFI16 (interferon-inducible protein 16), AIM2 (absent in melanoma 2) inflammasomes, and so on.

2.5.2 PTM of cGAS Signaling Pathways

Triggered by viral dsDNA, the mammalian cGAS enzyme with a single active site converts GTP and ATP into a mixed dinucleotide species G[2′-5′]pA[3′-5′]p (cGAMP) containing a 2′-5′ phosphodiester bond [87]. cGAMP then binds to STING resulting in the translocation of the adaptor protein from the endoplasmic reticulum to the Golgi and perinuclear sites, where activated STING dimerizes and binds to TBK1 and IRF3 and the STING-TBK1-IRF3 complex that finally triggers the IFN and ISG gene expression [21]. Here we introduce the posttranslational modification (PTM) of the relevant molecules in this pathway (Fig. 2.4).

2.5.2.1 Phosphorylation

Phosphorylation is one of the most common modifications to regulate the target protein functions by inducing their conformational changes or recruiting of other proteins. Akt kinase, one of the most critical and versatile protein kinases in higher eukaryotes, negatively regulates cGAS by phosphorylating cGAS at S291 of mouse and S305 of human [21]. Phosphorylation is also critical for activation and translocation of STING. ULK1/ATG (UCN-51-like kinase) can phosphorylate STING at S366 that suppresses the IRF3 function, while ULK1/ATG itself is activated via dissociation from unphosphorylated AMPK which is typically elicited by cGAMP [88]. In addition, viral interferon regulatory factor (vIRF1) of Kaposi's sarcoma-associated herpesvirus (KSHV/HHV8) inhibits the phosphorylation and activation of STING by inhibiting its interaction with TBK1 [89].

2.5.2.2 Ubiquitination

Protein ubiquitination is inversely regulated by E1, E2, and E3 enzymes or DUB which regulates degradation and activation of a wide variety of molecules in different immunological processes. TRIM (tripartite motif) is one of the most dominant E3 ligases families. TRIM14 not only recruits USP14 (ubiquitin-specific protease

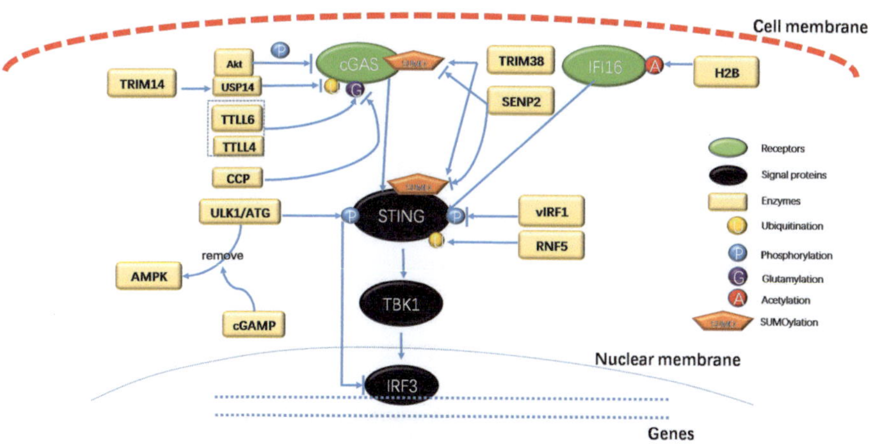

Fig. 2.4 Inflammatory signaling transduction and PTMs of intracellular DNA sensor pathways

14) that cleaves K48-linked ubiquitin chains of cGAS at K414 to prevent it from degradation but also blocks p62-mediated selective autophagic degradation of cGAS [43]. Moreover, STING is also a target for ubiquitination, of which K48-linked ubiquitination leading to degradation while K11-linked and K63-linked ubiquitination facilitating activation [43].

2.5.2.3 Others

Besides phosphorylation and ubiquitination, there are also other essential modifications that regulate the cGAS signaling pathway. Glutamylation of the cGAS suppresses its activation via two ways, in one is polyglutamylated at Glu272 by TTLL6 (tubulin tyrosine ligase-like glutamylases 6) and the other one is monoglutamylated at Glu302 by TTLL4. Conversely, CCP inverses this progress by CCP6 (cytosolic carboxypeptidases 6) via removing the polyglutamylation of cGAS whereas CCP5 hydrolyzing the monoglutamylation of cGAS [90]. What's more, sumoylation of cGAS at K217 and K464 and of STING at K337 by TRIM38 promotes their protein stability and prevents their ubiquitination and degradation, while in the late phase of viral infection, desumoylation of both cGAS and STING mediated by SENP2 avoids the sustained inflammatory response, and SENP2 is recruited by phosphorylated STING at S365 [91].

2.5.3 PTM of IFI16 Signaling Pathways

Interferon-inducible protein 16 (IFI16) is a critical DNA sensor binding to ssDNA, albeit more weakly than to dsDNA in the cytoplasm [92]. After triggered by viral DNA, IFI16 interacts with other adaptor proteins to form BRCAI-IFI16-ASC-procaspase-1 complex, which induces IL-1β and IFN-β production and STING activation. Acetylation of IFI16 complex by H2B, a histone acetyltransferase, enhances its interaction with STING that induces the redistribution of STING from cytoplasm to nucleus and the interaction with cGAS promoting cGAMP production and generation of IL-1β and IFN-β [93].

2.5.4 PTM of AIM2 Signaling Pathway

Located in the cytoplasm, absent in melanoma 2 (AIM2) inflammasome directly interacts with dsDNA driving activation of caspase 1, which promotes pyroptosis and the release of the pro-inflammatory cytokines. This progress requires the type I IFN-potentiated expression of GBPs (guanylate-binding proteins) via the transcriptional factor IRF1 [94]. AIM2 indirectly inhibits the phosphorylation of STAT1 (the signal transducers and activators of transcription 1) which impedes the phosphorylation of NF-κB p65 at S536 and acetylation of it at L310, the modification facilitating NF-κB translocation and activation, while deacetylation of it by SIRT1/HDAC3 blocks the activation pathway [95].

2.6 Intracellular RNA Sensors

RNA sensors sense atypical RNAs associated with virus infection, recognizing the 555p' RNA and 55p' RNA not protected by a 5′-cap. The typical RNA sensors RIG-1 and MDA5 share the similar functional domains including CARD domain of N-terminal that not only recruits downstream proteins like MAVS, IRF3, and TBK1 but also functions in polyubiquitination and C-terminal domain (CTD) binding to exogenous RNA, consistent with their similarity in immune response. PTM plays an important role in regulation of RNA sensor signaling pathway (Fig. 2.2).

2.6.1 PTM of RIG-I Signaling Pathways

As mentioned in Sect. **2.3**

2.6.2 PTM of MDA5 Signaling Pathways

MDA5 with 2CARD domains in high degree of heterogeneity can self-assemble into a defined oligomeric arrangement, while CARD domains form discrete patches of oligomers along MDA5 filament about 11 monomers per patch [96]. The phosphorylation of MDA5 at S88 in the first CARD and S828 in the CTD dampen its ability to interact with MAVS in uninfected cells [97, 98]. TRIM13, one member of the TRIM family, an E3 ligase family, inhibits the activation of MDA5 and the subsequent MDA5-mediated production of IFN-β but regulates RIG-I pathway in a positive way, whose role in the RNA-sensing signaling network needs to be further investigated [99]. In addition, RNF125 also facilitates the K48-linked polyubiquitination as the proteasome-dependent degradation of MDA5. What's more, PIAS2β, an E3 ligase of SUMOylation, SUMOylates at the C-terminal region of MDA5, which promotes the MDA5-mediated IFN induction, while the SUMOylation has no influence on the K48-linked ubiquitination of MDA5 [100].

2.7 Inflammasomes

Inflammasomes are protein complexes assemble on recognition of exogenous and endogenous danger signals and serve as platforms for activation of canonical caspase 1 or noncanonical caspase11 and secretion of pro-inflammatory cytokines: such as IL-1β, IL-18, and HMGB1 (high-mobility group box 1). Different types of inflammasomes vary in types of activation and downstream molecules. The most widely studied inflammasome up to now is the NLRP3 inflammasome, while NLRC4, AIM2, NLRP1, and other inflammasomes require more experimental investigations. As their effective influence on immune response, the activity of inflammasomes is tightly regulated to avoid from the generation of excessive inflammatory functions. Generally speaking, there are mainly two aspects of regulation in host cells. (A) The expression of inflammasome-relevant proteins is limited at a low level. For example, the level of NLRP3 in many cell types is particularly low with a priming signal to be triggered [101]. (B) The variants of inflammatory components alter in distinct, even controversial functions. Splice variants of ASC have been identified with different activities, with one variant even inhibiting the signaling pathway but not facilitating it as usual [102].

Adaptor protein ASC bridges the sensor protein (NLRP3 and AIM2) and caspase 1 to form ternary inflammasome complexes by its PYD and CARD domains. Then NLRP3 or NLRC4 inflammasomes recruit ASC through their PYD domains, while ASC recruits caspase 1 by the CARD/CARD interaction. Activated ASC trends to form filament, of which the PYD domain is nuclear for ASC^{PYD} forming filament, required for ASC^{PYD} removing His-MBP from the fusion protein [103].

2.7.1 *PTM of NRLP3 Signaling Pathway*

NLRP3 inflammasome senses a wide range of pathogens including bacteria, virus, fungi, and parasites through both DAMP and PAMP pathways. In addition, NLRP3 can be also activated by the reactive oxygen species (ROS) and lysosome destabilization [104]. In majority of cases, NLRP3 does not interact with pathogens directly. Excessive ROS are sensed by a complex of thioredoxin and thioredoxin-interacting protein (TXNIP), which leads to the dissociation of the complex. Then subsequent binding of TXNIP to NLRP3 triggers the recruitment of ASC and pro-caspase-1, together forming the active inflammasome complex. In addition, NLRP1, AIM2, NAIP, and NLRP4 that can bind to pathogens directly and proteases caused by lysosomal destruction facilitate the interaction between ASC, pro-caspase 1, and NLRP3 [102]. PTM plays an important role in regulating the whole progress (Fig. 2.5).

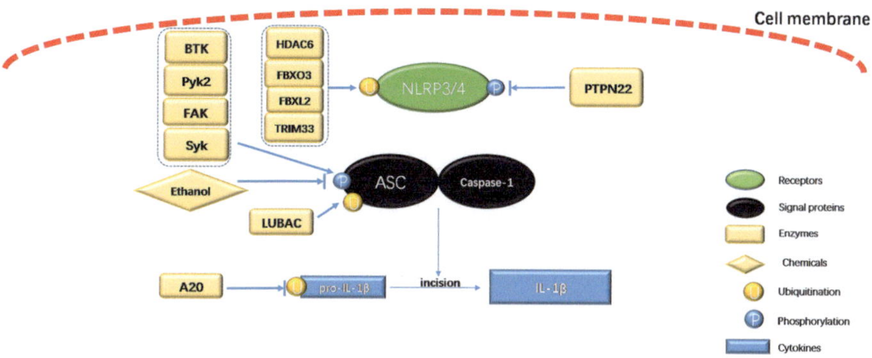

Fig. 2.5 Inflammatory signaling transduction and PTMs of inflammasome pathways

2.7.1.1 Phosphorylation

PKR (double-stranded RNA-dependent protein kinase) physically binds to NLRP3 as a PKR-NLRP3 complex and promotes the inflammasome formation. PKR also plays a critical role in activation of the NLRP1, AIM2, and NLRC4 [105]. Interacting with NLRP3 directly, PTPN22 interferes the tyrosine phosphorylation at Tyr86 of NLRP3, while phosphorylation of NLRP3 negatively regulates the activation of the complex. ASC is necessary for NLRP3 activation [106]. Ethanol inhibits NLRP3 inflammasome by impeding the phosphorylation of ASC at Y146 of human [107]. Syk (spleen tyrosine kinase) interacts with ASC and NLRP3 directly by its kinase domain but without procaspase-1, which phosphorylates ASC at Y146 and Y187 to promote the NLRP3-dependent caspase-1 activation [108]. As downstream proteins of Syk, Pyk2 (proline-rich tyrosine kinase) and FAK (focal adhesion kinase) also phosphorylate ASC at Y146 required for ASC oligomerization, caspase-1 activation, and IL-1β secretion. And only the depletion of FAK leads to the significant inhibition of AIM2-mediated IL-1β secretion [109]. In addition, BTK (Bruton's tyrosine kinase) phosphorylates ASC facilitating its oligomerization and distribution in macrophages required for NLRP3 activation, which also phosphorylates PLCγ to enhance Ca^{2+} influx as an assistant factor to NLRP3 inflammasome formation [110]. Conversely, AMPK (adenosine monophosphate-activated protein kinase) phosphorylation negatively regulates the activation of NLRP3 inflammasome [111]. Mechanism different from the kinases mentioned above, though Nek-7 interacts with NLRP3 by its NOD, LRRs and N-terminal (amino acid residues 34–212) directly, it does not function in phosphorylation but improves the K^+ efflux to activate NLRP3 inflammasome [112].

2.7.1.2 Ubiquitination

LUBAC plays a critical role in ASC foci formation and NLRP3-ASC inflammatory assembly required for linear ubiquitination of ASC and ASC-dependent NLRP3 inflammasome activation [113]. Two members of F-box protein as ubiquitin E3 ligases with a RING domain, FBXO3 (f-box O3) and FBXL2 (f-box L2), function in regulating the ubiquitination of NLRP3. FBXL2 mediates directly degradation of NLRP3 at K689 which impedes the activation of the inflammasome, while FBXO3 targeting on FBXL2 facilitates the ubiquitin-dependent degradation of FBXL2 promoted by the increased level of intracellular LPS [114]. TRIM33 directly interacts with NLRP3 at PYD domain of NLRP3 by its second C-terminal coiled-coil domain to facilitate the K48-linked proteasomal degradation of formation and activation via binding to the HA2-DUF region of DHX33 which induces the K63-linked ubiquitination at K218 [115]. DUBs (deubiquitinases) like OTUB1, UCH-L1, USP7, or CYLD have been shown to play a crucial role in ubiquitin-mediated cellular processes. It is also required for inflammasome-dependent IL-1β processing via promoting oligomerization of ASC to improve the activation of AIM2/NLRP3 inflammasomes; activation of the latter is more evident [116]. For example, UCH-L5 (ubiquitin C-terminal hydrolase 37, UCH37) knockdown leads to decrease in inflammasome-dependent IL-1β release [117]. In addition, upon LPS stimulation, A20, a ubiquitin-modifying enzyme, binds to pro-IL-1β to restrict the K63-linked ubiquitination of pro-IL-1β at K133 that promotes its proteolytic processing [118]. Interestingly, a deacetylase mentioned before, HDAC6 negatively regulates the NLRP3 pathway independent of deacetylation but by facilitating the ubiquitin-dependent degradation of NLRP3 with the Buz (binder of ubiquitin zinc finger) domain of HDAC6 [119].

2.7.1.3 Others

JAK/STAT1 is required for LPS- and IFN-β-induced HMGB1 acetylation catalyzed by the two NLS (nuclear localization sequence) domains of the acetylase, key to efficient HMGB1 release [120]. In addition, UBC9, a sole SUMO-conjugating enzyme, mediates the SUMO-1 SUMOylation to prevent IkBα from degradation, which inhibits the activation of NF-κB. Upon LPS stimuli, CDK1 phosphorylates UBC9 that promotes this progress [121].

2.7.2 PTM of Other Inflammasome Signaling Pathway

NLRC4 Different from NLRP3, NLRC4 is only activated by specific bacterial proteins but not binding directly [122]. NAIP proteins sense bacterial proteins like *S. typhimurium flagellin* causing the recruitment of NLRC4 and assembly of the NLRC4 inflammasome [104]. PKCζ and PAK2 phosphorylate NLRC4 at S553, but the depletion of PKCζ only reduces the phosphorylation of NLRC4 resulting in ASC focus formation, caspase-1 activation, and pyroptosis. Upon certain stimuli, PKCζ itself is activated via its phosphorylation at Y311 by PMA [123].

Pyrin Encoded by MEFV gene, pyrin recognizes the Rho GTPases-modified bacterial toxins like *C. difficile* cytotoxin TcdA/B, rather than microbial products like other inflammasomes. After phosphorylated at both S205 and S241, pi-pyrin recruits 14-3-3 protein which assists with the activation of pyrin upon specific Rho-modifying toxin stimuli, which triggers the dephosphorylation, ASC recruitment, and assembly of pyrin inflammasome [124].

2.8 PTM-Related Diseases and Immune Disorders

Since PTMs exert a wide range of influence on our immune system, any incapacity in regulations of posttranslational modifications can be enervating factors to immune homeostasis, which is responsible for a series of disease directly and indirectly. Up to now, several disorders in modulation of PTMs are identified and proved to be relative with certain diseases, such as Crohn's disease, a disease caused mainly by immune disorders, as well as exogenous factors that are able to effect the PTM patterns and lead to acquired disorders in immune systems, which cause the upregulation or repression in inflammation. A clear study of the factors, whatever endogenous or exogenous, is helpful for a precise remedy against the diseases derived from them.

Several notorious diseases are actually relative with the profound influence caused by PTM disorders. The Crohn's disease, marked by symptoms including severe stomachache, diarrhea, fever, and chronic anemia, is demonstrated related to the PTM disorders in the immune system. It is proved that NOD-2-involved inflammatory signal pathways are responsible for the onset of Crohn's disease. The mutation of NOD-2 including *L1007fsinsC* which removes the last 33 amino acids of its polypeptide is detected common in Crohn's patients [125, 126], and this specific mutation results in the defect in inducing RIP2 tyrosine autophosphorylation, which is discussed above in NLR pathways that are crucial for the activation of NOD-2 signaling pathways [75].With this certain deficiency, the immune systems of patients with Crohn's diseases are failed to eliminate the infective microbe colonized in our intestines leading to the overbudden of bacteria in gut cells. Also, the ITCH can recognize the phosphorylated RIP2 and thus deactivates it for a moderate NF-κB activation, which serves as an inhibitory role of NOD-2 pathway. As a result, the antagonists, gefitinib and erlotinib, against the autophosphorylation of RIP2 are able to repress the overreaction of NOD-2 pathways downregulating the inflammatory conditions [75].

Some pathogens are capable to induce enzymes responsible for PTMs that negatively regulate our immune response in order to escape the elimination executed by inflammatory conditions. RIG-1, serving both as sensors in RLR pathways and

RNA sensors, is essential for RNA virus defense. However, it is reported that some RNA virus are able to recruit lectin family number Siglec-G which induces and E3 ubiquitin ligase c-Cbl, and this enzyme is responsible for the degradation of RIG-I via K48-linked ubiquitination at Lys813. Sequentially, the degradation of RIG-I dampen the RNA sensor pathway activations as expected, and thus RNA virus makes their own escape [22]. Actually, this feedback-like relationship between virus and our immune systems represents a widespread rivalry.

Not all of PTMs that suppress the inflammatory signaling pathways are unwelcomed. In fact, under most circumstances, our immune system remains inactivated state to avoid unexpected inflammation, which is harmful to tissues. In RLR pathways, K48-linked ubiquitination displays dual roles in regulation. While the K48-linked ubiquitination modified by RNF135 is a positive regulation, those modified by RNF122/125 are responsible for degradation of RIG-I, MDA5, and MAVS, which abrogate the activation of RLR pathways [33, 34, 54]. Also, the K48-linked ubiquitination of TRIM25, a crucial modifier of RLR pathways, displays similar role of inducing degradation of TRIM25 [49]. For MAVS, Smurf1/2, RNF5, AIP4, and MARCH5 are responsible for its K48-linked ubiquitination. In NLR pathways, the ubiquitination modified by ITCH of the RIP2 is an inhibitory factor for activation, which serves as a negative feedback mechanism to limit the inflammation. Similarly, every negative modification mentioned in this article is set to protect cells from dysbiosis brought by improper or overrated inflammation.

To clinical aspect, PTMs that are involved in disease generating are new molecular sites for pharmaceutical design. Although the complete acknowledge of the PTMs is required for a more precise and safer remedy targeting specific disease, well-designed studies of such drugs are promising.

References

1. Akira S (2006) TLR signaling. Curr Top Microbiol Immunol 311:1–16
2. Jimenez-Dalmaroni MJ, Gerswhin ME, Adamopoulos IE (2016) The critical role of toll-like receptors–from microbial recognition to autoimmunity: a comprehensive review. Autoimmun Rev 15(1):1–8
3. Liu S et al (2015) Phosphorylation of innate immune adaptor proteins MAVS, STING, and TRIF induces IRF3 activation. Science 347(6227):aaa2630
4. Lei (2010) Glycogen synthase kinase 3b regulates IRF3 transcription factor-mediated antiviral response via activation of the kinase TBK1. Immunity 33:878–899
5. An H et al (2008) Phosphatase SHP-1 promotes TLR- and RIG-I-activated production of type I interferon by inhibiting the kinase IRAK1. Nat Immunol 9(5):542–550
6. Jiao S et al (2015) The kinase MST4 limits inflammatory responses through direct phosphorylation of the adaptor TRAF6. Nat Immunol 16(3):246–257
7. Medunjanin S et al (2016) GSK-3beta controls NF-kappaB activity via IKKgamma/NEMO. Sci Rep 6:38553
8. Jiang XACZ (2012) The role of ubiquitylation in immune defence and pathogen evasion. Nat Rev Immunol 12:35–48
9. Liu J, Qian C, Cao X (2016) Post-translational modification control of innate immunity. Immunity 45(1):15–30

10. Ye R et al (2017) Herpes simplex virus 1 ubiquitin-specific protease UL36 abrogates NF-kappaB activation in DNA sensing signal pathway. J Virol:91(5)
11. Tokunaga FSS (2009) Involvement of linear polyubiquitylation of NEMO in NF-kappaB activation. Nat Cell Biol 11:123–132
12. Mansell A et al (2006) Suppressor of cytokine signaling 1 negatively regulates toll-like receptor signaling by mediating mal degradation. Nat Immunol 7(2):148–155
13. Wang C et al (2009) The E3 ubiquitin ligase Nrdp1 'preferentially' promotes TLR-mediated production of type I interferon. Nat Immunol 10(7):744–752
14. Moon G et al (2015) Phosphoinositide-dependent kinase-1 inhibits TRAF6 ubiquitination by interrupting the formation of TAK1-TAB2 complex in TLR4 signaling. Cell Signal 27(12):2524–2533
15. Boone DL et al (2004) The ubiquitin-modifying enzyme A20 is required for termination of toll-like receptor responses. Nat Immunol 5(10):1052–1060
16. Kawagoe T et al (2009) TANK is a negative regulator of toll-like receptor signaling and is critical for the prevention of autoimmune nephritis. Nat Immunol 10(9):965–972
17. Wang Y et al (2006) Association of beta-arrestin and TRAF6 negatively regulates toll-like receptor-interleukin 1 receptor signaling. Nat Immunol 7(2):139–147
18. Yuk JM et al (2011) The orphan nuclear receptor SHP acts as a negative regulator in inflammatory signaling triggered by toll-like receptors. Nat Immunol 12(8):742–751
19. Schneider M et al (2012) The innate immune sensor NLRC3 attenuates toll-like receptor signaling via modification of the signaling adaptor TRAF6 and transcription factor NF-kappaB. Nat Immunol 13(9):823–831
20. Liu X et al (2013) Negative regulation of TLR inflammatory signaling by the SUMO-deconjugating enzyme SENP6. PLoS Pathog 9(6):e1003480
21. Chiang C, Gack MU (2017) Post-translational control of intracellular pathogen sensing pathways. Trends Immunol 38(1):39–52
22. Chen W et al (2013) Induction of Siglec-G by RNA viruses inhibits the innate immune response by promoting RIG-I degradation. Cell 152(3):467–478
23. Lim YJ et al (2015) A Src-family-tyrosine kinase, Lyn, is required for efficient IFN-beta expression in pattern recognition receptor, RIG-I, signal pathway by interacting with IPS-1. Cytokine 72(1):63–70
24. Gack MU et al (2010) Phosphorylation-mediated negative regulation of RIG-I antiviral activity. J Virol 84(7):3220–3229
25. Nistal-Villan E et al (2010) Negative role of RIG-I serine 8 phosphorylation in the regulation of interferon-beta production. J Biol Chem 285(26):20252–20261
26. Cao Z et al (2016) Pyruvate carboxylase activates the RIG-I-like receptor-mediated antiviral immune response by targeting the MAVS signalosome. Sci Rep 6:22002
27. Zhao Y et al (2012) PPM1B negatively regulates antiviral response via dephosphorylating TBK1. Cell Signal 24(11):2197–2204
28. An H et al (2006) SHP-2 phosphatase negatively regulates the TRIF adaptor protein-dependent type I interferon and proinflammatory cytokine production. Immunity 25(6):919–928
29. Li S et al (2017) Zebrafish STAT6 negatively regulates IFNphi1 production by attenuating the kinase activity of TANK-binding kinase 1. Dev Comp Immunol 67:189–201
30. Qin Y et al (2016) TRIM9 short isoform preferentially promotes DNA and RNA virus-induced production of type I interferon by recruiting GSK3beta to TBK1. Cell Res 26(5):613–628
31. Gu M et al (2014) Protein phosphatase PP1 negatively regulates the toll-like receptor- and RIG-I-like receptor-triggered production of type I interferon by inhibiting IRF3 phosphorylation at serines 396 and 385 in macrophage. Cell Signal 26(12):2930–2939
32. Lin L et al (2014) HSPD1 interacts with IRF3 to facilitate interferon-beta induction. PLoS One 9(12):e114874
33. Arimoto K et al (2007) Negative regulation of the RIG-I signaling by the ubiquitin ligase RNF125. Proc Natl Acad Sci U S A 104(18):7500–7505

34. Wang W et al (2016) RNF122 suppresses antiviral type I interferon production by targeting RIG-I CARDs to mediate RIG-I degradation. Proc Natl Acad Sci U S A 113(34):9581–9586
35. Oshiumi H et al (2009) Riplet/RNF135, a RING finger protein, ubiquitinates RIG-I to promote interferon-beta induction during the early phase of viral infection. J Biol Chem 284(2):807–817
36. Zemirli N et al (2014) The E3 ubiquitin ligase RNF121 is a positive regulator of NF-kappaB activation. Cell Commun Signal 12:72
37. Zhong B et al (2010) The E3 ubiquitin ligase RNF5 targets virus-induced signaling adaptor for ubiquitination and degradation. J Immunol 184(11):6249–6255
38. Gack MU et al (2007) TRIM25 RING-finger E3 ubiquitin ligase is essential for RIG-I-mediated antiviral activity. Nature 446(7138):916–920
39. Sanchez JG et al (2016) Mechanism of TRIM25 catalytic activation in the antiviral RIG-I pathway. Cell Rep 16(5):1315–1325
40. Jiang X et al (2012) Ubiquitin-induced oligomerization of the RNA sensors RIG-I and MDA5 activates antiviral innate immune response. Immunity 36(6):959–973
41. Gack MU et al (2009) Influenza A virus NS1 targets the ubiquitin ligase TRIM25 to evade recognition by the host viral RNA sensor RIG-I. Cell Host Microbe 5(5):439–449
42. Tsuchida T et al (2010) The ubiquitin ligase TRIM56 regulates innate immune responses to intracellular double-stranded DNA. Immunity 33(5):765–776
43. Zhang J et al (2012) TRIM32 protein modulates type I interferon induction and cellular antiviral response by targeting MITA/STING protein for K63-linked ubiquitination. J Biol Chem 287(34):28646–28655
44. Ashida H et al (2010) A bacterial E3 ubiquitin ligase IpaH9.8 targets NEMO/IKKgamma to dampen the host NF-kappaB-mediated inflammatory response. Nat Cell Biol 12(1):66–73. sup pp 1-9
45. McEwan WA et al (2013) Intracellular antibody-bound pathogens stimulate immune signaling via the Fc receptor TRIM21. Nat Immunol 14(4):327–336
46. Yan J et al (2014) TRIM4 modulates type I interferon induction and cellular antiviral response by targeting RIG-I for K63-linked ubiquitination. J Mol Cell Biol 6(2):154–163
47. Kuniyoshi K et al (2014) Pivotal role of RNA-binding E3 ubiquitin ligase MEX3C in RIG-I-mediated antiviral innate immunity. Proc Natl Acad Sci U S A 111(15):5646–5651
48. Zhao K et al (2016) Cytoplasmic STAT4 promotes antiviral type I IFN production by blocking CHIP-mediated degradation of RIG-I. J Immunol 196(3):1209–1217
49. Inn KS et al (2011) Linear ubiquitin assembly complex negatively regulates RIG-I- and TRIM25-mediated type I interferon induction. Mol Cell 41(3):354–365
50. Belgnaoui SM et al (2012) Linear ubiquitination of NEMO negatively regulates the interferon antiviral response through disruption of the MAVS-TRAF3 complex. Cell Host Microbe 12(2):211–222
51. Zhao X et al (2016) c-Cbl-mediated ubiquitination of IRF3 negatively regulates IFN-beta production and cellular antiviral response. Cell Signal 28(11):1683–1693
52. Fan Y et al (2014) USP21 negatively regulates antiviral response by acting as a RIG-I deubiquitinase. J Exp Med 211(2):313–328
53. Zhang H (2015) Ubiquitin-specific protease 15 negatively regulates virus induced type I interferon signaling via catalytically-dependent and -independent mechanisms. Sci Rep. doi:10.1038/srep11220
54. Hao Q et al (2015) A non-canonical role of the p97 complex in RIG-I antiviral signaling. EMBO J 34(23):2903–2920
55. Pan Y et al (2014) Smurf2 negatively modulates RIG-I-dependent antiviral response by targeting VISA/MAVS for ubiquitination and degradation. J Immunol 192(10):4758–4764
56. Zhou X et al (2012) Poly(C)-binding protein 1 (PCBP1) mediates housekeeping degradation of mitochondrial antiviral signaling (MAVS). Cell Res 22(4):717–727
57. Yu Y, Hayward GS (2010) The ubiquitin E3 ligase RAUL negatively regulates type i interferon through ubiquitination of the transcription factors IRF7 and IRF3. Immunity 33(6):863–877

58. Nguyen NT et al (2016) Ubiquitin-like modifier FAT10 attenuates RIG-I mediated antiviral signaling by segregating activated RIG-I from its signaling platform. Sci Rep 6:23377
59. Cui J et al (2014) USP3 inhibits type I interferon signaling by deubiquitinating RIG-I-like receptors. Cell Res 24(4):400–416
60. Friedman CS et al (2008) The tumour suppressor CYLD is a negative regulator of RIG-I-mediated antiviral response. EMBO Rep 9(9):930–936
61. Kim MJ et al (2008) Negative feedback regulation of RIG-I-mediated antiviral signaling by interferon-induced ISG15 conjugation. J Virol 82(3):1474–1483
62. Hay RT (2005) SUMO: a history of modification. Mol Cell 18(1):1–12
63. Mi Z et al (2010) SUMOylation of RIG-I positively regulates the type I interferon signaling. Protein Cell 1(3):275–283
64. Xia P et al (2015) IRTKS negatively regulates antiviral immunity through PCBP2 sumoylation-mediated MAVS degradation. Nat Commun 6:8132
65. Wang Y, Tong X, Ye X (2012) Ndfip1 negatively regulates RIG-I-dependent immune signaling by enhancing E3 ligase Smurf1-mediated MAVS degradation. J Immunol 189(11):5304–5313
66. Sun T et al (2014) alpha-Lipoic acid (alpha-LA) inhibits the transcriptional activity of interferon regulatory factor 1 (IRF-1) via SUMOylation. Toxicol In Vitro 28(7):1242–1248
67. Kubota T et al (2008) Virus infection triggers SUMOylation of IRF3 and IRF7, leading to the negative regulation of type I interferon gene expression. J Biol Chem 283(37):25660–25670
68. He S et al (2015) Viral pseudo-enzymes activate RIG-I via deamidation to evade cytokine production. Mol Cell 58(1):134–146
69. Choi SJ et al (2016) HDAC6 regulates cellular viral RNA sensing by deacetylation of RIG-I. EMBO J 35(4):429–442
70. Liu HM et al (2016) Regulation of retinoic acid inducible gene-I (RIG-I) activation by the histone deacetylase 6. EBioMedicine 9:195–206
71. Harton JA et al (2002) Cutting edge: CATERPILLER: a large family of mammalian genes containing CARD, pyrin, nucleotide-binding, and leucine-rich repeat domains. J Immunol 169(8):4088–4093
72. Carneiro LA et al (2008) Nod-like proteins in inflammation and disease. J Pathol 214(2):136–148
73. Chamaillard M et al (2003) An essential role for NOD1 in host recognition of bacterial peptidoglycan containing diaminopimelic acid. Nat Immunol 4(7):702–707
74. Kobayashi KS et al (2005) Nod2-dependent regulation of innate and adaptive immunity in the intestinal tract. Science 307(5710):731–734
75. Tigno-Aranjuez JT, Asara JM, Abbott DW (2010) Inhibition of RIP2's tyrosine kinase activity limits NOD2-driven cytokine responses. Genes Dev 24(23):2666–2677
76. Dorsch M et al (2006) Identification of a regulatory autophosphorylation site in the serine-threonine kinase RIP2. Cell Signal 18(12):2223–2229
77. Oppermann FS et al (2009) Large-scale proteomics analysis of the human kinome. Mol Cell Proteomics 8(7):1751–1764
78. Dephoure N et al (2008) A quantitative atlas of mitotic phosphorylation. Proc Natl Acad Sci U S A 105(31):10762–10767
79. Olsen JV et al (2010) Quantitative phosphoproteomics reveals widespread full phosphorylation site occupancy during mitosis. Sci Signal 3(104):ra3
80. Abbott DW et al (2004) The Crohn's disease protein, NOD2, requires RIP2 in order to induce ubiquitinylation of a novel site on NEMO. Curr Biol 14(24):2217–2227
81. Yang S et al (2013) Pellino3 ubiquitinates RIP2 and mediates Nod2-induced signaling and protective effects in colitis. Nat Immunol 14(9):927–936
82. Bertrand MJ et al (2009) Cellular inhibitors of apoptosis cIAP1 and cIAP2 are required for innate immunity signaling by the pattern recognition receptors NOD1 and NOD2. Immunity 30(6):789–801
83. Damgaard RB et al (2012) The ubiquitin ligase XIAP recruits LUBAC for NOD2 signaling in inflammation and innate immunity. Mol Cell 46(6):746–758

84. Tao M et al (2009) ITCH K63-ubiquitinates the NOD2 binding protein, RIP2, to influence inflammatory signaling pathways. Curr Biol 19(15):1255–1263
85. Lee KH et al (2012) Proteasomal degradation of Nod2 protein mediates tolerance to bacterial cell wall components. J Biol Chem 287(47):39800–39811
86. Zurek B et al (2012) TRIM27 negatively regulates NOD2 by ubiquitination and proteasomal degradation. PLoS One 7(7):e41255
87. Kranzusch PJ et al (2014) Structure-guided reprogramming of human cGAS dinucleotide linkage specificity. Cell 158(5):1011–1021
88. Konno H, Konno K, Barber GN (2013) Cyclic dinucleotides trigger ULK1 (ATG1) phosphorylation of STING to prevent sustained innate immune signaling. Cell 155(3):688–698
89. Ma Z et al (2015) Modulation of the cGAS-STING DNA sensing pathway by gammaherpesviruses. Proc Natl Acad Sci U S A 112(31):E4306–E4315
90. Xia P et al (2016) Glutamylation of the DNA sensor cGAS regulates its binding and synthase activity in antiviral immunity. Nat Immunol 17(4):369–378
91. Hu MM et al (2016) Sumoylation promotes the stability of the DNA sensor cGAS and the adaptor STING to regulate the kinetics of response to DNA virus. Immunity 45(3):555–569
92. Monroe KM et al (2014) IFI16 DNA sensor is required for death of lymphoid CD4 T cells abortively infected with HIV. Science 343(6169):428–432
93. Iqbal J et al (2016) Histone H2B-IFI16 recognition of nuclear Herpesviral genome induces cytoplasmic interferon-beta responses. PLoS Pathog 12(10):e1005967
94. Meunier E et al (2015) Guanylate-binding proteins promote activation of the AIM2 inflammasome during infection with Francisella novicida. Nat Immunol 16(5):476–484
95. Furrer A, Hottiger MO, Valaperti A (2016) Absent in Melanoma 2 (AIM2) limits pro-inflammatory cytokine transcription in cardiomyocytes by inhibiting STAT1 phosphorylation. Mol Immunol 74:47–58
96. Wu B et al (2013) Structural basis for dsRNA recognition, filament formation, and antiviral signal activation by MDA5. Cell 152(1–2):276–289
97. Wies E et al (2013) Dephosphorylation of the RNA sensors RIG-I and MDA5 by the phosphatase PP1 is essential for innate immune signaling. Immunity 38(3):437–449
98. Takashima K et al (2015) RIOK3-mediated phosphorylation of MDA5 interferes with its assembly and attenuates the innate immune response. Cell Rep 11(2):192–200
99. Narayan K et al (2014) TRIM13 is a negative regulator of MDA5-mediated type I interferon production. J Virol 88(18):10748–10757
100. Fu J et al (2011) MDA5 is SUMOylated by PIAS2beta in the upregulation of type I interferon signaling. Mol Immunol 48(4):415–422
101. Hornung V, Latz E (2010) Critical functions of priming and lysosomal damage for NLRP3 activation. Eur J Immunol 40(3):620–623
102. Bryan NB et al (2010) Differential splicing of the apoptosis-associated speck like protein containing a caspase recruitment domain (ASC) regulates inflammasomes. J Inflamm (Lond) 7:23
103. Lu A et al (2014) Unified polymerization mechanism for the assembly of ASC-dependent inflammasomes. Cell 156(6):1193–1206
104. Strowig T et al (2012) Inflammasomes in health and disease. Nature 481(7381):278–286
105. Lu B et al (2012) Novel role of PKR in inflammasome activation and HMGB1 release. Nature 488(7413):670–674
106. Spalinger MR et al (2016) NLRP3 tyrosine phosphorylation is controlled by protein tyrosine phosphatase PTPN22. J Clin Invest 126(5):1783–1800
107. Hoyt LR et al (2016) Ethanol and other short-chain alcohols inhibit NLRP3 Inflammasome activation through protein tyrosine phosphatase stimulation. J Immunol 197(4):1322–1334
108. Lin YC et al (2015) Syk is involved in NLRP3 inflammasome-mediated caspase-1 activation through adaptor ASC phosphorylation and enhanced oligomerization. J Leukoc Biol 97(5):825–835

109. Chung IC et al (2016) Pyk2 activates the NLRP3 inflammasome by directly phosphorylating ASC and contributes to inflammasome-dependent peritonitis. Sci Rep 6:36214
110. Ito M et al (2015) Bruton's tyrosine kinase is essential for NLRP3 inflammasome activation and contributes to ischaemic brain injury. Nat Commun 6:7360
111. Bullon P et al (2016) AMPK phosphorylation modulates pain by activation of NLRP3 Inflammasome. Antioxid Redox Signal 24(3):157–170
112. He Y et al (2016) NEK7 is an essential mediator of NLRP3 activation downstream of potassium efflux. Nature 530(7590):354–357
113. Rodgers MA et al (2014) The linear ubiquitin assembly complex (LUBAC) is essential for NLRP3 inflammasome activation. J Exp Med 211(7):1333–1347
114. Han S et al (2015) Lipopolysaccharide primes the NALP3 Inflammasome by inhibiting its Ubiquitination and degradation mediated by the SCFFBXL2 E3 ligase. J Biol Chem 290(29):18124–18133
115. Weng L et al (2014) The E3 ubiquitin ligase tripartite motif 33 is essential for cytosolic RNA-induced NLRP3 inflammasome activation. J Immunol 193(7):3676–3682
116. Lopez-Castejon G et al (2013) Deubiquitinases regulate the activity of caspase-1 and interleukin-1beta secretion via assembly of the inflammasome. J Biol Chem 288(4):2721–2733
117. Kummari E et al (2015) Activity-based proteomic profiling of Deubiquitinating enzymes in salmonella-infected macrophages leads to identification of putative function of UCH-L5 in Inflammasome regulation. PLoS One 10(8):e0135531
118. Duong BH et al (2015) A20 restricts ubiquitination of pro-interleukin-1beta protein complexes and suppresses NLRP3 inflammasome activity. Immunity 42(1):55–67
119. Hwang I et al (2015) Histone deacetylase 6 negatively regulates NLRP3 inflammasome activation. Biochem Biophys Res Commun 467(4):973–978
120. Lu B et al (2014) JAK/STAT1 signaling promotes HMGB1 hyperacetylation and nuclear translocation. Proc Natl Acad Sci U S A 111(8):3068–3073
121. Tomasi ML, Ramani K, Ryoo M (2016) Ubiquitin-conjugating enzyme 9 phosphorylation as a novel mechanism for potentiation of the inflammatory response. Am J Pathol 186(9):2326–2336
122. Roberts TL et al (2009) HIN-200 proteins regulate caspase activation in response to foreign cytoplasmic DNA. Science 323(5917):1057–1060
123. Qu Y et al (2012) Phosphorylation of NLRC4 is critical for inflammasome activation. Nature 490(7421):539–542
124. Inohara et al (2005) NOD-LRR proteins: role in host-microbial interactions and inflammatory disease. Annu Rev Biochem 74:355–383
125. Hugot JP et al (2001) Association of NOD2 leucine-rich repeat variants with susceptibility to Crohn's disease. Nature 411(6837):599–603
126. Gao W et al (2016) Site-specific phosphorylation and microtubule dynamics control Pyrin inflammasome activation. Proc Natl Acad Sci U S A 113(33):E4857–E4866

Chapter 3
Emerging Roles for Epigenetic Programming in the Control of Inflammatory Signaling Integration in Heath and Disease

Yiqun Hu, Feng Yan, Le Ying, and Dakang Xu

Abstract Macrophages and dendritic cells initiate the innate immune response to infection and injury and contribute to inflammatory signaling to maintain the homeostasis of various tissues, which includes resident macrophages for the elimination of invading microorganisms and tissue damage. Inappropriate inflammatory signaling can lead to persistent inflammation and further develop into autoimmune and inflammation-associated diseases. Inflammatory signaling pathways have been well characterized, but how these signaling pathways are converted into sustained and diverse patterns of expression of cytokines, chemokines, and other genes in response to environmental challenges is unclear. Emerging evidence suggests the important role of epigenetic mechanisms in finely tuning the outcome of the host innate immune response. An understanding of epigenetic regulation of innate immune cell identity and function will enable the identification of the mechanism

Y. Hu
Faculty of Medical Laboratory Science, Ruijin Hospital, Shanghai Jiao Tong University School of Medicine, Shanghai, China

F. Yan
Faculty of Medical Laboratory Science, Ruijin Hospital, Shanghai Jiao Tong University School of Medicine, Shanghai, China

Institute of Ageing Research, Hangzhou Normal University School of Medicine, Hangzhou, China

L. Ying
Faculty of Medical Laboratory Science, Ruijin Hospital, Shanghai Jiao Tong University School of Medicine, Shanghai, China

Department of Tea Science, Zhejiang University, Hangzhou, China

D. Xu (✉)
Faculty of Medical Laboratory Science, Ruijin Hospital, Shanghai Jiao Tong University School of Medicine, Shanghai, China
e-mail: dakang.xu@monash.edu

D. Xu (ed.), *Regulation of Inflammatory Signaling in Health and Disease*, Advances in Experimental Medicine and Biology 1024,
DOI 10.1007/978-981-10-5987-2_3

between gene-specific host defenses and inflammatory disease and will also allow for exploration of the program of innate immune memory in health and disease. This information could be used to develop therapeutic agents to enhance the host response, preventing chronic inflammation through preserving tissues and signaling integrity.

Keywords Epigenetic • Inflammatory signaling • Innate immune memory • Inflammatory diseases

3.1 Introduction

The immune system is composed of an innate (non-specific) system with the same protective response regardless of the initiating infection and an adaptive (specific) response with specificity that can generate immunological memory [1]. Innate immunity, which is constitutively present, is the first line of defense and is immediately mobilized following infection within a few minutes to hours in contrast to the adaptive immune system, which is slower and requires a few days to weeks to mount a response [2]. The primary component of innate immunity is inflammation [3]. Inflammation is part of the complex biological response of body tissues to harmful stimuli from pathogens and damaged cells that provide a protective process. The function of inflammation is to eliminate the initial cause of cell injury, clear out necrotic cells and tissues damaged by the original insult during the inflammatory process, and initiate tissue repair [4]. Injured cells release pro- and anti-inflammatory cytokines and other factors to limit the spread of infection and promote healing. At the same time, the innate response induces phagocytes and neutrophils that subsequently attract leukocytes and lymphocytes through these proinflammatory mediators [5]. Activation of the accompanying cascade enhances the innate response, which is an important consequence of complement cascade activation through inflammatory signaling, including Toll-like receptors (TLRs), RIG-I-like receptors (RLRs), and NOD-like receptors (NLRs). Through pattern recognition receptors (PRRs), these pathogen-associated molecular patterns (PAMPs) play a critical role in innate immune responses, and PAMP signaling prompts the induction of cytokines and other subsets of genes, the ingestion of complement-tagged pathogens, and the destruction of microorganisms by phagocytes [6]. The innate response involves various cell types, including neutrophils, macrophages, and dendritic cells. TLRs recognize PAMPs in the extracellular milieu and endosomes, while NLRs patrol the cytoplasm. A set of Nod-like receptors that include NLRP1b, NLRP3, and NLRC4 assemble multi-protein complexes termed inflammasomes. Inflammasome assembly is critical for the activation of caspase-1, which ultimately cleaves pro-IL-1β and pro-IL-18 into their mature bioactive forms, IL-1β and IL-18. Current studies have demonstrated inflammasome involvement during infection, autoimmune disease, and injury [7–15]. A comprehensive review of the inflammasome can be found in this book by Guangxun Meng. Another important function of the innate immune system is to stimulate the adaptive immune response via cross-presentation [16]. TLRs are one of the largest and best studied families of PRRs and are activated following recognition of specific, conserved PAMPs present in microbial proteins, bacteria, and nucleic acid variants that are normally associated with viruses, lipids, or carbohydrates. Recognition of these molecules by TLRs triggers signal transduction cascades that activate the transcription factors nuclear factor (NF-κB) and interferon-regulatory factors (IRFs), which induce the expression of proinflammatory cytokines and interferons or interferon-stimulated gene (ISG) and dictate the outcome of innate immune responses [17, 18]. Activation of these pathways is controlled by several molecular events, the basis of which depends on the interplay of a variety of elements, including transcription factors, epigenetic regulation, and post-transcriptional control mechanisms [18]. A growing number of results have documented the control of signaling pathways by transcription factors and epigenetic factors/chromatin-based mechanisms, which have provided critical evidence for context-specific gene expression in diverse innate immune cell types and chronic inflammatory autoimmune disease [19–22]. Epigenetic regulation comprises the posttranslational modification (PTM) of histones (histone methylation and histone acetylation and deacetylation), DNA methylation, and noncoding RNA. Epigenetic regulation is not only coupled with transcription factor-mediated regulation but is also linked with upstream signaling pathways that connect external signals and gene function to shape the identity and function of innate immune cells. In light of the recent data from high-throughput epigenomic techniques and human macrophage biology [23, 24], we are able to build an understanding of these regulatory mechanisms from detailed information to connect how innate immune cells develop functions in association with epigenetic changes.

In this chapter, we will discuss the role of epigenetic regulation in innate immune cell functional responses to insulting stimuli, the inflammatory process, and disease. In particular, we focus on the prominent role of DNA demethylation, histone modifications, promoter–enhancer interactions, and noncoding RNA-mediated regulation. Finally, we will describe newly emerging data to emphasize the important role of epigenetic mechanisms in the control of memory-type responses in innate immune cells. We will also summarize the future of innate immunity in epigenetics and medical translation.

3.2 Epigenetic Control of Gene Expression

Epigenetic changes in gene activity or expression occur without alteration of the underlying DNA sequence, which results in a change in phenotype without a change in genotype [25]. Genomic DNA is packaged by histones to form protein/DNA complexes termed chromatin in eukaryotic cells; chromatin alterations include DNA methylation, histone modifications, promoter–enhancer interactions, and non-coding RNA-mediated regulation. Epigenetic modulations are determined by the influence of chromatin organization/compaction and function and the critical regulation of gene expression mechanisms at the molecular level through cellular, tissue, and organ levels [26, 27]. The coordinated actions of these multifaceted epigenetic modulations determine cell fate, cell cycle regulation, and development and ultimately control responses in health and disease [28, 29].

In the genome, negatively charged, linear DNA is highly compacted and organized into three-dimensional (3D) chromosomes. DNA is coiled around histones (the main proteins of chromosomes) to form nucleosomes, the basic structural units of chromosomes. Histones are positively charged in the N-terminus with abundant lysine and arginine residues and can thus bind tightly to DNA to constrain its accessibility [30]. Each nucleosome contains a histone octamer that consists of two of each histone monomer (H2A, H2B, H3, and H4) [31]. These highly regulated organization mechanisms not only compact linear DNA into chromosomes but also allow the selective accessibility of transcription machinery [e.g., transcription factors (TFs) and cofactors] to specific genomic elements, including enhancers (facilitate transcription), promoters (initiate transcription), gene body/open reading frames (transcribe and translate into proteins), silencers (suppress transcription), and dielectric (block promoter–enhancer interactions). Epigenetic mechanisms chemically or structurally alter chromatin through modifications of DNA and histones or through chromatin remodeling and inter-/intrachromosomal DNA–DNA interaction, respectively [32] (Fig. 3.1). In DNA methylation, a methyl (CH3) group is enzymatically added onto the cytosine rings of DNA with temporal and spatial precision. In the human genome, DNA methylation occurs in most CpG dinucleotides [33], mainly to form 5-methylcytosine (5-mC). Methylation was once thought to be a stable heritable genetic trait, but recent studies have indicated that methyl groups can be dynamically added or erased [34]. DNA methylation occurs in promoter regions, such as transcription start sites (TSSs), and usually suppresses the expression of downstream genes through the recruitment of DNA-binding proteins and histone modifiers that repress transcription [35, 36]. Methylation of other genomic regions, such as enhancers and dielectric, has also been identified, but its functional importance remains to be investigated.

DNA methylation is a dynamic process in which methylation patterns with methylation marks in genomic regions can be synthesized de novo, maintained, or removed. DNA methylation is processed by an intricate balance between DNA methyltransferases (DNMTs) and DNA demethylases. The genome encodes three DNMTs (DNMT1, DNMT3A, and DNMT3B) to catalyze DNA methylation.

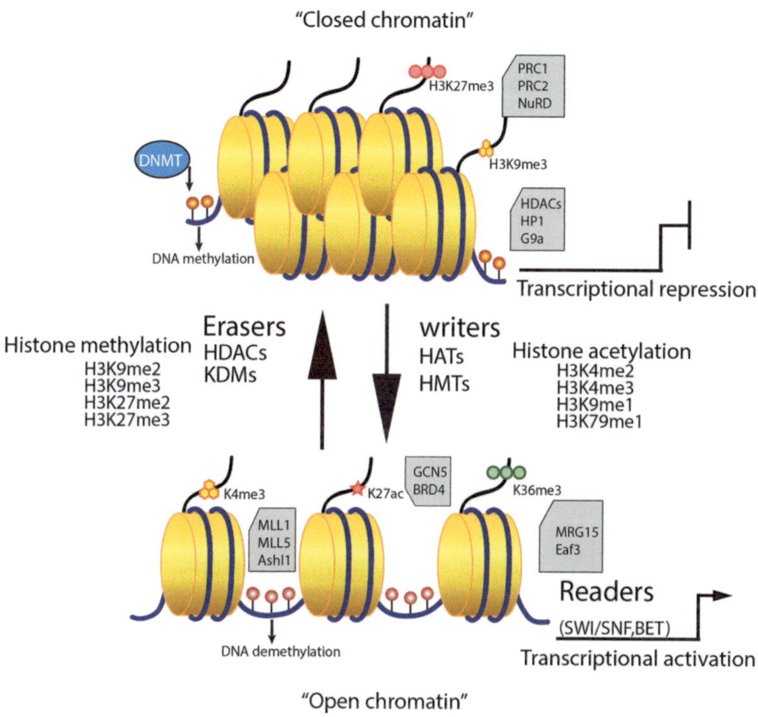

Fig. 3.1 Chromatins can exist as euchromatin (open) or as heterochromatin (compact). The chromatin state is marked (open) by promoter DNA hypomethylation and various posttranslational histone (H) modifications, especially H3, including H3K4me3, H3K27ac, and H3K36me3. In this state, DNA is loosely wrapped around nucleosomes and enables active (on) transcription. The heterochromatin state is marked (compact) by promoter DNA hypermethylation, H3K27me3, and H3K9me3, where compact DNA nucleosomes repress (off) transcription
The on and off states of transcription can be switched by the epigenetic markers writers, readers, and erasers, for example, "writers" by DNA methyltransferases [(DNMTs), histone acetyltransferases (HATs), and histone methyltransferases (HMTs)] and epigenetic "erasers" by DNA demethylases [(TETs), histone deacetylases (HDACs), and histone-lysine demethylases (KDMs)]. "Readers" by epigenetic marks on DNA or histones appear in *gray boxes*

DNMT3A and DNMT3B play major roles in de novo DNA methylation [37]. The major pathway mediating DNA demethylation is catalyzed by 11 translocation methylcytosine dioxygenase (TET) family proteins, with the methyl group of 5-mC oxidized to yield 5-hydroxymethyl cytosine (5-hmC), which can be further oxidized to 5-formylcytosine and 5-carboxylcytosine [38]. These derivatives of 5-mC are novel epigenetic marks that are linked to further biological functions.

Histone modification is subjected to posttranslational modifications (PTMs), including acetylation, methylation, phosphorylation, sumoylation, and ubiquitination [39] (Fig. 3.1). Histone PTMs are widely distributed throughout the whole genome to form the histone code, which can control the accessibility of DNA and recruit transcription factors and coactivators/cosuppressors to result in active, poised, or silenced transcription [40–42]. In conjunction with DNA methylation, the histone code affects the transcriptional states of genomic regions of promoters, enhancers, and gene bodies, which can determine transcriptional programs. Distinct histone modifications can lead to activation or inactivation of adjacent genes by recruiting chromatin remodeling complexes, TFs, and transcriptional coactivators/cosuppressors. In particular, H3K27ac [acetyl-histone H3 (Lys-27)], H3K4me3 [trimethyl-histone H3 (Lys-4)], H3K4me1 [methyl-histone H3 (Lys-4)], and H3K36me3 [tri-methyl-histone H3 (Lys-36)] are associated with the active transcription region. In contrast, H3K27me3 [tri-methyl-histone (Lys-27)] and H3K9me3 [tri-methyl-histone H3 (Lys-9)] are mainly distributed in the inactive gene locus [43, 44]. These histone PTMs are illustrated in Fig. 3.1. The three main classes of regulatory models are presented as follows: (1) "writers" that create covalent modifications for DNA methylation and histone methyltransferases and acetyltransferases, (2)"erasers" that eliminate modifications for demethylases and histone deacetylases, and (3)"readers" that recognize modification through specific protein domains for bromodomain-containing proteins and mating-type switching (SWI) and sucrose fermentation (sucrose nonfermenting – SNF) SWI/SNF nucleosome remodelers, which are the major players that govern dynamic changes in chromatin structure for development or response to environmental stimuli [45]. Histone modifications can affect transcriptional activities by the two following major mechanisms: (1) histone PTMs can alter chromatin structure and conformation. For example, H3K27 acetylation inhibits the positive charge of histones, thus decreasing their binding to DNA and increasing their accessibility for positive regulation of gene expression [46]. (2) Histone PTMs can provide signals for the "reader" enzymes to further recruit transcriptional activators/suppressors. For instance, H3K27 trimethylation is recognized by the polycomb repressive complex, and it can ubiquitinate histone H2A, leading to transcriptional repression [47]. For each histone PTM, there are specific enzymes that catalyze the dynamic "writing," "reading," or "erasing" of these modifications (Fig. 3.1). The "writers" include histone acetyltransferase (HAT), histone methyltransferase (HMT), and protein arginine methyltransferase, whereas the "erasers" are histone deacetylase (HDAC) and lysine demethylase (KDM). The "reader" proteins containing bromodomain-containing proteins or SWI/SNF nucleosome remodelers can recognize differentially modified histones. These enzymes are also subjected to PTMs elicited by upstream signal transduction. A study has shown that sumoylation of HDAC1 increases its transcriptional repression activity [48] and that phosphorylation of HDAC1/2 renders them enzymatically active [49].

Noncoding RNAs (ncRNAs) are RNAs that are not translated into proteins. Although 30% of the human genome can be transcribed, only a small portion of genes are translated into proteins [50]. ncRNAs can be classified on the basis of their size into small ncRNAs [small interfering RNAs (siRNAs), microRNAs (miRNAs), Piwi-interacting RNAs (piRNAs), transfer RNAs (tRNAs), and small nucleolar RNAs (snRNAs)] and long ncRNAs (lncRNAs)]. According to their modes of function, lncRNAs have been divided into four major types (signals, decoys, guides, and scaffolds) [51]. lncRNAs with functions related to signals or decoys contribute to gene activation or suppression, respectively. lncRNAs that act as guides recruit chromatin-modifying enzymes to regulate gene expression in *cis* or *trans*. lncRNAs act as scaffolds and recruit multiple proteins to assemble ribonucleoprotein complexes, which usually function on chromatin and/or modulate histone marks.

Understanding the dynamic changes in the 3D conformation of chromatin is very challenging; thus, precise mapping of the epigenome, such as genome-wide epigenetic modifications that require high-throughput profiling technologies at nucleotide resolution, has been carried out. With the achievement of the Human Genome Project and the development of next-generation sequencing (NGS) at high-throughput levels, a large number of RNA-seq profiling studies have been performed in the past few years. FANTOM5 provides novel insights from a large network into human macrophage biology [52, 53], and with more advanced bioinformatics tools, we could build the foundation and principles to allow for investigations of epigenetics by different disciplines. Histone codes are comprehensibly presented in the Encyclopedia of DNA Elements (ENCODE), which was developed to identify all functional elements of the human genome [54, 55]. ENCODE has collected a large amount of data on epigenetic regulation from various cell types and tissues and made this information publicly accessible for us. The archived large-scale data sets, including functional DNA elements, transcriptional machinery, and histone modifications, can be easily uploaded to the human genome browser, which allows us to perform further integrative analyses.

3.3 DNA Methylation and Innate Immune Cell Responses

A growing number of data sets have shown that epigenetic modification mechanisms, including DNA methylation, histone modifications, and noncoding RNAs, contribute to the regulation of inflammatory signaling for innate immune responses [18, 56] (Fig. 3.1). DNA methylation occurs in several regions, including intragenic, intergenic, and promoters with CpG islands in mammals [57]. Methylation of promoters leads to gene expression silencing, while methylation of intragenic regions can result in alternative transcripts and expression, which may lead to tissue- and cell-specific changes [57]. DNA methylation of CpG islands impacts negative regulation of gene expression. Previous research has reported that the switching of M1 and M2 phenotypes occurs by inactivating state-specific signature genes through DNA methylation of CpG islands in gene promoters. The influence of global methylation on gene expression in macrophages is evidenced by numerous hypomethylated regions in intragenic and intergenic regions in genes [58]. DNA methylation is a heritable DNA modification, which refers to the addition of a methyl group to 5-cytosine (C) to DNA catalyzed by DNA methyltransferase, such as DNMT3a and DNMT3b, to form 5-methylcytosine (5mC). DNA methylation, which is a transcriptional repression mark, plays an important role in many biological processes, such as development, tumorigenesis, and immune responses [34]. In Dnmt1 and Dnmt3a double knockout (DKO) neurons, upregulation of MHC class I gene and transcription factor signal transducer and activator of transcription 1 (STAT1) was found to subsequently lead to the expression of ISGs that are known to contribute to synaptic plasticity, and this regulation was associated with a significant decrease in DNA methylation, indicating that ISGs are potential targets of DNMTs [59]. Hypermethylation of the suppressor of cytokine signaling 1 (SOSC1) promoter under DNMT1 silencing causes overproduction of the proinflammatory cytokines TNF- and IL-6 during lipopolysaccharide (LPS)-induced activation of macrophages [60]. DNMT3b expression was also significantly elevated in isolated tissue resident macrophages, specifically adipose tissue macrophages from *ob*/*ob* mice. In obesity, elevated saturated fatty acids enhance DNMT3b expression, leading to DNA methylation of CpG sites on the peroxisome proliferator-activated receptor γ1(PPARγ1) promoter, a key transcriptional factor that regulates adipose tissue macrophage polarization, inflammation, and insulin resistance [61]. Genome-wide DNA methylation sequencing with isolated tissue macrophages and endothelial cells from ischemic muscle of hyperlipidemic and type 2 diabetes mellitus mice demonstrated significant promoter hypomethylation of genes typical of active proinflammatory M1 macrophage gene promoters, including Cfb, Serping1, and Tnfsf15, and simultaneous hypermethylation of anti-inflammatory, alternatively activated M2-Mϕ gene promoters, including Nrp1, Cxcr4, Plxnd1, Arg1, Cdk18, and Fes. This alternative activation skewed macrophage phenotypes toward M1 lineage [62].

Epigenetic modifications triggered by environmental factors are an important mechanism that leads to altered gene expression. DNA methylation has an important role in the regulation of inflammatory signaling, while aberrations in DNA methylation are involved in autoimmune and inflammatory diseases. Periodontitis is a common chronic inflammatory disease that poses significant global health and financial burdens; the initiation and maintenance of chronic inflammatory periodontitis occurs through a complex microbial biofilm in which the predominant pathogen is *Porphyromonas gingivalis* [63]. TLRs play an important role in periodontitis by recognizing *Porphyromonas gingivalis* pathogens and maintaining tissue homeostasis [64]. In an analysis of the TLR2 promoter CpG island, increased DNA methylation was observed in the gingiva of mice infected with *P. gingivalis* in a periodontitis oral gavage model. Infection was correlated with higher CpG methylation in dysregulated gingiva. Furthermore, tissues obtained from periodontitis patients also exhibited differential TLR2 promoter methylation through bisulfite DNA sequencing analysis. TLR2 promoter DNA methylation creates innate immune dysbiosis [65, 66]. DNMTs may also affect innate immune signaling.

3.4 Histone Methylation and Innate Immune Cell Responses

Increasingly, studies have indicated that histone methylation has an important role in the regulation of inflammatory triggered immune responses by controlling the transcriptional regulation of target-related genes [67]. Several histone methyltransferases (HMTs) and demethylases have been listed as key regulators that control the innate immune response genes, which include transcription factors, cytokines, chemokine, IFNs, and encoding enzymes [68]. H3 histone-lysine 4 trimethylation (H3K4me3), a modification localized to the transcription start site of active genes, is controlled by mixed lineage leukemia (MLL) enzymes. MLL contains a central zinc finger domain homologous to that of the *Drosophila* protein trithorax (Trx) with transactivation activity and a carboxyl terminal SET domain homologous to regions of the suppression of variegation 3–9 [Su(var)3–9], enhancer of zeste [E(z)], and trithorax *Drosophila* proteins. Trx and E(z) are encoded by members of the *trx* and *Polycomb* groups (*trx-G* and *Pc-G*) of developmental regulatory genes, respectively [69–71]. SET and MLL proteins have been associated with a wide variety of diseases [72–75]. MLL4 (the mouse ortholog is Wbp7) has a key role in the innate immune response. Wbp7 is required for robust activation of pathogen-mediated gene responses. Severe response defects were observed in Wbp7-deficient macrophages after LPS challenge, resulting in dysregulation of a very specific subset of genes, such as *Pigp*. The H3K4me3 levels of phosphatidylinositol-glycan biosynthesis class P protein (PIGP) were directly downregulated by Wbp7 depletion. PIGP regulates the glycosylphosphatidylinositol (GPI)-anchor synthesis pathway, which directs CD14 to the extracellular membrane. Membrane-tethered CD14 serves as an important accessory receptor for TLR4 and is critical for innate immune responses [76]. Nuclear translocation of another H3K4 methyltransferase, MLL1 (also known as KMT2A), to chromatin was found to function in innate immunity by selectively regulating the activation of genes downstream of NF-κB-mediated genes, such as interleukin-6 (IL-6) by tumor necrosis factor-α (TNF-α) and LPS. [77]. Ash1l with a SET (Su[var]3–9, E[z] and trithorax) domain, a H3K4 methyltransferase, suppressed IL-6 and TNF-α production in TLR-triggered macrophages, protecting mice from sepsis. The mechanism involved Ash1l promoting A20 (Tnfaip3) expression by increasing H3K4 methylation at the Tnfaip3 promoter that in turn facilitates A20-mediated deubiquitination of NF-κB signaling pathways and the production of proinflammatory cytokines. Ash1l-deficient mice were more susceptible to sepsis with *Escherichia coli* challenge. Ash1l-deficient mice also had increased susceptibility to collagen-II-induced arthritis, and compared with wild-type mice, more Ash1l-deficient mice suffered from spontaneous systemic autoimmune disease [78]. In addition, Polycomb group (PcG) proteins mediate gene silencing and repress transdifferentiation in a manner dependent on histone H3 lysine 27 trimethylation (H3K27me3). JmjC-domain protein Jmjd3 (also known as KDM6B) is a H3K27me3 demethylase expressed in macrophages; Jmjd3 binds PcG target genes and regulates their H3K27me3 levels and transcriptional activity. Research has shown that macrophage terminal differentiation is associated with a sharp drop in Jmjd3 occupancy of both *HoxA7* and *HoxA11*, which correlated with an increase in H3K27me3 levels and gene inactivation. As an inducible enzyme, Jmjd3 erases a histone mark that controls differentiation and cell identity, which provides a link between inflammation and reprogramming of the epigenome [79]. In previous experiments, interleukin-4 (IL-4) treatment led to upregulated Jmjd3 expression and decreased H3K27me3 at the promoter of M2 marker genes, such as Ym1 (*Chi3l3*), FIZZ1 (*Retnla*), and Arginase 1 (*Arg1*), and a concomitant increase in Jmjd3 expression. IL-4 stimulation leads to increased activated transcription factor STAT6 and binds to consensus sites at the Jmjd3 promoter. Increased Jmjd3 contributes to the decrease in H3K27 dimethylation and trimethylation (H3K27me2/3) marks, as well as the transcriptional activation of specific M2 marker genes [80]. A later study also identified interferon-regulatory factor 4 (IRF4) as a key transcription factor that controls M2 macrophage polarization. Jmjd3-mediated H3K27 demethylation is crucial for regulating M2 macrophage polarization, and TLRs mediated innate immune responses against helminth infection by regulating the expression of *Irf4* [79, 81]. In this study, Jmjd3 is recruited to transcription start sites with RNA polymerase II complex to bind Jmjd3 to target genes by the presence of the activation marker H3K4me3; this action is a H3K27 demethylation-independent mechanism. This finding indicates that histone methylation state exchange in H3K4 and H3K27 methylation can be an important epigenetic process for the control of innate immune response genes [81]. The histone-lysine N-methyltransferase, enhancer of zeste homologue 1 (EZH1), is another H3K27 methyltransferase. Ezh1 was the most upregulated HMT during DC maturation, and silencing of Ezh1 significantly reduced TLR-triggered production of cytokines, including IL-6, TNF-α, and IFN-β, in DCs and macrophages by suppressing the TLR negative regulator Toll-interacting protein (TOLLIP); the regulation mechanism directly targeted the proximal promoter of tollip and maintained the high level of H3K27 methylation for repression [82]. Furthermore, the H3K9 methyltransferase G9A (also known as EHMT2) with SET domain-containing histone-lysine methyltransferase (HKMT), is crucial for the negative regulation of IFN and ISG for antiviral responses, genetic ablation, or pharmacological inactivation of lysine methyltransferase G9a, which is essential for the generation of H3K9me2, which resulted in highly potent IFN-producing cells and rendered these cells resistant to pathogenic RNA viruses [83]. Interestingly, recent data show that methyltransferase Setdb2 was the only protein lysine methyltransferase induced during infection with influenza virus and repressed the expression of CXC-chemokine ligand 1 (*Cxcl1*), coinciding with occupancy by Setdb2 at the Cxcl1 promoter, which displayed diminished trimethylation of histone H3 Lys9 (H3K9me3) in setdb2 genetrap mice. Mice with a hypomorphic gene trap construct of Setdb2 exhibited increased infiltration of neutrophils during sterile lung inflammation and were less sensitive to bacterial superinfection after infection with influenza virus. This finding indicated cross talk between virus-induced susceptibility and bacterial superinfections [84]. TET2, an enzyme that modifies the DNA base methylcytosine to 5-hydroxymethylcytosine (5HMC), is upregulated in macrophages after LPS stimulation, and TET2 and TET2-induced 5HMC have a feedback loop to prevent persistently high transcription of IL-6 during an innate immune response. In a previous study, the authors found that IkBζ targets TET2 to the IL-6 promoter to indirectly recruit HDAC2, which deacetylates H3 and H4 histones and suppresses transcription [85]. Another recent study provides a unique glimpse of yet another aspect of coordinated DNA methylation and protein acetylation in the host response to pathogenic stimuli. Dnmt3a selectively impaired the production of type I interferons triggered by PRRs through an epigenetic mechanism by maintaining high expression of HDAC9. HDAC9 directly maintained the deacetylation status of the key PRR signaling molecule TBK1 and enhanced its kinase activity. Understanding how the complex coordination between methyltransferases and histones impacts signaling intermediates and transcriptional regulators at different levels of gene expression is necessary [86]. Overall, these findings indicated that epigenetic modification enzymes are recruited by multi-protein transcriptional complexes or corepressors with histone markers to target gene promoters, where they either directly or indirectly coordinate with DNA or histone modification for transcriptional regulation control.

3.5 Histone Acetylation, Deacetylation, and Innate Immune Cell Responses

The regulation of histone acetylation and deacetylation, which are essential processes for gene regulation, is controlled by HDACs and HATs [87]. HATs are enzymes that transfer acetyl groups to core histones, which have subsequent effects on chromatin remodeling and gene expression. HATs have the two following families [88]: (1) p300/CBP, Taf1, and nuclear receptor coactivators that also possess catalytic acetyltransferase activities and (2) Gcn5 N-acetyltransferases (GNATs) and Morf, Ybf2, Sas2, and Tip60 HATs. HATs are involved in initiating gene expression in macrophages during inflammation [89, 90]. However, we only understand global histone acetylation and its role in regulating gene expression. HATs may also interact with the opposing HDAC enzymes to enhance acetylation and eventually activation of antiviral gene promoters. HATs p300/CBP were recruited to the inactive IFN promoter upon transcription factor interferon regulator factor 5 (IRF5) phosphorylation and displaced the SMRT/Sin3a repressive complexes. IRF5 is subsequently acetylated by p300/CBP, facilitating H3 histone acetylation of target genes, histone phosphorylation by MAPK cascades, and recruitment of HATs p300/CBP by signaling transcription factors such as NF-κB and STATs, but little is known about how TLR-induced signals are propagated to chromatin and histones, including TNF-α and IL-6 [54]. A limited number of reports have shown that HATs catalyze GNATs and Morf, Ybf2, Sas2, and Tip60 HATs in the regulation of the expression of specific M1- or M2-associated genes.

The enzymes that oppose HAT functions are HDACs. Histone deacetylation is a dynamic process and may be the result of other posttranslational modifications. HDAC functions may induce further epigenetic changes and alternative gene expression. Eighteen mammalian HDACs have been identified, which are classified into the five following groups: class I (HDAC1, HDAC2, HDAC3, and HDAC8), class IIa (HDAC4, HDAC5, HDAC7, and HDAC9), class IIb (HDAC6 and HDAC10), and class III (consists of the NAD + −dependent HDACs) and HDAC11, which constitutes a class of its own [118]. The effects of HDACs during macrophage activation include HDAC sirtuin 6 repressing the transcription of NF-κB target genes through binding to NF-κB p65 and deacetylating H3K9 at the regulated gene promoters [91]. In addition, promyelocytic leukemia zinc finger (PLZF) protein, also known as Zbtb16 or Zfp145, suppressed the expression levels of inflammatory cytokines IL-6 and IL-12p40 by forming a corepressor complex that has HDAC activity to modify chromatin in a subset of transcription factor NF-κB regulatory elements [40, 92]. Using small-molecule inhibitors of HDAC classes I and II in several macrophage populations results in reduced levels of TLR receptors, cytokines, and chemokines. These inhibitors also impact cellular processes, such as chemotaxis, phagocytosis, apoptosis, and cellular metabolism [30, 55–57]. Inhibition of HDACs may also be beneficial in a complex inflammatory environment due to interactions between macrophage populations and other resident cells that occur for host defense. For example, HDAC inhibition rescued oligodendrocytes during traumatic brain injury via induction of the M2 phenotype in resident microglia [59]. There are many probable explanations for the HDAC class I and II effects on macrophage activation status. It has been reported that HDAC inhibition increase the recruitment of the repressive complex to the promoters of M1 activation state genes like IL-6 [55]. Another possibility is that these effects are a result of the decline in PU.1 levels in macrophages treated with TSA; PU.1 transcription factor is known to be important for the transcriptional control of macrophage development and function [60, 61]. Interestingly, Serrat et al. proposed that TSA induces acetylation-mediated repression of C/EBP, which binds with lower efficiency to the Arg1 promoter in macrophages [62]. A more comprehensive review of HDACs can be found in this book by Bandar Suliman.

3.6 Noncoding RNA and Innate Immune Responses

3.6.1 miRNAs

miRNAs are small noncoding RNAs of ~22 nucleotides in length that are produced by two RNases – Drosha and Dicer. The main function of miRNAs is RNA silencing by targeting protein coding transcripts. miRNAs regulate TLR signaling pathways at several levels, including regulation of TLR expression, TLR-associated signaling proteins and regulatory molecules, and TLR-induced transcription factors and functional cytokines [93, 94]. Therefore, miRNAs greatly contribute to inflammatory signaling and have been extensively studied. A comprehensive review of miRNAs can be found in this book by Claire McCoy.

3.6.2 lncRNAs

Thousands of lncRNAs have been discovered by high-throughput transcriptome sequencing [95, 96]. lncRNAs interact with small RNAs and RNA-binding protein and play important roles in controlling chromatin structure, gene expression, and mRNA translation via posttranscriptional mechanisms targeting splicing and stability [97]. Hundreds of lncRNAs are induced by innate immune activation through TLR ligation stimulation experiments. lnc-DC was expressed exclusively in DCs, which has been recently identified as a requirement of LPS-induced functional maturation and activation of DCs [98]. lnc-DC directly bound to STAT3 in the cytoplasm, which promoted STAT3 phosphorylation on tyrosine-705 by interfering with STAT3 binding and dephosphorylation by SHP1, which regulates DC differentiation. RNA-seq analysis revealed that 72 lincRNAs were significantly upregulated in mouse bone marrow-derived macrophages after treatment with bacterial lipoprotein Pam3CSK4 (TLR2) [99]. One of these lincRNAs, (lincRNA)-*Cox2*, was shown to act as a regulator of a MYD88- and NF-κB-mediated program in macrophages, and both activate and repress distinct subsets of proinflammatory genes. A later study showed that lincRNA-Cox2 is assembled into the SWI/SNF complex after TLR treatment in macrophages. The lincRNA-Cox2/SWI/SNF complex can modulate the assembly of transcription factor NF-κB subunits p65/p50 to the SWI/SNF complex, leading to chromatin remodeling and transactivation of the late primary inflammatory response genes in response to microbial challenge [100]. lincRNA-Cox2 is highly induced by multiple inflammatory triggers, such as TLR ligands (LPS) and Pam3CSK4 and microbial pathogens (*Sendai virus* and *Listeria monocytogenes*) [101]. Lethe, a functional pseudogene (Rps15a-ps4) lncRNA, is also highly inducible by TNF-α and IL-1β treatment. In addition, after treatment with an anti-inflammatory glucocorticoid receptor agonist, dexamethasone, Lethe expression is induced and impacts IL6, IL-8, and superoxide dismutase 2 (SOD2) expression levels through inhibiting transcription factor p65 (RelA) occupancy at its target gene promoter [102]. Furthermore, THRIL is another immunoregulatory lincRNA (TNF- and HNRNPL-related immunoregulatory lincRNA) that is induced by TNF stimulation in the human monocyte-like THP-1 cell line. THRIL was also differentially expressed in response to Pam3CSK4 (TLR2). Knockdown of nine lincRNAs, including THRIL, led to impairment of IL-6 and/or TNF-α induction [103]. lincRNA-Tnfaip3 is also required for the transactivation of NF-κB-regulated inflammatory response genes after LPS treatment. lincRNA-Tnfaip3 physically interacts with the high-mobility group box 1 (Hmgb1), assembling a NF-κB/Hmgb1/lincRNA-Tnfaip3 complex after LPS treatment. The resulting NF-κB/Hmgb1/lincRNA-Tnfaip3 complex can modulate Hmgb1-associated histone modification [104]. More recently, lincRNA-EPS was found to be regulated in macrophages and impacted control of the expression of immune response genes from transcriptome profiles of macrophages in lincRNA-EPS-deficient mice. Additionally, a gain-of-function study and rescue experiments indicated that lincRNA-EPS can functionally restrain inflammatory gene expression, which occurs through lincRNA-EPS interacting with heterogeneous nuclear ribonucleoprotein L through a CANACA motif located in its 3′ end [105]. Overall, lncRNA was highlighted as an emerging and important regulator that controls gene expression in the innate immune cell response.

3.7 Epigenetic and Innate Immune Cell Memory

The innate immune defense through macrophages is tightly controlled by timing and space. Macrophages have the ability to respond readily and balance inflammatory signaling for the host defense against pathogens through the process of resistance and tolerance [106]. Resistance will reduce pathogens, whereas tolerance will limit tissue damage. These processes achieve effective host protection [107]. More recent evidence has shown that macrophages display long-term changes in their functional programs after infection or vaccination, and those changes affect its responsiveness to cytokines and inflammatory mediators to eliminate infection upon secondary pathogen stimulation [108]. This finding revealed the existence of innate immune cell memory in addition to classical adaptive immune system memory. Emerging studies have shown that chromatin structure and epigenetic modifications play an essential role in the behavior of trained immunity and endotoxin tolerance for the macrophage memory phenotype [109].

Priming macrophages with an initial infection or vaccine challenge results in enhanced macrophage responsiveness to a secondary challenge [110, 111]. Mice lacking functional T and B lymphocytes are protected against reinfection with non-lethal infection with *Candida albicans* in a monocyte-dependent manner, which lead to enhanced cytokine production both in vivo and in vitro. *C. albicans* and fungal cell wall β-glucans can induce functional reprogramming of monocytes [112]. This finding also applies to human monocytes treated with *C. albicans*, which increased the inflammatory response following subsequent stimulation. Monocyte training by β-glucans was associated with H3K4me3, which suggests the involvement of epigenetic mechanisms in this phenomenon [112]. H3K4 methylation (H3K4me3) and H3K27 acetylation (H3K27ac) occupied the regulatory element, which is in the promoter during the first infection stimulus in some cases. Recent studies have termed regulatory elements latent enhancers; after stimulation, transcription is terminated, and H3K4 methylation and H3K27 acetylation also disappear, but H3K4 methylation persists at latent enhancers and constitutes the basal level occupying the region for a faster and enhanced response until restimulation [113, 114] (Fig. 3.2). In β-glucan-trained macrophage cells, we can see the alteration of glycolysis enzymes, which form a link between metabolism and inflammation, and more detail has shown that increased glycolysis and suppressed oxidative phosphorylation under sufficient oxygen supply occur through the Akt-mTOR-HIF-1α pathway [115]. After training with *C. albicans*-derived β-glucan, macrophages regulated TNF-α, IL-6, and IL-18 under stable changes in H3K4me3, showing that both histone methylation and acetylation patterns change with envi-

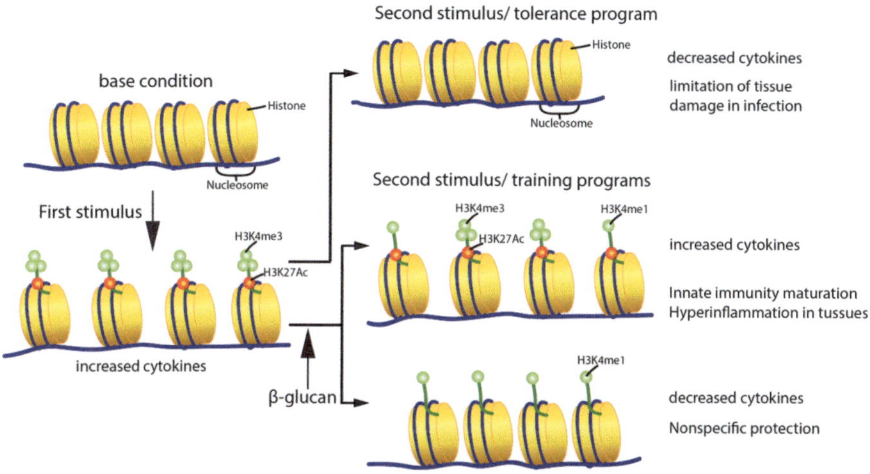

Fig. 3.2 The figure represents innate immune memory with endotoxin tolerance and trained immunity after pathogen-associated molecular patterns (PAMPs) challenge; innate immune cell recognition of pattern recognition receptors (PRRs) leads to functional reprogramming of the cell. The model of trained immunity with β-glucan stimuli from *Candida albicans* leads to metabolic changes with increased glycolysis and suppressed oxidative phosphorylation that links epigenetic reprogramming to the innate immune response through histone modifications (methylation and acetylation). Subsequent PRR stimulation may cause decreased or increased responsiveness of these cells, which depends on the initial type of stimulus. Initial PRR stimulation leads to H3K4me3 and H3K27Ac on the promoters of proinflammatory genes. In the endotoxin tolerance model, the removal of the stimulus causes the disappearance of H3K4me, and gene expression returns to basal levels. Following subsequent stimulus, some tolerized genes will lose H3K4me3 or H3K27Ac and will remain silent in response to a second stimulation (*top figure*). In the trained immunity model, proinflammatory genes will maintain enhancers marked with H3K4me1. Subsequent stimulus will lead transcription factors to bind to both the enhancers and the promoters of innate immune response genes and recruit a chromatin modifier complex, including histone acetyltransferase p300 and elevated H3K4me3, for enhancing the expression of a subset trained genes (*middle figure*). On the other hand, some genes will prevent H3K4me3 and have less gene expression induction (*bottom figure*)

ronmental challenges, which is most likely the metabolic switch for macrophage training and activation [109, 115].

Endotoxin tolerance is another form of memory that contrasts with trained macrophages; the underlying mechanism first involves stimulation with TLR4 ligands that causes the cell refractory state, which leads to tolerization from gene expression after restimulation [116]. Tolerized genes will be diminished, and non-tolerized genes will be increased or unchanged. This pattern reflects clinical situations such as sepsis in which patients enter a suppressed response state [117] (Fig. 3.2). The transcription active mark H3K4 methylation disappears after first stimulation, which is associated with the re-induction of tolerized inflammatory cytokines [19, 118]. The contribution of DNA methylation changes to macrophage memory has also shown that high levels of TET2 in cells have the ability to demethylate DNA; we would like to explore the regulatory mechanism of macrophage functional plasticity that depends on stimuli and restimuli. DNMT3B affects the control of inflammatory signaling, which impacts macrophage polarized phenotypes for M1-like and M2-like states. Therefore, DNA methylation is another factor that contributes to macrophage memory [61, 119]. A recent study has indicated that environmental factors, such as circulating glucose levels and gut microflora, may alter innate immune responsiveness through modification of the epigenetic states of peripheral blood monocytes and resident tissue macrophages. Increasing expression levels of TLR1, TLR2, and TLR4 by high glucose through circulating glucose and insulin levels have acute effects on peripheral blood mononuclear cell (PBMC) receptor levels, indicating a direct impact on innate immune receptor expression [120]. High glucose drives epigenetic changes in monocytes that result in increased expression of proinflammatory stressors, promoting epigenetic changes and facilitated binding of transcription factor NF-κB subunits in monocytes that result in increased expression of the inflammatory cytokines IL-6 and TNF-α [121]. Under high-glucose conditions, enhanced binding of NF-κB to its target genes occurs through epigenetic changes with reduced HDAC2 and increased HAT activity [122].

Tissue macrophages are also regulated by environmental cues, and this relationship is evidenced by disease phenotypes, such as colitis models. There are correlations between gut microbiota and the immune responsiveness of resident macrophages; in humans, alteration of commensal bacteria populations is associated with inflammatory bowel disease (IBD), and in mice, enhanced cellular responses to dextran sodium sulfate (DSS)-induced colitis are observed in germ-free mice [123]. Regarding the deletion of TLR2 in multidrug resistance gene (MDR1/ABCB1) deficiency, TLR2/MDR1A double-knockout mice exhibited a fulminant pancolitis phenotype with early expansion of CD11b(+) myeloid cells and rapid changes in Th1 immune responses in the lamina propria during colitis models [124]. The double-knockout CD11b(+) myeloid cells expressed MD-2/TLR4 and hyperresponded to LPS or nonpathogenic *Escherichia coli* with reactive oxygen species and caspase-1 activation and release of proinflammatory IL-1β, leading to pyroptosis. Compared with active ulcerative colitis (US) with no polymorphisms, active ulcerative colitis (UC) with TLR2-R753Q and MDR1-C3435T polymorphisms was associated with increased expression of caspase-1 protein and cell death in acutely inflamed tissues. The results demonstrate that the restricted TLR signaling network is important in maintaining healthy intestinal homeostasis [125]. Interestingly, "training" mouse or human hematopoietic progenitor cells by prior exposure of differentiated macrophages to a TLR2 agonist ("tolerance") suppress inflammatory cytokine production but elevate reactive oxygen species. Soluble factors produced following the exposure of progenitor cells to a TLR2 agonist can also act in a paracrine manner to influence the function of macrophages [126].

3.8 Epigenetics in Host Defense and Inflammatory Disease

The majority of disease-associated single nucleotide polymorphisms (SNPs) that were identified by genome-wide association studies (GWAS) are localized to regulatory regions of the genome, and less than 10% of SNPs are located in protein coding genes [127]. Regulatory regions of the genome, including regulatory elements of promoters, enhancers, and intergenic regions of the genome, are rich in miRNAs and lncRNAs [128]. Those genetic variations could affect the expression and/or function of RNAs and impact human diseases. Dysregulated expression of enhancers, miRNAs, and lncRNAs could lead to several immune-related diseases, such as diabetes, multiple sclerosis, and IBD [129]. The important, highly specific miR-150 was found in leukocytes and was upregulated in mucosal tissue of patients with IBD for both ulcerative colitis and Crohn's disease and murine models of experimental colitis; this upregulation may represent a colitis-associated increase in mucosal leukocyte infiltration as opposed to representing pathophysiological relevance. Investigations into miRNA function in IBD will continue to be challenging given the high rate of inflammatory cell infiltration related to fibrosis and turnover of epithelial cells [130]. An atherosclerosis-associated SNP located in the intron of lncRNA LINC00305 was recently identified by a search of the GWAS database. LINC00305 expression is enriched in atherosclerotic plaques and monocytes. Overexpression of LINC00305 promoted the expression of inflammation-associated genes in THP-1 cells and reduced the expression of contractile markers in co-cultured human aortic smooth muscle cells [131]. Higher mRNA expression levels of DNMT1 and MBD2 were found in PBMCs from rheumatoid arthritis (RA) patients who had lower levels of global DNA methylation, which indicated that DNA hypomethylation results in overexpression of autoimmune-related genes that contribute to RA [132]. The current challenge is to define the functional role of these regulators (miRNA, IncRNA, and SNPs in regulatory elements in genomic DNA) in the context of disease pathophysiology. Recent use of promoter capture high-throughput chromosome conformation capture (Hi-C) will enable identification of novel candidate genes and complex long-range interactions with related autoimmune risk loci. Long-range interactions between regulatory elements and gene promoters are uncharted and will provide us with a complete picture of the link to understand genome control [133]. A study has identified interacting regions of 31,253 promoters in 17 human primary immune cell types. The data show that promoter interactions are highly cell type specific and enriched for links between epigenetically marked enhancers and active promoters [134]. The power of primary cell promoter interactomes to reveal insights into genomic regulatory mechanisms and rich resources to connect noncoding disease variants with putative target promoters can help us understand disease pathways and screen candidate genes. Genetic variants that affect gene expression can be identified using expression quantitative trait loci (eQTLs), which is a novel approach to analyzing regions of the genome containing DNA sequence variants that influence the expression level of one or more genes. This approach is crucial to addressing genes examined in the context of human inflammatory diseases. Performing eQTL mapping and parallel analysis in multiple leukocyte subsets from patients with active disease provides new insights into the genetic basis of immune-mediated diseases [135]. Overall, the aberrant expression of miRNA, IncRNA, and SNPs in promoter or enhancers may play a causal role or could be the consequence of disease pathology, which requires further investigation.

3.9 Respective of Innate Immunity in Epigenetics

NGS technology offers unique ways to profile genome-wide epigenetic changes, including genome methylation and histone modifications. Three-dimensional genome architecture reveals chromatin topology, including ChIA-PET (Chromatin Interaction Analysis by Paired-End Tag Sequencing), which was designed to detect genome-wide chromatin interactions mediated by specific protein factors [136], whereas Hi-C was developed to capture all chromatin contacts [137]. Hi-C has been proven effective for mapping large-scale structures, such as topologically associated domains [138, 139], based on the methods described in the previous sections [24]. Bioinformatics tools have been developed for the analysis of these NGS data to infer precise methylation changes, TF binding, and topological interactions. Publicly accessible databases have been established to enable researchers. We can use various methods to conduct profiling and analysis but can still limit the investigations to conduct epigenetic analyses with bioinformatics data and tools [140–142]. Understanding epigenetics requires a systems biology approach with concern regarding the kinetics of various processes, such as methylation, chromatin modifications, topological reorganization, TF activation, gene transcription, translation, and cellular processes, that span a range of timescales from milliseconds to hours. To understand how to uncover causal information, we have to analyze the sequence of time points, which provides a dynamic perspective of epigenetic changes. Therefore, we can ultimately build directed networks that regulate homeostasis and disease stage. There have been several recent reports of integrative analyses involving epigenetics, transcription, and, in some cases, cellular signaling. For example, we used TLR-induced macrophages as a model and combined time series transcriptomic profiling, cluster analysis, ChIP-seq, and network modeling to identify a prominent role of activating transcription factor 3(ATF3) in high-density lipoprotein-mediated reprogramming of macrophages in anti-inflammatory signaling on TLR responses [143].

Systems models provide a powerful way to guide designs for perturbation experiments, as well as therapeutic interventions. Existing methods used in epigenetics research, including NGS, are largely bulk assays, which provide global information averaged from a large number of cells. However, current approaches are moving to studies at the single-cell level. For example, by using the methyl-CpG-binding domain of methyl-CpG-binding domain protein 1 tagged with enhanced green fluorescent protein, we can achieve single-cell visualization of methylated DNA [144], and FRET-based sensors have been used for single-cell detection of histone acetylation states [145]. Chromatin is a recently developed technique, which is a quantitative high-resolution imaging approach combining fluorescence in situ hybridization, immunostaining, and array tomography imaging that enables us to investigate chromatin organization in tissue specimens at the cellular level [146]. This excellent novel real-time imaging technology can be used in living cells with high sensitivity and specificity at locus-level resolution. The CRISPR (clustered regularly interspaced short palindromic repeats) bacterial immune system and derived Cas9 protein have provided unprecedented opportunities for genomic editing by introducing sequence-specific gene mutations [146]. Although CRISPR/Cas9 represents a valuable way to manipulate both the *cis* and *trans* elements involved in epigenetics both in vitro and in vivo, researchers are still improving undesirable off-target mutations, the lack of tissue specificity, and the efficacy of in vivo delivery, which will be expected in the future.

A growing understanding of how epigenetic mechanisms control gene expression patterns has highlighted the potential for targeting epigenetic mechanisms for the treatment of inflammation and related diseases [147]. Numerous HDAC inhibitors have been investigated for their effects on innate immune responses to PAMP stimulation. For example, vorinostat has been shown to reduce inflammatory cytokine production elicited both by LPS stimulation of human PBMCs and after LPS injection into mice [148, 149]; HDAC inhibitors have also been tested for their effects on human disease tissues, with studies primarily focusing on RA. In this context, TSA, vorinostat, and sodium phenylbutyrate were all shown to inhibit cytokine production by macrophages derived from the inflamed joints of patients with RA and cytokine production from RA synovial explants [150]. In addition, TSA and an HDAC3 inhibitor (MI192) were reported to inhibit IL-6 production from RA PBMCs stimulated with LPS [151]. Bromodomain and extraterminal (BET) proteins recognized acetylated lysine residues in the histone tail, which can also contribute to the transcription of inflammatory cytokine gene expression by forming a complex with positive transcription elongation factor-b and RNA polymerase II at the transcription start site [152, 153]. BRD-containing proteins bound to chromatin loci such as enhancers are often deregulated in disease, leading to aberrant expression of proinflammatory cytokines and growth-promoting genes. Recent developments targeting the BET subset of BRD proteins demonstrated remarkable efficacy in murine models, providing a compelling rationale for drug development and translation to the clinic [154]. Using I-BET (a synthetic compound) to inhibit inflammatory cytokine genes in macrophages after treatment LPS, the mechanism of this drug appears to comprise interfering with the binding of BET to acetylated histones, which can prevent inflammatory disorders [155, 156]. Jmjd3 could be another therapeutic target for inflammatory diseases. Indeed, a selective jumonji H3K27 demethylase (catalytic site inhibitor) [157], GSK-J4 (GlaxoSmithKline corporate compound collection), was recently reported to inhibit LPS-induced production of inflammatory cytokines in human macrophages from healthy volunteers.

Epigenetic research has had a profound impact on personalized medicine. Advances in NGS technology have enabled the sequencing of the human genome within days with reduced cost. Collectively, DNA-, RNA-, and ChIP-seq data are anticipated to provide genetic and epigenetic information regarding inflammatory signaling at the homeostasis stage or inflammatory disease stage, which will allow us to understand how epigenetic mechanisms finely tune the outcome of host innate immune response and will also provide valuable information or guidance on human inflammatory disease etiology, diagnosis, treatment, and prognosis. The integration of NGS technology and data analysis by bioinformaticians and immunologists, using clinical information from physicians, will revolutionize translational medicine.

Acknowledgements This work was supported by grants from the National Natural Science Foundation of China (81273247, 81472655 and 31670905) and the Shanghai Municipal Education Commission key discipline support project 2015-10101001-1.

References

1. Asgari E, Zhou W, Sacks S (2010) Complement in organ transplantation. Curr Opin Organ Transplant 15(4):486–491
2. Iwasaki A, Medzhitov R (2010) Regulation of adaptive immunity by the innate immune system. Science 327(5963):291–295
3. Barton GM (2008) A calculated response: control of inflammation by the innate immune system. J Clin Invest 118(2):413–420
4. Markiewski MM, Lambris JD (2007) The role of complement in inflammatory diseases from behind the scenes into the spotlight. Am J Pathol 171(3):715–727
5. Kolaczkowska E, Kubes P (2013) Neutrophil recruitment and function in health and inflammation. Nat Rev Immunol 13(3):159–175
6. O'Neill LA, Golenbock D, Bowie AG (2013) The history of toll-like receptors – redefining innate immunity. Nat Rev Immunol 13(6):453–460
7. Chen KW, Schroder K (2013) Antimicrobial functions of inflammasomes. Curr Opin Microbiol 16(3):311–318
8. Schroder K, Tschopp J (2010) The inflammasomes. Cell 140(6):821–832
9. Franklin BS, Mangan MS, Latz E (2016) Crystal formation in inflammation. Annu Rev Immunol 34:173–202
10. Heneka MT, Golenbock DT, Latz E (2015) Innate immunity in Alzheimer's disease. Nat Immunol 16(3):229–236
11. Zimmer S, Grebe A, Latz E (2015) Danger signaling in atherosclerosis. Circ Res 116(2):323–340
12. Broderick L, De Nardo D, Franklin BS, Hoffman HM, Latz E (2015) The inflammasomes and autoinflammatory syndromes. Annu Rev Pathol 10:395–424
13. Kono H, Kimura Y, Latz E (2014) Inflammasome activation in response to dead cells and their metabolites. Curr Opin Immunol 30:91–98
14. Heneka MT, Kummer MP, Latz E (2014) Innate immune activation in neurodegenerative disease. Nat Rev Immunol 14(7):463–477
15. Meng G, Zhang F, Fuss I, Kitani A, Strober W (2009) A mutation in the Nlrp3 gene causing inflammasome hyperactivation potentiates Th17 cell-dominant immune responses. Immunity 30(6):860–874
16. Iwasaki A, Medzhitov R (2004) Toll-like receptor control of the adaptive immune responses. Nat Immunol 5(10):987–995

17. Kawai T, Akira S (2011) Toll-like receptors and their crosstalk with other innate receptors in infection and immunity. Immunity 34(5):637–650
18. Cao X (2016) Self-regulation and cross-regulation of pattern-recognition receptor signalling in health and disease. Nat Rev Immunol 16(1):35–50
19. Foster SL, Hargreaves DC, Medzhitov R (2007) Gene-specific control of inflammation by TLR-induced chromatin modifications. Nature 447(7147):972–978
20. Stender JD, Glass CK (2013) Epigenomic control of the innate immune response. Curr Opin Pharmacol 13(4):582–587
21. Maciejewska Rodrigues H, Jungel A, Gay RE, Gay S (2009) Innate immunity, epigenetics and autoimmunity in rheumatoid arthritis. Mol Immunol 47(1):12–18
22. Biggar KK, Li SS (2015) Non-histone protein methylation as a regulator of cellular signalling and function. Nat Rev Mol Cell Biol 16(1):5–17
23. Xue J, Schmidt SV, Sander J, Draffehn A, Krebs W, Quester I, De Nardo D, Gohel TD, Emde M, Schmidleithner L, Ganesan H, Nino-Castro A, Mallmann MR, Labzin L, Theis H, Kraut M et al (2014) Transcriptome-based network analysis reveals a spectrum model of human macrophage activation. Immunity 40(2):274–288
24. Winter DR, Jung S, Amit I (2015) Making the case for chromatin profiling: a new tool to investigate the immune-regulatory landscape. Nat Rev Immunol 15(9):585–594
25. Baylin SB, Jones PA (2016, September 1) Epigenetic determinants of cancer. Cold Spring Harb Perspect Biol 8(9). pii: a019505 doi:10.1101/cshperspect.a019505
26. Kouzarides T (2007) Chromatin modifications and their function. Cell 128(4):693–705
27. Kouzarides T (2007) SnapShot: histone-modifying enzymes. Cell 131(4):822
28. Stricker SH, Koferle A, Beck S (2017) From profiles to function in epigenomics. Nat Rev Genet 18(1):51–66
29. Liu F, Wang L, Perna F, Nimer SD (2016) Beyond transcription factors: how oncogenic signalling reshapes the epigenetic landscape. Nat Rev Cancer 16(6):359–372
30. Bell O, Tiwari VK, Thoma NH, Schubeler D (2011) Determinants and dynamics of genome accessibility. Nat Rev Genet 12(8):554–564
31. Luger K, Mader AW, Richmond RK, Sargent DF, Richmond TJ (1997) Crystal structure of the nucleosome core particle at 2.8 A resolution. Nature 389(6648):251–260
32. Jiang C, Pugh BF (2009) Nucleosome positioning and gene regulation: advances through genomics. Nat Rev Genet 10(3):161–172
33. Ehrlich M (2003) Expression of various genes is controlled by DNA methylation during mammalian development. J Cell Biochem 88(5):899–910
34. Wu H, Zhang Y (2014) Reversing DNA methylation: mechanisms, genomics, and biological functions. Cell 156(1–2):45–68
35. Lin JC, Jeong S, Liang G, Takai D, Fatemi M, Tsai YC, Egger G, Gal-Yam EN, Jones PA (2007) Role of nucleosomal occupancy in the epigenetic silencing of the MLH1 CpG island. Cancer Cell 12(5):432–444
36. Wade PA (2001) Methyl CpG-binding proteins and transcriptional repression. BioEssays 23(12):1131–1137
37. Okano M, Bell DW, Haber DA, Li E (1999) DNA methyltransferases Dnmt3a and Dnmt3b are essential for de novo methylation and mammalian development. Cell 99(3):247–257
38. Ito S, Shen L, Dai Q, Wu SC, Collins LB, Swenberg JA, He C, Zhang Y (2011) Tet proteins can convert 5-methylcytosine to 5-formylcytosine and 5-carboxylcytosine. Science 333(6047):1300–1303
39. Rivera CM, Ren B (2013) Mapping human epigenomes. Cell 155(1):39–55
40. Sadler AJ, Rossello FJ, Yu L, Deane JA, Yuan X, Wang D, Irving AT, Kaparakis-Liaskos M, Gantier MP, Ying H, Yim HC, Hartland EL, Notini AJ, de Boer S, White SJ, Mansell A et al (2015) BTB-ZF transcriptional regulator PLZF modifies chromatin to restrain inflammatory signaling programs. Proc Natl Acad Sci U S A 112(5):1535–1540

41. Schmidt SV, Krebs W, Ulas T, Xue J, Bassler K, Gunther P, Hardt AL, Schultze H, Sander J, Klee K, Theis H, Kraut M, Beyer M, Schultze JL (2016) The transcriptional regulator network of human inflammatory macrophages is defined by open chromatin. Cell Res 26(2):151–170
42. Waszak SM, Delaneau O, Gschwind AR, Kilpinen H, Raghav SK, Witwicki RM, Orioli A, Wiederkehr M, Panousis NI, Yurovsky A, Romano-Palumbo L, Planchon A, Bielser D, Padioleau I, Udin G, Thurnheer S et al (2015) Population variation and genetic control of modular chromatin architecture in humans. Cell 162(5):1039–1050
43. Calo E, Wysocka J (2013) Modification of enhancer chromatin: what, how, and why? Mol Cell 49(5):825–837
44. Nagy G, Daniel B, Jonas D, Nagy L, Barta E (2013) A novel method to predict regulatory regions based on histone mark landscapes in macrophages. Immunobiology 218(11):1416–1427
45. Falkenberg KJ, Johnstone RW (2014) Histone deacetylases and their inhibitors in cancer, neurological diseases and immune disorders. Nat Rev Drug Discov 13(9):673–691
46. Bannister AJ, Kouzarides T (2011) Regulation of chromatin by histone modifications. Cell Res 21(3):381–395
47. Pasini D, Hansen KH, Christensen J, Agger K, Cloos PA, Helin K (2008) Coordinated regulation of transcriptional repression by the RBP2 H3K4 demethylase and polycomb-repressive complex 2. Genes Dev 22(10):1345–1355
48. Yang SH, Sharrocks AD (2004) SUMO promotes HDAC-mediated transcriptional repression. Mol Cell 13(4):611–617
49. Pflum MK, Tong JK, Lane WS, Schreiber SL (2001) Histone deacetylase 1 phosphorylation promotes enzymatic activity and complex formation. J Biol Chem 276(50):47733–47741
50. Kung JT, Colognori D, Lee JT (2013) Long noncoding RNAs: past, present, and future. Genetics 193(3):651–669
51. Wang KC, Chang HY (2011) Molecular mechanisms of long noncoding RNAs. Mol Cell 43(6):904–914
52. Andersson R, Gebhard C, Miguel-Escalada I, Hoof I, Bornholdt J, Boyd M, Chen Y, Zhao X, Schmidl C, Suzuki T, Ntini E, Arner E, Valen E, Li K, Schwarzfischer L, Glatz D et al (2014) An atlas of active enhancers across human cell types and tissues. Nature 507(7493):455–461
53. FANTOM Consortium and the RIKEN PMI and CLST (DGT), Forrest AR, Kawaji H, Rehli M, Baillie JK, de Hoon MJ, Haberle V, Lassmann T, Kulakovskiy IV, Lizio M, Itoh M, Andersson R, Mungall CJ, Meehan TF et al (2014) A promoter-level mammalian expression atlas. Nature 507(7493):462–470
54. Boguski MS (2004) ENCODE and ChIP-chip in the genome era. Genomics 83(3):347–348
55. Consortium EP (2004) The ENCODE (ENCyclopedia Of DNA elements) project. Science 306(5696):636–640
56. Alvarez-Errico D, Vento-Tormo R, Sieweke M, Ballestar E (2015) Epigenetic control of myeloid cell differentiation, identity and function. Nat Rev Immunol 15(1):7–17
57. Maunakea AK, Nagarajan RP, Bilenky M, Ballinger TJ, D'Souza C, Fouse SD, Johnson BE, Hong C, Nielsen C, Zhao Y, Turecki G, Delaney A, Varhol R, Thiessen N, Shchors K, Heine VM et al (2010) Conserved role of intragenic DNA methylation in regulating alternative promoters. Nature 466(7303):253–257
58. Zilbauer M, Rayner TF, Clark C, Coffey AJ, Joyce CJ, Palta P, Palotie A, Lyons PA, Smith KG (2013) Genome-wide methylation analyses of primary human leukocyte subsets identifies functionally important cell-type-specific hypomethylated regions. Blood 122(25):e52–e60
59. Feng J, Zhou Y, Campbell SL, Le T, Li E, Sweatt JD, Silva AJ, Fan G (2010) Dnmt1 and Dnmt3a maintain DNA methylation and regulate synaptic function in adult forebrain neurons. Nat Neurosci 13(4):423–430
60. Cheng C, Huang C, Ma TT, Bian EB, He Y, Zhang L, Li J (2014) SOCS1 hypermethylation mediated by DNMT1 is associated with lipopolysaccharide-induced inflammatory cytokines in macrophages. Toxicol Lett 225(3):488–497

61. Yang X, Wang X, Liu D, Yu L, Xue B, Shi H (2014) Epigenetic regulation of macrophage polarization by DNA methyltransferase 3b. Mol Endocrinol 28(4):565–574
62. Babu M, Durga Devi T, Makinen P, Kaikkonen M, Lesch HP, Junttila S, Laiho A, Ghimire B, Gyenesei A, Yla-Herttuala S (2015) Differential promoter methylation of macrophage genes is associated with impaired vascular growth in ischemic muscles of hyperlipidemic and type 2 diabetic mice: genome-wide promoter methylation study. Circ Res 117(3):289–299
63. Teles FR, Teles RP, Uzel NG, Song XQ, Torresyap G, Socransky SS, Haffajee AD (2012) Early microbial succession in redeveloping dental biofilms in periodontal health and disease. J Periodontal Res 47(1):95–104
64. Burns E, Bachrach G, Shapira L, Nussbaum G (2006) Cutting Edge: TLR2 is required for the innate response to Porphyromonas gingivalis: activation leads to bacterial persistence and TLR2 deficiency attenuates induced alveolar bone resorption. J Immunol 177(12):8296–8300
65. Benakanakere M, Abdolhosseini M, Hosur K, Finoti LS, Kinane DF (2015) TLR2 promoter hypermethylation creates innate immune dysbiosis. J Dent Res 94(1):183–191
66. de Faria Amormino SA, Arao TC, Saraiva AM, Gomez RS, Dutra WO, da Costa JE, de Fatima Correia Silva J, Moreira PR (2013) Hypermethylation and low transcription of TLR2 gene in chronic periodontitis. Hum Immunol 74(9):1231–1236
67. Feinberg AP (2007) Phenotypic plasticity and the epigenetics of human disease. Nature 447(7143):433–440
68. Ruthenburg AJ, Li H, Patel DJ, Allis CD (2007) Multivalent engagement of chromatin modifications by linked binding modules. Nat Rev Mol Cell Biol 8(12):983–994
69. Del Rizzo PA, Trievel RC (2011) Substrate and product specificities of SET domain methyltransferases. Epigenetics 6(9):1059–1067
70. Vedadi M, Blazer L, Eram MS, Barsyte-Lovejoy D, Arrowsmith CH, Hajian T (2017) Targeting human SET1/MLL family of proteins. Protein Sci 26(4):662–676
71. Zhang Y, Mittal A, Reid J, Reich S, Gamblin SJ, Wilson JR (2015) Evolving catalytic properties of the MLL family SET domain. Structure 23(10):1921–1933
72. Krivtsov AV, Hoshii T, Armstrong SA (2017, February 27) Mixed-lineage leukemia fusions and chromatin in leukemia. Cold Spring Harb Perspect Med. pii: a026658 doi:10.1101/cshperspect.a026658
73. Shilatifard A (2012) The COMPASS family of histone H3K4 methylases: mechanisms of regulation in development and disease pathogenesis. Annu Rev Biochem 81:65–95
74. Rousseau M, Ferraiuolo MA, Crutchley JL, Wang XQ, Miura H, Blanchette M, Dostie J (2014) Classifying leukemia types with chromatin conformation data. Genome Biol 15(4):R60
75. Krivtsov AV, Armstrong SA (2007) MLL translocations, histone modifications and leukaemia stem-cell development. Nat Rev Cancer 7(11):823–833
76. Austenaa L, Barozzi I, Chronowska A, Termanini A, Ostuni R, Prosperini E, Stewart AF, Testa G, Natoli G (2012) The histone methyltransferase Wbp7 controls macrophage function through GPI glycolipid anchor synthesis. Immunity 36(4):572–585
77. Wang X, Zhu K, Li S, Liao Y, Du R, Zhang X, Shu HB, Guo AY, Li L, Wu M (2012) MLL1, a H3K4 methyltransferase, regulates the TNFalpha-stimulated activation of genes downstream of NF-kappaB. J Cell Sci 125(Pt 17):4058–4066
78. Xia M, Liu J, Wu X, Liu S, Li G, Han C, Song L, Li Z, Wang Q, Wang J, Xu T, Cao X (2013) Histone methyltransferase Ash1l suppresses interleukin-6 production and inflammatory autoimmune diseases by inducing the ubiquitin-editing enzyme A20. Immunity 39(3):470–481
79. De Santa F, Totaro MG, Prosperini E, Notarbartolo S, Testa G, Natoli G (2007) The histone H3 lysine-27 demethylase Jmjd3 links inflammation to inhibition of polycomb-mediated gene silencing. Cell 130(6):1083–1094
80. Ishii M, Wen H, Corsa CA, Liu T, Coelho AL, Allen RM, Carson WF, Cavassani KA, Li X, Lukacs NW, Hogaboam CM, Dou Y, Kunkel SL (2009) Epigenetic regulation of the alternatively activated macrophage phenotype. Blood 114(15):3244–3254

81. Satoh T, Takeuchi O, Vandenbon A, Yasuda K, Tanaka Y, Kumagai Y, Miyake T, Matsushita K, Okazaki T, Saitoh T, Honma K, Matsuyama T, Yui K, Tsujimura T, Standley DM, Nakanishi K et al (2010) The Jmjd3-Irf4 axis regulates M2 macrophage polarization and host responses against helminth infection. Nat Immunol 11(10):936–944
82. Liu Y, Zhang Q, Ding Y, Li X, Zhao D, Zhao K, Guo Z, Cao X (2015) Histone lysine methyltransferase Ezh1 promotes TLR-triggered inflammatory cytokine production by suppressing Tollip. J Immunol 194(6):2838–2846
83. Fang TC, Schaefer U, Mecklenbrauker I, Stienen A, Dewell S, Chen MS, Rioja I, Parravicini V, Prinjha RK, Chandwani R, MacDonald MR, Lee K, Rice CM, Tarakhovsky A (2012) Histone H3 lysine 9 di-methylation as an epigenetic signature of the interferon response. J Exp Med 209(4):661–669
84. Schliehe C, Flynn EK, Vilagos B, Richson U, Swaminathan S, Bosnjak B, Bauer L, Kandasamy RK, Griesshammer IM, Kosack L, Schmitz F, Litvak V, Sissons J, Lercher A, Bhattacharya A, Khamina K et al (2015) The methyltransferase Setdb2 mediates virus-induced susceptibility to bacterial superinfection. Nat Immunol 16(1):67–74
85. Zhang Q, Zhao K, Shen Q, Han Y, Gu Y, Li X, Zhao D, Liu Y, Wang C, Zhang X, Su X, Liu J, Ge W, Levine RL, Li N, Cao X (2015) Tet2 is required to resolve inflammation by recruiting Hdac2 to specifically repress IL-6. Nature 525(7569):389–393
86. Li X, Zhang Q, Ding Y, Liu Y, Zhao D, Zhao K, Shen Q, Liu X, Zhu X, Li N, Cheng Z, Fan G, Wang Q, Cao X (2016) Methyltransferase Dnmt3a upregulates HDAC9 to deacetylate the kinase TBK1 for activation of antiviral innate immunity. Nat Immunol 17(7):806–815
87. Li Y, Seto E (2016) HDACs and HDAC inhibitors in cancer development and therapy. Cold Spring Harb Perspect Med 6(10):a026831
88. Lee KK, Workman JL (2007) Histone acetyltransferase complexes: one size doesn't fit all. Nat Rev Mol Cell Biol 8(4):284–295
89. Chauhan A, Quenum FZ, Abbas A, Bradley DS, Nechaev S, Singh BB, Sharma J, Mishra BB (2015) Epigenetic modulation of microglial inflammatory gene loci in helminth-induced immune suppression: implications for immune regulation in neurocysticercosis. ASN Neuro 7(4):1–12
90. Feng D, Sangster-Guity N, Stone R, Korczeniewska J, Mancl ME, Fitzgerald-Bocarsly P, Barnes BJ (2010) Differential requirement of histone acetylase and deacetylase activities for IRF5-mediated proinflammatory cytokine expression. J Immunol 185(10):6003–6012
91. Kawahara TL, Michishita E, Adler AS, Damian M, Berber E, Lin M, McCord RA, Ongaigui KC, Boxer LD, Chang HY, Chua KF (2009) SIRT6 links histone H3 lysine 9 deacetylation to NF-kappaB-dependent gene expression and organismal life span. Cell 136(1):62–74
92. Sadler AJ, Suliman BA, Yu L, Yuan X, Wang D, Irving AT, Sarvestani ST, Banerjee A, Mansell AS, Liu JP, Gerondakis S, Williams BR, Xu D (2015) The acetyltransferase HAT1 moderates the NF-kappaB response by regulating the transcription factor PLZF. Nat Commun 6:6795
93. Alam MM, O'Neill LA (2011) MicroRNAs and the resolution phase of inflammation in macrophages. Eur J Immunol 41(9):2482–2485
94. O'Neill LA, Sheedy FJ, McCoy CE (2011) MicroRNAs: the fine-tuners of Toll-like receptor signalling. Nat Rev Immunol 11(3):163–175
95. Hangauer MJ, Vaughn IW, McManus MT (2013) Pervasive transcription of the human genome produces thousands of previously unidentified long intergenic noncoding RNAs. PLoS Genet 9(6):e1003569
96. Cabili MN, Trapnell C, Goff L, Koziol M, Tazon-Vega B, Regev A, Rinn JL (2011) Integrative annotation of human large intergenic noncoding RNAs reveals global properties and specific subclasses. Genes Dev 25(18):1915–1927
97. Guttman M, Rinn JL (2012) Modular regulatory principles of large non-coding RNAs. Nature 482(7385):339–346
98. Wang P, Xue Y, Han Y, Lin L, Wu C, Xu S, Jiang Z, Xu J, Liu Q, Cao X (2014) The STAT3-binding long noncoding RNA lnc-DC controls human dendritic cell differentiation. Science 344(6181):310–313

99. Carpenter S, Aiello D, Atianand MK, Ricci EP, Gandhi P, Hall LL, Byron M, Monks B, Henry-Bezy M, Lawrence JB, O'Neill LA, Moore MJ, Caffrey DR, Fitzgerald KA (2013) A long noncoding RNA mediates both activation and repression of immune response genes. Science 341(6147):789–792
100. Hu G, Gong AY, Wang Y, Ma S, Chen X, Chen J, Su CJ, Shibata A, Strauss-Soukup JK, Drescher KM, Chen XM (2016) LincRNA-Cox2 promotes late inflammatory gene transcription in macrophages through modulating SWI/SNF-mediated chromatin remodeling. J Immunol 196(6):2799–2808
101. Tong Q, Gong AY, Zhang XT, Lin C, Ma S, Chen J, Hu G, Chen XM (2016) LincRNA-Cox2 modulates TNF-alpha-induced transcription of Il12b gene in intestinal epithelial cells through regulation of Mi-2/NuRD-mediated epigenetic histone modifications. FASEB J 30(3):1187–1197
102. Rapicavoli NA, Qu K, Zhang J, Mikhail M, Laberge RM, Chang HY (2013) A mammalian pseudogene lncRNA at the interface of inflammation and anti-inflammatory therapeutics. eLife 2:e00762
103. Li Z, Chao TC, Chang KY, Lin N, Patil VS, Shimizu C, Head SR, Burns JC, Rana TM (2014) The long noncoding RNA THRIL regulates TNFalpha expression through its interaction with hnRNPL. Proc Natl Acad Sci U S A 111(3):1002–1007
104. Ma S, Ming Z, Gong AY, Wang Y, Chen X, Hu G, Zhou R, Shibata A, Swanson PC, Chen XM (2017) A long noncoding RNA, lincRNA-tnfaip3, acts as a coregulator of NF-kappaB to modulate inflammatory gene transcription in mouse macrophages. FASEB J 31(3):1215–1225
105. Atianand MK, Hu W, Satpathy AT, Shen Y, Ricci EP, Alvarez-Dominguez JR, Bhatta A, Schattgen SA, McGowan JD, Blin J, Braun JE, Gandhi P, Moore MJ, Chang HY, Lodish HF, Caffrey DR et al (2016) A Long Noncoding RNA lincRNA-EPS Acts as a transcriptional brake to restrain inflammation. Cell 165(7):1672–1685
106. Medzhitov R, Horng T (2009) Transcriptional control of the inflammatory response. Nat Rev Immunol 9(10):692–703
107. Medzhitov R, Schneider DS, Soares MP (2012) Disease tolerance as a defense strategy. Science 335(6071):936–941
108. Witteveldt J, Cifuentes CC, Vlak JM, van Hulten MC (2004) Protection of Penaeus monodon against white spot syndrome virus by oral vaccination. J Virol 78(4):2057–2061
109. Saeed S, Quintin J, Kerstens HH, Rao NA, Aghajanirefah A, Matarese F, Cheng SC, Ratter J, Berentsen K, van der Ent MA, Sharifi N, Janssen-Megens EM, Ter Huurne M, Mandoli A, van Schaik T, Ng A et al (2014) Epigenetic programming of monocyte-to-macrophage differentiation and trained innate immunity. Science 345(6204):1251086
110. Zhang X, Morrison DC (1993) Lipopolysaccharide-induced selective priming effects on tumor necrosis factor alpha and nitric oxide production in mouse peritoneal macrophages. J Exp Med 177(2):511–516
111. Hirohashi N, Morrison DC (1996) Low-dose lipopolysaccharide (LPS) pretreatment of mouse macrophages modulates LPS-dependent interleukin-6 production in vitro. Infect Immun 64(3):1011–1015
112. Quintin J, Saeed S, Martens JH, Giamarellos-Bourboulis EJ, Ifrim DC, Logie C, Jacobs L, Jansen T, Kullberg BJ, Wijmenga C, Joosten LA, Xavier RJ, van der Meer JW, Stunnenberg HG, Netea MG (2012) Candida albicans infection affords protection against reinfection via functional reprogramming of monocytes. Cell Host Microbe 12(2):223–232
113. Ostuni R, Piccolo V, Barozzi I, Polletti S, Termanini A, Bonifacio S, Curina A, Prosperini E, Ghisletti S, Natoli G (2013) Latent enhancers activated by stimulation in differentiated cells. Cell 152(1–2):157–171
114. Bhatt D, Ghosh S (2014) Regulation of the NF-kappaB-mediated transcription of inflammatory genes. Front Immunol 5:71
115. Cheng SC, Quintin J, Cramer RA, Shepardson KM, Saeed S, Kumar V, Giamarellos-Bourboulis EJ, Martens JH, Rao NA, Aghajanirefah A, Manjeri GR, Li Y, Ifrim DC, Arts RJ,

van der Veer BM, Deen PM et al (2014) mTOR- and HIF-1alpha-mediated aerobic glycolysis as metabolic basis for trained immunity. Science 345(6204):1250684
116. Rajaiah R, Perkins DJ, Polumuri SK, Zhao A, Keegan AD, Vogel SN (2013) Dissociation of endotoxin tolerance and differentiation of alternatively activated macrophages. J Immunol 190(9):4763–4772
117. Adib-Conquy M, Cavaillon JM (2009) Compensatory anti-inflammatory response syndrome. Thromb Haemost 101(1):36–47
118. Ivashkiv LB, Park SH (2016, June) Epigenetic regulation of myeloid cells. Microbiol Spectr 4(3). doi:10.1128/microbiolspec.MCHD-0010-2015
119. Takeuch O, Akira S (2011) Epigenetic control of macrophage polarization. Eur J Immunol 41(9):2490–2493
120. Dandona P, Ghanim H, Green K, Sia CL, Abuaysheh S, Kuhadiya N, Batra M, Dhindsa S, Chaudhuri A (2013) Insulin infusion suppresses while glucose infusion induces Toll-like receptors and high-mobility group-B1 protein expression in mononuclear cells of type 1 diabetes patients. Am J Physiol Endocrinol Metab 304(8):E810–E818
121. Shanmugam N, Reddy MA, Guha M, Natarajan R (2003) High glucose-induced expression of proinflammatory cytokine and chemokine genes in monocytic cells. Diabetes 52(5):1256–1264
122. Yun JM, Jialal I, Devaraj S (2011) Epigenetic regulation of high glucose-induced proinflammatory cytokine production in monocytes by curcumin. J Nutr Biochem 22(5):450–458
123. Maslowski KM, Vieira AT, Ng A, Kranich J, Sierro F, Yu D, Schilter HC, Rolph MS, Mackay F, Artis D, Xavier RJ, Teixeira MM, Mackay CR (2009) Regulation of inflammatory responses by gut microbiota and chemoattractant receptor GPR43. Nature 461(7268):1282–1286
124. Ey B, Eyking A, Klepak M, Salzman NH, Gothert JR, Runzi M, Schmid KW, Gerken G, Podolsky DK, Cario E (2013) Loss of TLR2 worsens spontaneous colitis in MDR1A deficiency through commensally induced pyroptosis. J Immunol 190(11):5676–5688
125. Huang E, Wells CA (2014) The ground state of innate immune responsiveness is determined at the interface of genetic, epigenetic, and environmental influences. J Immunol 193(1):13–19
126. Yanez A, Hassanzadeh-Kiabi N, Ng MY, Megias J, Subramanian A, Liu GY, Underhill DM, Gil ML, Goodridge HS (2013) Detection of a TLR2 agonist by hematopoietic stem and progenitor cells impacts the function of the macrophages they produce. Eur J Immunol 43(8):2114–2125
127. Consortium EP (2012) An integrated encyclopedia of DNA elements in the human genome. Nature 489(7414):57–74
128. Ricano-Ponce I, Wijmenga C (2013) Mapping of immune-mediated disease genes. Annu Rev Genomics Hum Genet 14:325–353
129. Kumar V, Westra HJ, Karjalainen J, Zhernakova DV, Esko T, Hrdlickova B, Almeida R, Zhernakova A, Reinmaa E, Vosa U, Hofker MH, Fehrmann RS, Fu J, Withoff S, Metspalu A, Franke L et al (2013) Human disease-associated genetic variation impacts large intergenic non-coding RNA expression. PLoS Genet 9(1):e1003201
130. Kent OA, McCall MN, Cornish TC, Halushka MK (2014) Lessons from miR-143/145: the importance of cell-type localization of miRNAs. Nucleic Acids Res 42(12):7528–7538
131. Zhang DD, Wang WT, Xiong J, Xie XM, Cui SS, Zhao ZG, Li MJ, Zhang ZQ, Hao DL, Zhao X, Li YJ, Wang J, Chen HZ, Lv X, Liu DP (2017) Long noncoding RNA LINC00305 promotes inflammation by activating the AHRR-NF-kappaB pathway in human monocytes. Sci Rep 7:46204
132. Liu CC, Fang TJ, Ou TT, Wu CC, Li RN, Lin YC, Lin CH, Tsai WC, Liu HW, Yen JH (2011) Global DNA methylation, DNMT1, and MBD2 in patients with rheumatoid arthritis. Immunol Lett 135(1–2):96–99
133. Martin P, McGovern A, Orozco G, Duffus K, Yarwood A, Schoenfelder S, Cooper NJ, Barton A, Wallace C, Fraser P, Worthington J, Eyre S (2015) Capture Hi-C reveals novel candidate genes and complex long-range interactions with related autoimmune risk loci. Nat Commun 6:10069

134. Javierre BM, Burren OS, Wilder SP, Kreuzhuber R, Hill SM, Sewitz S, Cairns J, Wingett SW, Varnai C, Thiecke MJ, Burden F, Farrow S, Cutler AJ, Rehnstrom K, Downes K, Grassi L et al (2016) Lineage-specific genome architecture links enhancers and non-coding disease variants to target gene promoters. Cell 167(5):1369–1384. e1319
135. Peters JE, Lyons PA, Lee JC, Richard AC, Fortune MD, Newcombe PJ, Richardson S, Smith KG (2016) Insight into genotype-phenotype associations through eQTL mapping in multiple cell types in health and immune-mediated disease. PLoS Genet 12(3):e1005908
136. Fullwood MJ, Liu MH, Pan YF, Liu J, Xu H, Mohamed YB, Orlov YL, Velkov S, Ho A, Mei PH, Chew EG, Huang PY, Welboren WJ, Han Y, Ooi HS, Ariyaratne PN et al (2009) An oestrogen-receptor-alpha-bound human chromatin interactome. Nature 462(7269):58–64
137. Lieberman-Aiden E, van Berkum NL, Williams L, Imakaev M, Ragoczy T, Telling A, Amit I, Lajoie BR, Sabo PJ, Dorschner MO, Sandstrom R, Bernstein B, Bender MA, Groudine M, Gnirke A, Stamatoyannopoulos J et al (2009) Comprehensive mapping of long-range interactions reveals folding principles of the human genome. Science 326(5950):289–293
138. Dixon JR, Selvaraj S, Yue F, Kim A, Li Y, Shen Y, Hu M, Liu JS, Ren B (2012) Topological domains in mammalian genomes identified by analysis of chromatin interactions. Nature 485(7398):376–380
139. Tang Z, Luo OJ, Li X, Zheng M, Zhu JJ, Szalaj P, Trzaskoma P, Magalska A, Wlodarczyk J, Ruszczycki B, Michalski P, Piecuch E, Wang P, Wang D, Tian SZ, Penrad-Mobayed M et al (2015) CTCF-mediated human 3D genome architecture reveals chromatin topology for transcription. Cell 163(7):1611–1627
140. Laird PW (2010) Principles and challenges of genomewide DNA methylation analysis. Nat Rev Genet 11(3):191–203
141. Park PJ (2009) ChIP-seq: advantages and challenges of a maturing technology. Nat Rev Genet 10(10):669–680
142. Yong WS, Hsu FM, Chen PY (2016) Profiling genome-wide DNA methylation. Epigenetics Chromatin 9:26
143. De Nardo D, Labzin LI, Kono H, Seki R, Schmidt SV, Beyer M, Xu D, Zimmer S, Lahrmann C, Schildberg FA, Vogelhuber J, Kraut M, Ulas T, Kerksiek A, Krebs W, Bode N et al (2014) High-density lipoprotein mediates anti-inflammatory reprogramming of macrophages via the transcriptional regulator ATF3. Nat Immunol 15(2):152–160
144. Stelzer Y, Shivalila CS, Soldner F, Markoulaki S, Jaenisch R (2015) Tracing dynamic changes of DNA methylation at single-cell resolution. Cell 163(1):218–229
145. Sasaki K, Ito T, Nishino N, Khochbin S, Yoshida M (2009) Real-time imaging of histone H4 hyperacetylation in living cells. Proc Natl Acad Sci U S A 106(38):16257–16262
146. Sander JD, Joung JK (2014) CRISPR-Cas systems for editing, regulating and targeting genomes. Nat Biotechnol 32(4):347–355
147. Tough DF, Tak PP, Tarakhovsky A, Prinjha RK (2016) Epigenetic drug discovery: breaking through the immune barrier. Nat Rev Drug Discov 15(12):835–853
148. Leoni F, Zaliani A, Bertolini G, Porro G, Pagani P, Pozzi P, Dona G, Fossati G, Sozzani S, Azam T, Bufler P, Fantuzzi G, Goncharov I, Kim SH, Pomerantz BJ, Reznikov LL et al (2002) The antitumor histone deacetylase inhibitor suberoylanilide hydroxamic acid exhibits antiinflammatory properties via suppression of cytokines. Proc Natl Acad Sci U S A 99(5):2995–3000
149. Leoni F, Fossati G, Lewis EC, Lee JK, Porro G, Pagani P, Modena D, Moras ML, Pozzi P, Reznikov LL, Siegmund B, Fantuzzi G, Dinarello CA, Mascagni P (2005) The histone deacetylase inhibitor ITF2357 reduces production of pro-inflammatory cytokines in vitro and systemic inflammation in vivo. Mol Med 11(1–12):1–15
150. Grabiec AM, Krausz S, de Jager W, Burakowski T, Groot D, Sanders ME, Prakken BJ, Maslinski W, Eldering E, Tak PP, Reedquist KA (2010) Histone deacetylase inhibitors suppress inflammatory activation of rheumatoid arthritis patient synovial macrophages and tissue. J Immunol 184(5):2718–2728

151. Gillespie J, Savic S, Wong C, Hempshall A, Inman M, Emery P, Grigg R, McDermott MF (2012) Histone deacetylases are dysregulated in rheumatoid arthritis and a novel histone deacetylase 3-selective inhibitor reduces interleukin-6 production by peripheral blood mononuclear cells from rheumatoid arthritis patients. Arthritis Rheum 64(2):418–422
152. Belkina AC, Denis GV (2012) BET domain co-regulators in obesity, inflammation and cancer. Nat Rev Cancer 12(7):465–477
153. Khan YM, Kirkham P, Barnes PJ, Adcock IM (2014) Brd4 is essential for IL-1beta-induced inflammation in human airway epithelial cells. PLoS One 9(4):e95051
154. Wang CY, Filippakopoulos P (2015) Beating the odds: BETs in disease. Trends Biochem Sci 40(8):468–479
155. Nicodeme E, Jeffrey KL, Schaefer U, Beinke S, Dewell S, Chung CW, Chandwani R, Marazzi I, Wilson P, Coste H, White J, Kirilovsky J, Rice CM, Lora JM, Prinjha RK, Lee K et al (2010) Suppression of inflammation by a synthetic histone mimic. Nature 468(7327):1119–1123
156. Belkina AC, Nikolajczyk BS, Denis GV (2013) BET protein function is required for inflammation: Brd2 genetic disruption and BET inhibitor JQ1 impair mouse macrophage inflammatory responses. J Immunol 190(7):3670–3678
157. Kruidenier L, Chung CW, Cheng Z, Liddle J, Che K, Joberty G, Bantscheff M, Bountra C, Bridges A, Diallo H, Eberhard D, Hutchinson S, Jones E, Katso R, Leveridge M, Mander PK et al (2012) A selective jumonji H3K27 demethylase inhibitor modulates the proinflammatory macrophage response. Nature 488(7411):404–408

Chapter 4
Roles of HDACs in the Responses of Innate Immune Cells and as Targets in Inflammatory Diseases

Yiqun Hu and Bandar Ali Suliman

Abstract Histone deacetylases (HDACs) are an emerging class of molecules involved in the epigenetic regulation of innate immune responses through Toll-like receptor (TLR) and interferon (IFN) signaling pathways. HDACs are also key drivers of inflammatory diseases via epigenetic regulation through chromatin DNA and histone modification by methylation and acetylation, among other mechanisms, which control innate immune cell gene expression. Importantly, these epigenetic changes are reversible, and HDACs may also be targeted by small-molecule HDAC inhibitors, which have been used in clinical settings for cancer therapy. Here, we highlight HDACs as strong therapeutic molecules and explore HDAC-induced mechanisms regulating innate immune responses and inflammatory cytokine control, with the goal of developing personalized medicine for the treatment of human diseases, including inflammatory diseases and immune disorders. Currently, this novel class of immunomodulatory therapeutics is being evaluated in the laboratory, in preclinical models, and in the clinic.

Keywords Histone deacetylases (HDACs) • HDAC inhibitors • Inflammatory signaling • Inflammatory diseases

---

Y. Hu
Faculty of Medical Laboratory Science, Ruijin Hospital, Shanghai Jiao Tong University School of Medicine, Shanghai, China

B.A. Suliman (✉)
College of Applied Medical Sciences, Taibah University, Madinah, Saudi Arabia

Molecular Biomedicine Program, Research Center, King Faisal Specialist Hospital & Research Center, Riyadh, Saudi Arabia
e-mail: bsuliman@taibahu.edu.sa

D. Xu (ed.), *Regulation of Inflammatory Signaling in Health and Disease*,
Advances in Experimental Medicine and Biology 1024,
DOI 10.1007/978-981-10-5987-2_4

4.1 Introduction

4.1.1 The Innate Immune System

The innate immune system is a unique collection of immune cells and tissues that share a common goal: the nonspecific identification of pathogens [1]. Most innate immune cells are present at birth and harbor genetically conserved molecular receptors that are capable of identifying self from nonself molecules. These *innate immune receptors* are called *pattern recognition receptors* (*PRRs*) and are used by the immune system to recognize genetically conserved molecular structures, called *pathogen-associated molecular patterns* (*PAMPs*), which are shared by many microbial species. PAMP structures consist of combinations of lipid, sugar, protein, and nucleic acid sequences and constitute the outward-facing biological components of viruses, bacteria, fungi, and parasites. One of the hallmarks of the innate immune response is *speed*. The innate immune response is initiated by the interaction of PRRs with their ligands (i.e., PAMPs). Despite the molecular diversity of PAMPs in many microbial species, the immune responses initiated by innate immune cells are rapid and occur within seconds of successful recognition [2]. Receptor-ligand interactions are functionally translated into two distinct pathways: phagocytosis and the inflammatory response. Phagocytosis is the biological process by which professional innate immune cells, such as neutrophils, macrophages, and dendritic cells, engulf and destroy pathogenic molecules to extract specific protein structures called epitopes, which activate the adaptive arm of the immune system [3]. The inflammatory response, in contrast, consists of the production and subsequent release of regulatory chemical molecules aimed at orchestrating the functions of other immune cells [4].

This dependency of the adaptive immune response on the innate arm of the immune system demonstrates the crucial and important role of innate immune cells [5]. The human body requires this general and nonspecific immune response to achieve the full spectrum of adaptive immune recognition, either by supplying specific epitopes that are targeted by the adaptive immune system or by providing the necessary chemical modulators required for successful immune signaling.

4.2 Inflammatory Response Signaling

The immune system represents one of the largest signaling networks in the human body. Intertwined signaling cascades provide protection via millions of cells that circulate in the human body and defend it not only against external microorganisms but also internal defective and cancerous cells. There is an immaculate harmony between the two arms of the immune system. The innate immune cells, which are responsible for initiating the adaptive immune response, control the development and differentiation of adaptive immune cells. They also participate in the positive and negative regulation of the inflammatory response to fine-tune the involvement of different immune cells in the overall immunological reaction [6].

Many studies have investigated the inflammatory response and how each and every molecule plays a part in its execution. Innate immune cells such as macrophages, fibroblasts, and dendritic cells recognize pathogens and pathogenic epitopes and subsequently initiate complex immune signaling cascades by which they communicate their findings, the scope of the inflammatory reaction, and the exact level of the adaptive immune system response required to eliminate the threat [7]. To achieve such a delicate outcome, stimulated innate immune cells control both the spectrum and scope of the inflammatory reaction by controlling intermediary lipid molecules and enzymes that work together at the transcriptional level for each specific cell. Although many pathogenic molecules are recognized by the antigen-presenting cells of the innate immune system, genetic signatures at the level of transcriptional regulation are commonly shared among those cells.

Cytokine production is the hallmark and the ultimate outcome of the innate immune response [8]. When a macrophage encounters a virus, bacterium, or other microorganism, it produces cytokines known as interleukins that instruct other immune cells. The responses of these stimulated cells are not only directed toward other immune cells but also toward many other cells and tissues with essential roles in the immune system response to combat the infection. Some of the chemical mediators produced by innate immune cells control adjacent endothelial cells to enhance permeability, allowing necessary proteins to escape the blood circulation [9]. Other inflammatory mediators act on vascular smooth muscle cells to promote vasodilation, causing an increase in blood flow and helping other immune cells to enter the site of inflammation.

4.3 Transcriptional Regulation

Immune cells, like any other cell type, encompass a vast amount of DNA, which is tightly packed inside the nucleus. This DNA contains all genetic information needed by the cells to produce the necessary molecular structures for pathogenic identification, the enzymes and proteins involved in signal transduction, and the actual chemokines and cytokines produced as a result of immune activation. The DNA is contained in the nucleus in the form of supercoiled antiparallel strands of nucleotides wrapped around core histone proteins, ensuring the tight packing of DNA into structures called nucleosomes. Each nucleosome contains 146 base pairs of DNA wrapped twice around eight histone molecules, such as H2A, H2B, H3, and H4 [10]. This growing chain of nucleosomes along the DNA fiber forms the chromatin. The chromatin is maintained by the negative charge of the DNA strands together with the positive charge of the histone proteins, forming strong bonds that preserve the supercoiled structure. The arrangement of the DNA and histones closely regulates transcription by either allowing RNA polymerase complexes (the transcriptional machinery) to access the DNA strand to search for promoter regions or blocking

DNA access and thus precluding transcription. Euchromatin is the DNA state characterized by loose packaging because the core histone H1 has not yet been added; in contrast, heterochromatin is the DNA state characterized by tight packaging after the addition of H1 [10, 11]. Although transcriptional regulation involves many biological processes, modification of the chromatin state remains one of the most influential methods [12–14]. Many studies have established a direct link between transcriptional activation and altered chromatin states. Such alterations along DNA strands range from the simple addition of methyl groups to specific arrangements of the cytosine pyrimidine rings [15].

4.4 Histone Deacetylases (HDACs) and Transcriptional Regulation

The regulation of histone acetylation and deacetylation is controlled by enzymes, specifically HDACs and histone acetyltransferases (HATs), which are essential for gene regulation [16]. HATs are enzymes that transfer acetyl groups to core histones, with subsequent effects on chromatin remodeling and gene expression. HATs comprise two families [17]: (1) p300/CBP, Taf1, and nuclear receptor coactivators that also possess catalytic acetyltransferase activities and (2) Gcn5 N-acetyltransferases (GNATs) and Morf, Ybf2, Sas2, and Tip60 HATs. HATs are involved in initiating gene expression in macrophages during inflammation [18, 19]. HDACs are enzymes that oppose HAT functions and exert dynamic effects that may derive from other posttranslational modifications. HDAC functions potentially induce further epigenetic changes and alternative gene expression. Eighteen mammalian HDACs have been identified and classified into five groups: class I (HDAC1, HDAC2, HDAC3, and HDAC8), class IIa (HDAC4, HDAC5, HDAC7, and HDAC9), class IIb (HDAC6 and HDAC10), class III (consists of the NAD+-dependent HDACs), and HDAC11, which constitutes a class of its own (Fig. 4.1) [20]. A single DNA strand consists of a collection of histone molecules that bind to a long chain of nucleotides. The structural binding affinity used by histone molecules to control how tight the nucleotide chain is wrapped around them is dependent on the electrical charge difference between the phosphate groups in the DNA strand and the adjacent N-terminal tails of each histone molecule [21]. Since this electrical charge is manipulated by certain enzymes and chemicals, it is therefore possible to adjust the binding of histone

Fig. 4.1 Classification of the two common classes of HDACs

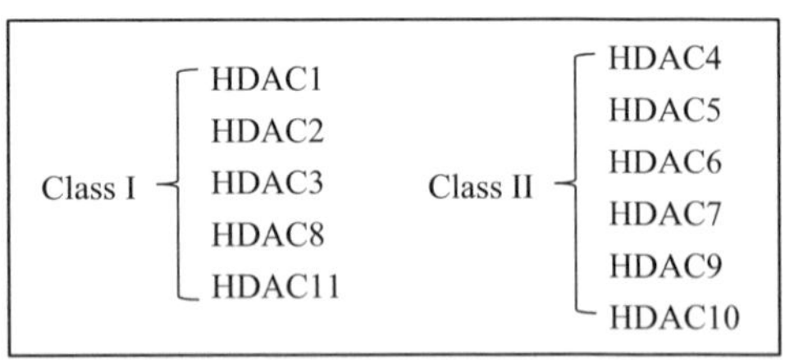

molecules to the DNA strand in individual nucleosomes through the process of acetylation [22]. To maintain electrical balance, the addition of an acetyl group to lysine residues in histone molecules neutralizes their positive charge and thus relaxes their grip on the DNA strand, allowing the nucleosome to be accessed by the transcriptional machinery and all necessary transcription factors, ultimately upregulating the transcription process and producing more mRNA transcripts. This process is very important for active genes for which the promoter regions are actively accessible to the regulatory molecules required for transcription. Alternatively, removal of acetyl groups restores the positive charge of histone molecules, tightening their grip on the DNA strand and rendering the nucleosome inaccessible to the transcriptional machinery. HDACs catalyze the removal of acetyl groups from histone molecules. Many cells harbor this powerful mechanism to regulate the expression of certain genes via specific, similar activities [23]. For example, the cell cycle, a universal biological process that is carried out by most cells in the human body, requires the availability and abundance of certain proteins in a specific time frame. To achieve this goal, cells employ different HDACs to alter the mRNA expression of genes of interest during different phases of the cell cycle [24].

Since the early days of experimental genetics, many scientists have performed experiments using *Drosophila* and *Saccharomyces cerevisiae*, which resulted in the discovery of a variety of biological activities that correlate with their human counterparts [25]. Therefore, eukaryotic and human HDACs have been classified to represent structural homology and catalytic activity. Based on this system, HDACs in humans are classified as class I and class II. Class I HDACs have a closer relationship with the transcriptional regulator RPD3 found in *Saccharomyces cerevisiae*, while class II HDACs share another domain that is more closely related to the HDAC1 protein in the same yeast [26]. Although this classification system has successfully distinguished between closely related HDACs in mammals to date, the overall DNA sequence similarities between eukaryotic enzymes and their prokaryotic counterparts are not sufficiently significant [27]. Eukaryotes are more complex and diverse, and therefore it is expected that different classes of enzymes have more prominent roles in specialized functions, such as cellular differentiation and basic developmental processes [28]. Consequently, some enzymes, such as class I HDACs, are found in almost all cells of the body, while class II HDACs are more restricted to a specialized set of cells [29]. *In vivo* experiments conducted in mouse models have shown that HDACs from both classes control the expression of specific subsets of DNA sequences in both a spatial and temporal manner. Thus, the functional acetylation or deacetylation of only a subpopulation of histones is achieved through the function of members of class I and class II HDACs, either individually or collectively [30, 31].

A very simple representation of HDAC classes shows the members of both class I and class II HDACs. There are more subclasses for some members, such as HDAC9a and HDAC9b. For a detailed and in-depth differentiation of HDAC classes and isoforms, please consult the review by De Ruijter and his colleagues [32].

4.5 Role of HDACs in the Innate Immune System

The cells of the innate immune system are the most crucial line of defense against specific pathogenic invaders as well as cancer cells. Because of their importance in maintaining a healthy state, the regulatory mechanisms controlling the functions and cellular interactions within this system are among the most active areas of medical research [33]. Cells of the innate immune system are controlled by several genetic and epigenetic mechanisms, in addition to external stimuli. This network of regulatory elements influences the overall inflammatory outcome and is essential for defining the epigenetic signatures of individual cells. For example, monozygotic twins may show variable susceptibility to pathological conditions, even while sharing the same environment [34–36]. This finding supports the impact of epigenetic regulation in defining the scope and overall outcome of the inflammatory responses caused by different stimuli. Current studies are focused on HDAC regulation of TLR and type I IFN-mediated signaling pathways.

4.6 HDAC Regulation of TLR-Mediated Signaling

TLRs comprise a very well-known group of genetically conserved receptors that are responsible for the identification of a wide range of pathogens. TLRs are found on the surfaces of many innate cells circulating in the bloodstream and are concentrated in various tissue types. Since their discovery in 2001, many molecular ligands have been identified for each specific TLR [7]. These ligands are molecular activators that are responsible for activating a particular TLR and initiating specific signaling cascades to produce cytokines targeting the pathogen or microbe harboring the activating molecular pattern. Although the contributions of different molecules to the responses of innate cells to different ligands have been established, the actual mechanism underlying this type of regulation remains poorly understood.

As discussed earlier in this chapter, many inflammatory signals activate certain TLRs to produce a required cytokine profile. Cigarette smoking, for instance, increases the production of a mixed group of inflammatory mediators, including MCP-1, TNF-α, and IL-8, which are directly implicated in the development of the classical clinical features of chronic obstructive pulmonary disease (COPD) [37, 38]. The continuous increase in the inflammatory signals that cause mucus accumulation during destruction of the lung parenchyma and ultimately fibrosis and obstruction of the airways has recently been linked to both TLR2 and TLR4 activation [39]. Many of the downstream genes activated by both of these TLRs contain acetylation sites in their promoter regions that attract HDAC1 binding [40]. In another study, HDAC1 was successfully immunoprecipitated along with ATF-3 from LPS-activated macrophages. ATF-3 is known for its function as a negative regulator of TLR4 activation that limits the production of IL-6 and IL-12 post-LPS treatment [41, 42]. The protein complex formed between ATF-3 and HDAC1 modifies histone structures and limits the accessibility of the transcriptional machinery to the IL-6 and IL-12 promoters. HDAC2 forms regulatory complexes with multiple proteins, enabling it to regulate the transcriptional activities of various targets [43, 44]. HDAC2 binds MTA1, a known chromatin modifier implicated in nucleosome remodeling, to form a repressor complex targeting the *MyD88* promoter after TLR4 activation of macrophages. MyD88 is an adaptor protein that is responsible for transferring molecular signals from TLRs to a group of interacting proteins known as nuclear factor-kappa-B (NF-κB), which in turn exerts its effects on multiple genes to determine the scope and amplitude of the immune response and inflammatory reaction. HDAC3 was recently associated with the direct functional inhibition of NF-κB. Nuclear co-localization of HDAC3 with inhibitor of NF-κB-like (IKBL) was observed after TLR stimulation, resulting in a decrease in NF-κB activity [45]. Similarly, the transcriptional inhibitor PLZF recruits HDAC3 to the promoter regions of many NF-κB target genes. These regions contain NF-κB-binding motifs, allowing HDAC3 to modify chromatin and NF-κB to access these DNA regions, ultimately controlling transcriptional activity [46, 47].

Similar to other HDACs, HDAC6 is responsible for the recruitment of the MyD88-GyrB complex after TLR2 and TLR4 activation. HDAC6 negatively regulates those TLR-activated pathways by inhibiting MyD88-TRAF6 formation and attenuating the resulting immune response [48]. HDAC7 is expressed at elevated levels in inflammatory macrophages (thioglycollate-elicited peritoneal macrophages) compared with its expression in bone marrow-derived macrophages and the RAW264 cell line. Overexpression of a specific, alternatively spliced isoform of HDAC7 lacking the N-terminal 22 amino acids promotes the LPS-inducible expression of HDAC-dependent genes (Edn1, IL-12p40, and IL-6) in RAW264 cells [49]. Such regulation is usually carried out by immune cells to limit the cytotoxic and destructive effects caused by strong stimulation of the signaling pathway under specific conditions [50, 51]. Although the exact effects of certain HDACs on various levels of immune signaling pathways have been dissected and elucidated, there is no clear understanding of how many HDACs control the final outcome of the immune response at the molecular level. For example, the effects of HDACs on the polarization and programming of CD4 T lymphocytes following TLR4 activation are not known. Both macrophages and dendritic cells are responsible for activating T lymphocytes to derive a specific adaptive immune response against certain pathogens. This activation may specify T-cell polarization: either Th1, which is pro-inflammatory, or Th2, which is anti-inflammatory [52, 53]. Polarization requires the delicate temporal activation of certain mediators, such as CXCL9, CXCL10, IL-12, IL-15, and MCP-1, to skew the programming toward a certain phenotype. The inhibition of HDAC activation and the subsequent modification of chromatin status change the molecular signatures of these mediators, leading to changes in the inflammatory environment [54].

Another example of how the modification of HDAC activity may influence the immune response against pathogens is seen in the ability of *Candida albicans* or *Saccharomyces cerevisiae* to regulate TLR1 and TLR6 downstream targets in the intestinal mucosa. Both of these fungal species are members of the normal commensal flora residing in the healthy human intestine. They positively prime the innate immune response against pathogens through the production of IL-8, which exerts a favorable effect by protecting the mucosal surface from intestinal pathogens [55]. The inhibition of HDAC activity, however, causes a marked increase in IL-8, which disturbs the immune balance and leads to the destruction of the commensal flora, creating a niche for pathogenic and opportunistic species to cause further infections [56]. Similarly, chronic and persistent activation of MAP kinases causes articular chondrocyte hypertrophy through the direct stimulation of a group of cytokines and chemokines that includes IL-8 and CXCL1 [57, 58]. Both of these cytokines are upregulated through the induction of TLR1 and TLR6, which increase after treatment with butyric acid, a known class I HDAC inhibitor [59].

4.7 HDAC Regulation of IFN-Mediated Signaling

When a virus infects a cell, the innate immune system employs a wide range of IFN molecules to prepare the surrounding tissues for viral invasion. During this viral invasion, innate immune cells respond in a general way to invading single-stranded or double-stranded DNA or RNA molecules. HDACs control this response by changing the genetic signature, either by amplifying or dampening the outcome. HDAC1 participates in the antiviral response by exerting its deacetylase activity to influence gene expression profiling. Its expression at both the transcriptional and translational levels is significantly reduced after influenza A viral infection. This decreased expression is accompanied by a decrease in the enzymatic activity of HDAC1 and subsequent downregulation of the expression of IFN-stimulated genes (ISGs) such as phosphorylated-STAT1 and IFITM3 [60]. Moreover, type I IFN stimulates the phosphorylation of STAT3, which in turn functions as a transcriptional regulator for many ISGs. The transcriptional activity of STAT3 itself may be modulated by the acetylation of many lysine residues as a direct response to cytokine stimulation [61, 62]. This cross talk between the regulation of STAT3 by IFN and the subsequent regulation of STAT3 by HDACs reveals the complicated regulation of the transcriptional activity underlying a simple viral infection.

The endogenous protein expression levels of HDAC7 are constitutively maintained in cytotoxic T cells (CTLs). Continuous phosphorylation of HDAC7 is actively maintained by CTLs to guarantee the cytosolic accumulation of HDAC7 [63]. Scientists investigating the cause of HDAC7 exclusion from the nucleus discovered that activated CTLs control functional capacity through the genetic signature of many intrinsic pathways [64]. These pathways are the targets of HDAC7 deacetylation [65]. Moreover, the expression of IFN-γ, which is an effector cytokine produced by CTLs in response to T-cell receptor activation, is hindered by the increased expression of nuclear HDAC7 [66] (Fig. 4.2).

The innate immune response is a rapid nonspecific response that is controlled by many molecules. HDACs play a crucial role in controlling the scope and amplitude

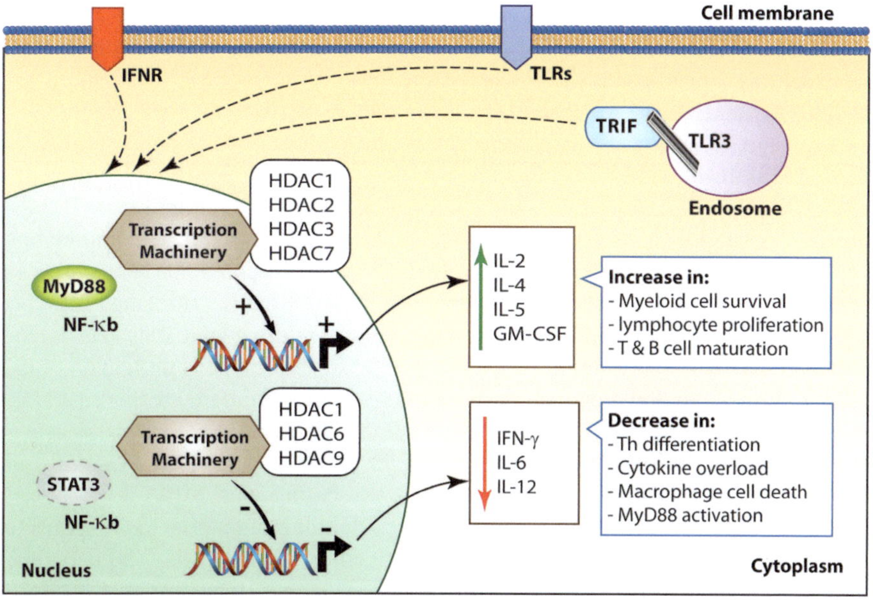

Fig. 4.2 The most important innate immune functions modulated by HDACs

of the reaction, leading to the production of precise amounts of inflammatory cytokines and interleukins.

4.8 HDAC Regulation of Other Immune Mediators

To control infections and inflammatory reactions, the immune system not only modulates the functions of its own cells but also affects many functional and biological aspects of neighboring cells and tissues. Both HDAC1 and HDAC2 regulate the transcription of many genes that are important for cell junction integrity, the overall permeability of the intestinal epithelium and the amount of mucin produced by goblet cells. Both HDACs are also involved in limiting the phosphorylation of the transcription factor STAT3, a major regulator of the inflammatory reaction in the intestinal epithelium. The dysregulation of HDAC1 and HDAC2 functions not only to disrupt the physiological structure of the intestine but also to modify the microbial environment through the effects of STAT3, causing major intestinal inflammatory complications [67].

Earlier in this chapter, we described the cytokine profile following an innate immune response as a major force underlying the cellular response to different immunological stimuli. These cytokines are commonly expressed by many innate immune cells through the transcriptional activation of their respective genes by a transcription factor that is a key player in the immune reaction, NF-kB. When

NF-kB is activated, it binds to promoter regions containing specific kB motifs with downstream signaling targets. Both HDAC1 [68] and HDAC2 [69] control the acetylation of the IL-6 promoter, which in turn controls the accessibility of the transcriptional machinery to chromatin fibers when an immune cell, such as a macrophage, is instructed by internal signaling pathways that such a protein is needed [68]. IL-6 is a key pro-inflammatory cytokine responsible for the differentiation of lymphocytes, monocytes, and B cells. It acts directly on fully mature B cells, T cells, hematopoietic progenitor cells, and many other tissues during inflammation [70]. Dysregulation of its expression during the immune response has been observed in many diseases, such as rheumatoid arthritis [71] and Kaposi's sarcoma [72]. This type of regulation, which affects the immune response during later stages, is not the only way HDACs control immune reactions. HDAC3 deacetylates protein 65 of the NF-kB complex to downregulate the duration of its transcriptional response and its nuclear expression levels [73]. Conversely, HDAC2 is responsible for the deacetylation of class II transactivator, a transcription factor that regulates the expression of the MHC-II complex on the surface of antigen-presenting cells. MHC-II complexes are essential for the recognition and subsequent binding of pathogenic epitopes to innate immune receptors and professional antigen-presenting cells. The deregulation of such proteins may affect the functional capacity of innate immune cells to execute their roles in the immune response [74].

4.9 HDAC Inhibitors and Inflammatory Cytokines

HDAC inhibitors, abbreviated as HDACis, are a class of compounds that specifically target HDACs to functionally inactivate their enzymatic activity or action and therefore prevent histone deacetylation. The collective function of HDACis augments the buildup of nucleosomes with acetylated lysine residues in their core histones, causing chromatin relaxation, attracting the transcriptional machinery, and promoting gene expression. The use of HDACis has been linked to the treatment of many tumors due to their ability to activate genes silenced by the oncogenic transformation of targeted cells and tissues. Many cells in our body are in a constant state of equilibrium between transcriptional activation and inhibition. Cellular proliferation and DNA replication require numerous housekeeping genes that are similar among different tissue types. However, some specialized cells only require certain proteins for selected biological functions, such as inflammatory reactions, or precise amounts of certain proteins during a certain period of time. The ability to change the transcriptional signatures of immune cells is not without precedent [75]. Few studies have delineated the exact mechanisms employed by HDACis to alter the scope of inflammation [76]. Current studies evaluating anti-inflammatory treatments are focused on blocking the signaling pathways of common and nonspecific immunoregulatory proteins. Alternatively, HDACis usually target specific transcription factors or cytokines to control the amplitude of the inflammatory reaction. Several classes of HDACis have been reported to regulate cytokine expression and immunological effects (Table 4.1).

Table 4.1 The functional effects of different HDACis on the activities of cytokines and immune cells

HDAC inhibitor	Immunological effects	Refs.
SAHA	SAHA or vorinostat is a class I inhibitor that also inhibits HDAC5 by directly blocking substrate access to the catalytic site of the enzymes	[87, 88]
	Inhibits the production of IL-6, TNF-α, IL-1-β, NO, and IFN-γ by macrophages in response to LPS stimulation in both mice and humans	
	Inhibits IFN-γ-dependent production of IL-18 and IL-12 by human PMNs	
TSA	TSA, or trichostatin A, is an antimicrobial gene that selectively inhibits class I and II HDACs through the removal of acetyl groups from core histones	[83]
	Inhibits the activity of many TLRs, NODs, RIG-I, Mda-5, and C-type lectins, leading to diminished microbial sensing	
	Inhibits the activity of many immunoregulatory mediators involved in chemotaxis, inflammation, the cell cycle, and tissue repair	
	Inhibits the production of TNF-α, IL-6, and IL-12p40 by macrophages	
LBH589	LBH589, or panobinostat, is a pan-HDAC inhibitor that was approved by the FDA as a treatment for patients with multiple myeloma	[89]
	Inhibits the maturation and immunological activity of dendritic cells, which are required to activate inflammatory T cells and polarization	
	Inhibits the production of IL-12, IL-23, IL-6, IL-10, and TNF-α by activated dendritic cells	
PXD-101	PDX-101, or belinostat, is a hydroxamic acid-type inhibitor that was approved by the FDA as a treatment for patients with peripheral T-cell lymphoma	[90]
	Enhances plasma levels of specific cytokines, such as IL-6, in patients with solid tumors	
MGCD0103	MGCD0103, or mocetinostat, is a selective class I and IV inhibitor that targets the programmed death ligand 1 (PD-L1) pathway (caspase cascade)	[91, 92]
	Induces the cleavage and mitochondrial shuttling of Bax, which results in caspase-9-dependent activation	
Sodium Butyrate	Sodium butyrate is a selective class I inhibitor that is naturally produced in the gut through the dietary consumption of legumes	[93–95]
	Inhibits the IFN-γ-dependent transcriptional activation of NF-κB, which in turn diminishes the production of NO, iNOS, TNF-α, and IL-12p40	
	Inhibits the proliferation of T cells and the motility of macrophages in response to LPS stimulation	

(continued)

Table 4.1 (continued)

HDAC inhibitor	Immunological effects	Refs.
MS-275	MS-275, SNDX-275, or entinostat is a selective HDAC1 and HDAC3 inhibitor that is undergoing clinical trials for the treatment of Hodgkin's lymphoma and breast cancer	[96, 97]
	Downregulates CD1a, CD11c, and CD14 after IL-4 treatment, limiting the differentiation of monocytes into mature dendritic cells	
	Inhibits the functional activity of mature dendritic cells by decreasing their production of TNF-α, IL-6, and IL-12p40	
	Controls CD4+ T helper cell polarization through the downregulation of IL-13 and IL-4 expression and the upregulation of both CXCL10 and IL-12	
FR276457	FR276457 is a selective class I inhibitor that exhibits inhibitory activity toward HDAC4	[98]
	Inhibits T-cell proliferation and prevents the maturation of monocytes into functional macrophages	
	Prevents the differentiation of CD8+ cytotoxic T cells	
Valproic acid	Valproic acid, or Depakene, is a class I inhibitor that was approved by the FDA due to its ability to promote the differentiation of hematopoietic progenitor cells and leukemic blasts	[99, 100]
	Inhibits the production of TNF-α and IL-6 in response to LPS stimulation through modulation of the p50 protein subunit that targets NF-κB activity	
LAQ824	LAQ824, or dacinostat, is a pan class I and II inhibitor	[54]
	Inhibits CD4+ T helper polarization by controlling IL-12, IL-15, CXCL9, and CXCL10 expression by macrophages and dendritic cells	
ITF2357	ITF2357, or givinostat, is a selective class I and II inhibitor	[101, 102]
	Inhibits TNFα, IL-12, IL-1α, IL-1β, and IFN-γ production by PMNs in response to LPS stimulation	

HDACis are a class of biological compounds that limit the activity of histone acetylation, leading to transcriptionally active chromatin. HDACis modulate the inflammatory response by regulating many cytokines and inflammatory mediators

4.10 HDACis and Inflammatory Disease

HDACis have been used to treat cancer in the clinic by functionally regulating anti-proliferation, the cell cycle, and apoptosis through gene expression mechanisms. Recently, HDACis have been implemented as an exciting approach to treat inflammatory disease [77]. Additionally, two HDACis (TSA and SAHA) have shown promising results for the treatment of diabetes, which has a substantial inflammatory component. TSA and SAHA suppress cytokines in pancreatic beta cells [78].

Trichostatin A (TSA) also reduces spinal cord inflammation, demyelination, and neuronal and axonal loss and ameliorates disability during the relapsing phase of experimental autoimmune encephalomyelitis (EAE), a model of multiple sclerosis (MS) [79]. The anti-inflammatory activity of inhibitors targeting HDAC1, such as NW-21 and MS-275, reduces the expression of monocyte chemotactic protein 1 (MCP-1), TNF-α, macrophage inflammatory protein 1α (MIP-1α), IL-1, and RANTES in human monocytes stimulated with either TNF-α or lipopolysaccharide for 24 h *in vitro*. Such inhibitors also reduce inflammation and bone loss in a model of arthritis [80]. As shown in animal models, HDACis may exert beneficial effects on colitis by boosting levels of Foxp3+ (forkhead box P3+) T-regulatory cells that dampen inflammation and may also have a potential role in the treatment of inflammatory bowel diseases (IBDs) [81]. Treatment with the HDACi SAHA improved survival by decreasing levels of the pro-inflammatory cytokine IL-6 and increasing levels of the anti-inflammatory cytokine IL-10 in an LPS-induced septic shock mouse model [82]. HDACis also impact host defenses against bacterial infections by reducing the levels of cytokines, chemokines, and, more importantly, microbial sensing molecules (c-type lectins, adhesion molecules) in the innate immune response to TLR treatment [83, 84]. The treatment of murine macrophages with the HDACis TSA or VPA resulted in the reduced killing of *E. coli* and *S. aureus* by impairing phagocytosis and reactive oxygen and nitrogen species generation [84].

Macrophage activation is tightly controlled by the reversible acetylation and deacetylation of histones, and the application of HDACis demonstrates therapeutic effects in animal models of chronic inflammatory disease, depressed macrophage HDAC activity in patients with asthma, chronic obstructive pulmonary (COPD) disease, and inflammatory-mediated rheumatoid arthritis (RA), potentially supporting the administration of therapeutic HDACis [85]. Some studies have directly examined the effects of HDACis on the activation of macrophages derived from the inflamed joints of patients with RA. The inhibition of class I/II HDACs or class III sirtuin HDACs potently blocked the production of IL-6 and TNF-a by macrophages from healthy donors and patients with RA [85]. Two HDACis, trichostatin A (TSA) and nicotinamide, selectively induced macrophage apoptosis by regulating the anti-apoptotic protein Bfl-1/A1, and inflammatory stimuli increased the sensitivity of macrophages to HDACi-induced apoptosis. Importantly, inflammatory cytokine and angiogenic factor induction in RA synovial biopsy explants were also suppressed by HDACis. Although studies have identified important roles for class I/II and sirtuin HDACs in promoting inflammation, angiogenesis, and cell survival in RA, potentially revealing some redundancy between class I and II HDACs, their therapeutic effects must be assessed.

4.11 HDACs and HDACis in Homeostasis and Inflammatory Diseases

Due to the dynamic nature of epigenetic programming, inflammatory signaling events may be reversed by targeting distinct epigenetic enzymes, including HDACs. To gain a better understanding of the roles of specific HDACs in inflammatory disease, we must define the molecular and biological consequences of HDAC inhibition. Although HDACis have been used for clinical trials in cancer, the effects of HDACis on immune regulation are also promising, and these compounds may be applied to treat inflammatory diseases in the future [77]. Various HDACi molecules define specific immune response signaling cascades in different disease models, from acute to chronic inflammation [86]. HDACi molecules primarily block the transcription of key inflammatory modulators to control the full capacity of the inflammatory response after TLR and IFN signaling. The inhibitory effects of HDACis prevent the activation and differentiation of many effector cells, including DCs and macrophages. Some pro-inflammatory cytokines, including TNF-α, IL-6, and IL-12p40, are suppressed by HDACi application. HDACi treatment also enhances the IFN-activated cascade via ISG induction to establish an antiviral state. This phenomenon reveals the potential anti-inflammatory functions of HDACi agents to control the inflammatory signals produced by M1 macrophages in a variety of inflammatory disease models, in addition to their antiviral potential. The successful therapeutic use of HDACi molecules, however, still requires much work given the lack of knowledge regarding the distinct HDACi targeting of pro-inflammatory cytokines or chemokines. Future HDACi therapeutic approaches should focus on the fundamental biological mechanisms underlying the HDACi-mediated inhibition of pro-inflammatory mediators, how to use current sequence and structural information to develop novel HDACi molecules capable of targeting certain HDAC enzymes, and elucidating the regulatory cross talk between HDACis that affects the same class of HDACs. It will also be important to enhance the strategic use of HDACis in combination with biomarkers to assess the response and resistance to an HDACi treatment to demonstrate its effectiveness and rational clinical application.

Acknowledgements This work was supported by grants from the National Natural Science Foundation of China (81273247, 81472655 and 31670905) and the Shanghai Municipal Education Commission key discipline support project 2015-10101001-1.

References

1. O'Neill LA, Golenbock D, Bowie AG (2013) The history of Toll-like receptors - redefining innate immunity. Nat Rev Immunol 13(6):453–460
2. Kawai T, Akira S (2011) Toll-like receptors and their crosstalk with other innate receptors in infection and immunity. Immunity 34(5):637–650

3. Kaufmann SH, Dorhoi A (2016) Molecular Determinants in Phagocyte-Bacteria Interactions. Immunity 44(3):476–491
4. Iwasaki A, Medzhitov R (2004) Toll-like receptor control of the adaptive immune responses. Nat Immunol 5(10):987–995
5. Iwasaki A, Medzhitov R (2010) Regulation of adaptive immunity by the innate immune system. Science 327(5963):291–295
6. Antonelli M, Kushner I (2017) It's time to redefine inflammation. FASEB J 31(5):1787–1791
7. Cao X (2016) Self-regulation and cross-regulation of pattern-recognition receptor signalling in health and disease. Nat Rev Immunol 16(1):35–50
8. Diacovich L, Gorvel JP (2010) Bacterial manipulation of innate immunity to promote infection. Nat Rev Microbiol 8(2):117–128
9. Pober JS, Sessa WC (2007) Evolving functions of endothelial cells in inflammation. Nat Rev Immunol 7(10):803–815
10. Lee DY, Hayes JJ, Pruss D, Wolffe AP (1993) A positive role for histone acetylation in transcription factor access to nucleosomal DNA. Cell 72(1):73–84
11. Vettese-Dadey M, Grant P, Hebbes T, Crane-Robinson C, Allis C, Workman J (1996) Acetylation of histone H4 plays a primary role in enhancing transcription factor binding to nucleosomal DNA in vitro. EMBO J 15(10):2508
12. Grunstein M (1997) Histone acetylation in chromatin structure and transcription. Nature 389(6649):349–352
13. Narlikar GJ, Fan HY, Kingston RE (2002) Cooperation between complexes that regulate chromatin structure and transcription. Cell 108(4):475–487
14. Paranjape SM, Kamakaka RT, Kadonaga JT (1994) Role of chromatin structure in the regulation of transcription by RNA polymerase II. Annu Rev Biochem 63(1):265–297
15. Bird A (2002) DNA methylation patterns and epigenetic memory. Genes Dev 16(1):6
16. Li Y, Seto E (2016) HDACs and HDAC inhibitors in cancer development and therapy. Cold Spring Harb Perspect Med 6(10) a026831
17. Lee KK, Workman JL (2007) Histone acetyltransferase complexes: one size doesn't fit all. Nat Rev Mol Cell Biol 8(4):284–295
18. Chauhan A, Quenum FZ, Abbas A, Bradley DS, Nechaev S, Singh BB, Sharma J, Mishra BB (2015) Epigenetic modulation of microglial inflammatory gene loci in helminth-induced immune suppression: implications for immune regulation in neurocysticercosis. ASN Neuro 7(4) 1–12
19. Feng D, Sangster-Guity N, Stone R, Korczeniewska J, Mancl ME, Fitzgerald-Bocarsly P, Barnes BJ (2010) Differential requirement of histone acetylase and deacetylase activities for IRF5-mediated proinflammatory cytokine expression. J Immunol 185(10):6003–6012
20. Takeuch O, Akira S (2011) Epigenetic control of macrophage polarization. Eur J Immunol 41(9):2490–2493
21. Davey CA, Sargent DF, Luger K, Maeder AW, Richmond TJ (2002) Solvent mediated interactions in the structure of the nucleosome core particle at 1.9 Å resolution+. J Mol Biol 319(5):1097–1113
22. Reinke H, Horz W (2003) Histones are first hyperacetylated and then lose contact with the activated PHO5 promoter. Mol Cell 11(6):1599–1607
23. Zhang HS, Gavin M, Dahiya A, Postigo AA, Ma D, Luo RX, Harbour JW, Dean DC (2000) Exit from G1 and S phase of the cell cycle is regulated by repressor complexes containing HDAC-Rb-hSWI/SNF and Rb-hSWI/SNF. Cell 101(1):79–89
24. Ferreira R, Naguibneva I, Mathieu M, Ait-Si-Ali S, Robin P, Pritchard LL, Harel-Bellan A (2001) Cell cycle-dependent recruitment of HDAC-1 correlates with deacetylation of histone H4 on an Rb–E2F target promoter. EMBO Rep 2(9):794–799
25. Kurdistani SK, Grunstein M (2003) Histone acetylation and deacetylation in yeast. Nat Rev Mol Cell Biol 4(4):276–284
26. Yang X-J, Seto E (2008) The Rpd3/Hda1 family of lysine deacetylases: from bacteria and yeast to mice and men. Nat Rev Mol Cell Biol 9(3):206–218

27. Gao L, Cueto MA, Asselbergs F, Atadja P (2002) Cloning and functional characterization of HDAC11, a novel member of the human histone deacetylase family. J Biol Chem 277(28):25748–25755
28. Galasinski SC, Resing KA, Goodrich JA, Ahn NG (2002) Phosphatase inhibition leads to histone deacetylases 1 and 2 phosphorylation and disruption of corepressor interactions. J Biol Chem 277(22):19618–19626
29. Joseph J, SIDERIS ML, Polly M, LORIMER DD, MCINTOSH B, CLARK JM (2000) Cloning and characterization of a novel human histone deacetylase, HDAC8. Biochem J 350(1):199–205
30. Lagger G, O'Carroll D, Rembold M, Khier H, Tischler J, Weitzer G, Schuettengruber B, Hauser C, Brunmeir R, Jenuwein T (2002) Essential function of histone deacetylase 1 in proliferation control and CDK inhibitor repression. EMBO J 21(11):2672–2681
31. Dovey OM, Foster CT, Cowley SM (2010) Histone deacetylase 1 (HDAC1), but not HDAC2, controls embryonic stem cell differentiation. Proc Natl Acad Sci 107(18):8242–8247
32. De Ruijter AJ, Van Gennip AH, Caron HN, Stephan K, Van Kuilenburg AB (2003) Histone deacetylases (HDACs): characterization of the classical HDAC family. Biochem J 370(3):737–749
33. Hackett CJ, Rotrosen D, Auchincloss H, Fauci AS (2007) Immunology research: challenges and opportunities in a time of budgetary constraint. Nature 8(2):114–117
34. Gervin K, Vigeland MD, Mattingsdal M, Hammerø M, Nygård H, Olsen AO, Brandt I, Harris JR, Undlien DE, Lyle R (2012) DNA methylation and gene expression changes in monozygotic twins discordant for psoriasis: identification of epigenetically dysregulated genes. PLoS Genet 8(1):e1002454
35. Höhler T, Reuss E, Evers N, Dietrich E, Rittner C, Freitag CM, Vollmar J, Schneider PM, Fimmers R (2002) Differential genetic determination of immune responsiveness to hepatitis B surface antigen and to hepatitis A virus: a vaccination study in twins. Lancet 360(9338):991–995
36. Feinberg AP (2007) Phenotypic plasticity and the epigenetics of human disease. Nature 447(7143):433–440
37. Weathington NM, van Houwelingen AH, Noerager BD, Jackson PL, Kraneveld AD, Galin FS, Folkerts G, Nijkamp FP, Blalock JE (2006) A novel peptide CXCR ligand derived from extracellular matrix degradation during airway inflammation. Nat Med 12(3):317–323
38. Mortaz E, Redegeld FA, Sarir H, Karimi K, Raats D, Nijkamp FP, Folkerts G (2008) Cigarette smoke stimulates the production of chemokines in mast cells. J Leukoc Biol 83(3):575–580
39. Sabroe I, Whyte M (2007) Toll-like receptor (TLR)-based networks regulate neutrophilic inflammation in respiratory disease. Portland Press Limited)
40. Chen X, Guan X-J, Peng X-H, Cui Z-L, Luan C-Y, Guo X-J (2015) Acetylation of lysine 9 on histone H3 is associated with increased pro-inflammatory cytokine release in a cigarette smoke-induced rat model through HDAC1 depression. Inflamm Res 64(7):513–526
41. Gilchrist M, Thorsson V, Li B, Rust AG, Korb M, Kennedy K, Hai T, Bolouri H, Aderem A (2006) Systems biology approaches identify ATF3 as a negative regulator of Toll-like receptor 4. Nature 441(7090):173–178
42. De Nardo D, Labzin LI, Kono H, Seki R, Schmidt SV, Beyer M, Xu D, Zimmer S, Lahrmann C, Schildberg FA, Vogelhuber J, Kraut M, Ulas T, Kerksiek A, Krebs W, Bode N et al (2014) High-density lipoprotein mediates anti-inflammatory reprogramming of macrophages via the transcriptional regulator ATF3. Nat Immunol 15(2):152–160
43. Rountree MR, Bachman KE, Baylin SB (2000) DNMT1 binds HDAC2 and a new co-repressor, DMAP1, to form a complex at replication foci. Nat Genet 25(3):269–277
44. Roopra A, Sharling L, Wood IC, Briggs T, Bachfischer U, Paquette AJ, Buckley NJ (2000) Transcriptional repression by neuron-restrictive silencer factor is mediated via the Sin3-histone deacetylase complex. Mol Cell Biol 20(6):2147–2157
45. Mankan AK, Daly J, Caraher E, Kelleher D, McManus R (2008) The IKBL protein inhibits TLR mediated activation of gene expression by NF kappa B. Ulster Med J 77(1) 69–70

46. Sadler AJ, Rossello FJ, Yu L, Deane JA, Yuan X, Wang D, Irving AT, Kaparakis-Liaskos M, Gantier MP, Ying H (2015) BTB-ZF transcriptional regulator PLZF modifies chromatin to restrain inflammatory signaling programs. Proc Natl Acad Sci 112(5):1535–1540
47. Sadler AJ, Suliman BA, Yu L, Yuan X, Wang D, Irving AT, Sarvestani ST, Banerjee A, Mansell AS, Liu JP, Gerondakis S, Williams BR, Xu D (2015) The acetyltransferase HAT1 moderates the NF-kappaB response by regulating the transcription factor PLZF. Nat Commun 6:6795
48. INTO T, INOMATA M, NIIDA S, MURAKAMI Y, K-I SHIBATA (2010) Regulation of MyD88 aggregation and the MyD88-dependent signaling pathway by sequestosome 1 and histone deacetylase 6. J Biol Chem 285(46):35759–35769
49. Shakespear MR, Hohenhaus DM, Kelly GM, Kamal NA, Gupta P, Labzin LI, Schroder K, Garceau V, Barbero S, Iyer A, Hume DA, Reid RC, Irvine KM, Fairlie DP, Sweet MJ (2013) Histone deacetylase 7 promotes Toll-like receptor 4-dependent proinflammatory gene expression in macrophages. J Biol Chem 288(35):25362–25374
50. Lehnardt S, Wennekamp J, Freyer D, Liedtke C, Krueger C, Nitsch R, Bechmann I, Weber JR, Henneke P (2007) TLR2 and caspase-8 are essential for group B Streptococcus-induced apoptosis in microglia. J Immunol 179(9):6134–6143
51. Peck-Palmer OM, Unsinger J, Chang KC, Davis CG, McDunn JE, Hotchkiss RS (2008) Deletion of MyD88 markedly attenuates sepsis-induced T and B lymphocyte apoptosis but worsens survival. J Leukoc Biol 83(4):1009–1018
52. Gu L, Tseng S, Horner RM, Tam C, Loda M, Rollins BJ (2000) Control of TH2 polarization by the chemokine monocyte chemoattractant protein-1. Nature 404(6776):407–411
53. Zou W, Borvak J, Marches F, Wei S, Galanaud P, Emilie D, Curiel TJ (2000) Macrophage-derived dendritic cells have strong Th1-polarizing potential mediated by β-chemokines rather than IL-12. J Immunol 165(8):4388–4396
54. Brogdon JL, Xu Y, Szabo SJ, An S, Buxton F, Cohen D, Huang Q (2007) Histone deacetylase activities are required for innate immune cell control of Th1 but not Th2 effector cell function. Blood 109(3):1123–1130
55. Saegusa S, Totsuka M, Kaminogawa S, Hosoi T (2004) Candida albicans and Saccharomyces cerevisiae induce interleukin-8 production from intestinal epithelial-like Caco-2 cells in the presence of butyric acid. FEMS Immunol Med Microbiol 41(3):227–235
56. Takami M, Terry V, Petruzzelli L (2002) Signaling pathways involved in IL-8-dependent activation of adhesion through Mac-1. J Immunol 168(9):4559–4566
57. Cecil DL, Johnson K, Rediske J, Lotz M, Schmidt AM, Terkeltaub R (2005) Inflammation-induced chondrocyte hypertrophy is driven by receptor for advanced glycation end products. J Immunol 175(12):8296–8302
58. Pulai JI, Chen H, Im H-J, Kumar S, Hanning C, Hegde PS, Loeser RF (2005) NF-κB mediates the stimulation of cytokine and chemokine expression by human articular chondrocytes in response to fibronectin fragments. J Immunol 174(9):5781–5788
59. Vaughan EE, Geiger RC, Miller AM, Loh-Marley PL, Suzuki T, Miyata N, Dean DA (2008) Microtubule acetylation through HDAC6 inhibition results in increased transfection efficiency. Mol Ther 16(11):1841–1847
60. Nagesh PT, Husain M (2016) Influenza A virus dysregulates host histone deacetylase 1 that inhibits viral infection in lung epithelial cells. J Virol 90(9):4614–4625
61. Hou T, Ray S, Lee C, Brasier AR (2008) The STAT3 NH2-terminal domain stabilizes enhanceosome assembly by interacting with the p300 bromodomain. J Biol Chem 283(45):30725–30734
62. Z-l Y, Y-j G, Chatterjee D, Chin YE (2005) Stat3 dimerization regulated by reversible acetylation of a single lysine residue. Science 307(5707):269–273
63. Martin M, Kettmann R, Dequiedt F (2009) Class IIa histone deacetylases: conducting development and differentiation. Int J Dev Biol 53(2–3):291–301
64. Kao H-Y, Verdel A, Tsai C-C, Simon C, Juguilon H, Khochbin S (2001) Mechanism for nucleocytoplasmic shuttling of histone deacetylase 7. J Biol Chem 276(50):47496–47507

65. Kasler HG, Verdin E (2007) Histone deacetylase 7 functions as a key regulator of genes involved in both positive and negative selection of thymocytes. Mol Cell Biol 27(14):5184–5200
66. Navarro MN, Goebel J, Feijoo-Carnero C, Morrice N, Cantrell DA (2011) Phosphoproteomic analysis reveals an intrinsic pathway for the regulation of histone deacetylase 7 that controls the function of cytotoxic T lymphocytes. Nat Immunol 12(4):352–361
67. Turgeon N, Blais M, Gagné J-M, Tardif V, Boudreau F, Perreault N, Asselin C (2013) HDAC1 and HDAC2 restrain the intestinal inflammatory response by regulating intestinal epithelial cell differentiation. PLoS One 8(9):e73785
68. Yao Z, Zhang Q, Li X, Zhao D, Liu Y, Zhao K, Liu Y, Wang C, Jiang M, Li N (2014) Death domain-associated protein 6 (Daxx) selectively represses IL-6 transcription through histone deacetylase 1 (HDAC1)-mediated histone deacetylation in macrophages. J Biol Chem 289(13):9372–9379
69. Zhang Q, Zhao K, Shen Q, Han Y, Gu Y, Li X, Zhao D, Liu Y, Wang C and Zhang X (2015) Tet2 is required to resolve inflammation by recruiting Hdac2 to specifically repress IL-6. Nature, 525(7569):389–393.
70. Hunter CA, Jones SA (2015) IL-6 as a keystone cytokine in health and disease. Nat Immunol 16(5):448–457
71. Fishman D, Faulds G, Jeffery R, Mohamed-Ali V, Yudkin JS, Humphries S, Woo P (1998) The effect of novel polymorphisms in the interleukin-6 (IL-6) gene on IL-6 transcription and plasma IL-6 levels, and an association with systemic-onset juvenile chronic arthritis. J Clin Investig 102(7):1369
72. Russo JJ, Bohenzky RA, Chien M-C, Chen J, Yan M, Maddalena D, Parry JP, Peruzzi D, Edelman IS, Chang Y (1996) Nucleotide sequence of the Kaposi sarcoma-associated herpesvirus (HHV8). Proc Natl Acad Sci 93(25):14862–14867
73. L-f C, Fischle W, Verdin E, Greene WC (2001) Duration of nuclear NF-κB action regulated by reversible acetylation. Science 293(5535):1653–1657
74. Kong X, Fang M, Li P, Fang F, Xu Y (2009) HDAC2 deacetylates class II transactivator and suppresses its activity in macrophages and smooth muscle cells. J Mol Cell Cardiol 46(3):292–299
75. Shakespear MR (2011) Halili MA. Irvine KM, Fairlie DP and Sweet MJ. Histone deacetylases as regulators of inflammation and immunity. Trends in Immunology
76. Adcock I (2007) HDAC inhibitors as anti inflammatory agents. Br J Pharmacol 150(7):829–831
77. Das Gupta K, Shakespear MR, Iyer A, Fairlie DP, Sweet MJ (2016) Histone deacetylases in monocyte/macrophage development, activation and metabolism: refining HDAC targets for inflammatory and infectious diseases. Clin Transl Immunol 5(1):e62
78. Larsen L, Tonnesen M, Ronn SG, Storling J, Jorgensen S, Mascagni P, Dinarello CA, Billestrup N, Mandrup-Poulsen T (2007) Inhibition of histone deacetylases prevents cytokine-induced toxicity in beta cells. Diabetologia 50(4):779–789
79. Camelo S, Iglesias AH, Hwang D, Due B, Ryu H, Smith K, Gray SG, Imitola J, Duran G, Assaf B, Langley B, Khoury SJ, Stephanopoulos G, De Girolami U, Ratan RR, Ferrante RJ et al (2005) Transcriptional therapy with the histone deacetylase inhibitor trichostatin A ameliorates experimental autoimmune encephalomyelitis. J Neuroimmunol 164(1–2):10–21
80. Cantley MD, Fairlie DP, Bartold PM, Marino V, Gupta PK, Haynes DR (2015) Inhibiting histone deacetylase 1 suppresses both inflammation and bone loss in arthritis. Rheumatology 54(9):1713–1723
81. Edwards AJ, Pender SL (2011) Histone deacetylase inhibitors and their potential role in inflammatory bowel diseases. Biochem Soc Trans 39(4):1092–1095
82. Finkelstein RA, Li Y, Liu B, Shuja F, Fukudome E, Velmahos GC, de Moya M, Alam HB (2010) Treatment with histone deacetylase inhibitor attenuates MAP kinase mediated liver injury in a lethal model of septic shock. J Surg Res 163(1):146–154
83. Roger T, Lugrin J, Le Roy D, Goy G, Mombelli M, Koessler T, Ding XC, Chanson AL, Reymond MK, Miconnet I, Schrenzel J, Francois P, Calandra T (2011) Histone deacetylase

inhibitors impair innate immune responses to Toll-like receptor agonists and to infection. Blood 117(4):1205–1217
84. Mombelli M, Lugrin J, Rubino I, Chanson AL, Giddey M, Calandra T, Roger T (2011) Histone deacetylase inhibitors impair antibacterial defenses of macrophages. J Infect Dis 204(9):1367–1374
85. Grabiec AM, Krausz S, de Jager W, Burakowski T, Groot D, Sanders ME, Prakken BJ, Maslinski W, Eldering E, Tak PP, Reedquist KA (2010) Histone deacetylase inhibitors suppress inflammatory activation of rheumatoid arthritis patient synovial macrophages and tissue. J Immunol 184(5):2718–2728
86. Suliman BA, Xu D, Williams BR (2012) HDACi: molecular mechanisms and therapeutic implications in the innate immune system. Immunol Cell Biol 90(1):23–32
87. Leoni F, Zaliani A, Bertolini G, Porro G, Pagani P, Pozzi P, Don√† G, Fossati G, Sozzani S, Azam T, Bufler P, Fantuzzi G, Goncharov I, Kim S-H, Pomerantz BJ, Reznikov LL et al (2002) The antitumor histone deacetylase inhibitor suberoylanilide hydroxamic acid exhibits antiinflammatory properties via suppression of cytokines. Proc Natl Acad Sci 99(5):2995–3000
88. Mishra N, Reilly CM, Brown DR, Ruiz P, Gilkeson GS (2003) Histone deacetylase inhibitors modulate renal disease in the MRL-lpr/lpr mouse. J Clin Investig 111(4):539–552
89. Song W, Tai YT, Tian Z, Hideshima T, Chauhan D, Nanjappa P, Exley MA, Anderson KC, Munshi NC (2011) HDAC inhibition by LBH589 affects the phenotype and function of human myeloid dendritic cells. Leukemia 25(1):161–168
90. Steele NL, Plumb JA, Vidal L, Tjørnelund J, Knoblauch P, Rasmussen A, Ooi CE, Buhl-Jensen P, Brown R, Evans TRJ, DeBono JS (2008) A phase 1 pharmacokinetic and pharmacodynamic study of the histone deacetylase inhibitor belinostat in patients with advanced solid tumors. Clin Cancer Res 14(3):804–810
91. Fournel M, Bonfils C, Hou Y, Yan PT, Trachy-Bourget MC, Kalita A, Liu J, Lu AH, Zhou NZ, Robert MF (2008) MGCD0103, a novel isotype-selective histone deacetylase inhibitor, has broad spectrum antitumor activity in vitro and in vivo. Mol Cancer 7(4):759
92. El-Khoury V, Moussay E, Janji B, Palissot VR, Aouali NR, NHC B, Van Moer K, Pierson S, Van Dyck E, Berchem G (2010) The histone deacetylase inhibitor MGCD0103 induces apoptosis in B-cell chronic lymphocytic leukemia cells through a mitochondria-mediated caspase Activation cascade. Mol Cancer 9(5):1349–1360
93. Kim H-S, Whang S-Y, Woo M-S, Park J-S, Kim W-K, Han I-O (2004) Sodium butyrate suppresses interferon-gamma-, but not lipopolysaccharide-mediated induction of nitric oxide and tumor necrosis factor-alpha in microglia. J Neuroimmunol 151(1–2):85–93
94. SÄEMANN MD, BÖHMIG GA, ÖSTERREICHER CH, BURTSCHER H, PAROLINI O, DIAKOS C, STÖCKL J, HÖRL WH, ZLABINGER GJ (2000) Anti-inflammatory effects of sodium butyrate on human monocytes: potent inhibition of IL-12 and up-regulation of IL-10 production. FASEB J 14(15):2380–2382
95. Maa M-C, Chang MY, Hsieh M-Y, Chen Y-J, Yang C-J, Chen Z-C, Li YK, Yen C-K, Wu R-R, Leu T-H (2010) Butyrate reduced lipopolysaccharide-mediated macrophage migration by suppression of Src enhancement and focal adhesion kinase activity. J Nutr Biochem 21(12):1186–1192
96. Nencioni A, Beck J, Werth D, Grünebach F, Patrone F, Ballestrero A, Brossart P (2007) Histone deacetylase inhibitors affect dendritic cell differentiation and immunogenicity. Clin Cancer Res 13(13):3933
97. Khaskhely N, Buglio D, Shafer J, Bollard C, Younes A (2009) The histone deacetylase (HDAC) inhibitor entinostat (SNDX-275) targets Hodgkin lymphoma through a dual mechanism of immune modulation and apoptosis induction. pp. 1562
98. Kinugasa F, Nagatomi I, Nakanishi T, Noto T, Mori H, Matsuoka H, Sudo Y, Mutoh S (2009) Effect of the immunosuppressant histone deacetylase inhibitor FR276457 in a canine renal transplant model. Transpl Immunol 21(4):198–202

99. Ichiyama T, Okada K, Lipton JM, Matsubara T, Hayashi T, Furukawa S (2000) Sodium valproate inhibits production of TNF-[alpha] and IL-6 and activation of NF-[kappa]B. Brain Res 857(1–2):246–251
100. Kochanek AR, Fukudome EY, Li Y, Smith EJ, Liu B, Velmahos GC, de Moya M, King D, Alam HB (2011) Histone deacetylase inhibitor treatment attenuates MAP kinase pathway activation and pulmonary inflammation following hemorrhagic shock in a rodent model. J Surg Res In Press, Corrected Proof
101. Golay J, Cuppini L, Leoni F, Mico C, Barbui V, Domenghini M, Lombardi L, Neri A, Barbui A, Salvi A (2007) The histone deacetylase inhibitor ITF2357 has anti-leukemic activity in vitro and in vivo and inhibits IL-6 and VEGF production by stromal cells. Leukemia 21(9):1892–1900
102. Leoni F, Fossati G, Lewis EC, Lee JK, Porro G, Pagani P, Modena D, Moras ML, Pozzi P, Reznikov LL (2005) The histone deacetylase inhibitor ITF2357 reduces production of proinflammatory cytokines in vitro and systemic inflammation in vivo. Mol Med 11(1–12):1

Chapter 5
miR-155 Dysregulation and Therapeutic Intervention in Multiple Sclerosis

Claire E. McCoy

Abstract microRNAs play a fundamental role in the immune system. One particular microRNA, miR-155 plays a critical role in hematopoietic cell development and tightly regulates innate and adaptive immune responses in response to infection. However, its dysregulation, more specifically its overexpression, is closely associated with various inflammatory disorders. The purpose of this review is to consolidate how miR-155 underpins a variety of processes that contribute to the pathology of multiple sclerosis (MS). In particular, the impact of miR-155 is discussed with respect to human pathology and animal models. How miR-155 contributes to the activation of pathogenic immune cells, the permeability of the blood-brain barrier, and neurodegeneration in relation to MS is described. Many environmental risk factors associated with MS susceptibility can cause upregulation of miR-155, while many of the current disease-modifying treatments may work by inhibiting miR-155. From this review, it is clear that miR-155 is a realistic and feasible diagnostic, prognostic, and therapeutic target for the treatment of MS.

Keywords miR-155 • MicroRNA • Multiple sclerosis • Experimental autoimmune encephalomyelitis • EAE • Immunopathology • Biomarker • Disease-modifying treatments • Blood-brain barrier • Environmental risk factors • Microglia • Astrocytes • Macrophages

---

C.E. McCoy (✉)
Hudson Institute of Medical Research, Department of Molecular and Translational Science, Faculty of Medicine, Nursing and Health Sciences, Monash University, Clayton, VIC 3168, Australia

Molecular and Cellular Therapeutics, Royal College of Surgeons in Ireland, 123 St Stephen's Green, Dublin 2, Ireland
e-mail: clairemccoy@rcsi.ie

D. Xu (ed.), *Regulation of Inflammatory Signaling in Health and Disease*, Advances in Experimental Medicine and Biology 1024,
DOI 10.1007/978-981-10-5987-2_5

5.1 Introduction

microRNAs (miRNAs) play a fundamental role in cellular biology. They are critical modulators for a wide range of cellular processes including development, differentiation, proliferation, metabolism, and apoptosis [2, 4]. miRNAs are small ~22 nucleotide RNA sequences that regulate the expression of protein-coding mRNA sequences, often targeting multiple proteins within a particular network or pathway. miRNA targeting is achieved when they are guided to the 3′ untranslated region (UTR) of mRNA sequences, where partial or exact complementary base pairing of the miRNA results in degradation or translational inhibition of the target mRNA molecules. Each miRNA is predicted to bind to more than 200 mRNA target sequences, and as a result, overexpression and/or inhibition of a single miRNA can have a profound effect on cellular function [12, 127]. Major efforts have focused on understanding how miRNAs are induced in the cell, with an overall aim to uncover their specific mRNA target genes. In particular, attention is drawn to understanding which miRNAs or panels of miRNAs are dysregulated in disease. With this in mind, miRNAs are thus emerging as valuable therapeutic targets. Indeed, basic and applied miRNA research has already made a vast contribution in bench-to-bedside application, where >10 miRNA inhibitors (antagomirs) are currently in clinical trials for the treatment of cancer, cardiovascular disease, and hepatitis C virus (HCV) infection [55].

Of significance, miRNAs are critical regulators of the immune system, and their dysregulation clearly impacts the pathogenesis of various inflammatory diseases [80, 81, 116]. One particular microRNA, miR-155, plays a remarkable role in the immune system [79]. In summary, it was first identified when the B-cell integration cluster (BIC) (the gene which encodes miR-155) was found to be highly overexpressed in B-cell-activated lymphomas [22, 52, 118]. Later studies illustrated that miR-155 was not restricted to B cells, and a large-scale sequencing study performed in 26 different organ systems identified that the expression of miR-155 is highly specific for hematopoietic cells [53]. Indeed, several groups noted that miR-155 is potently induced upon activation of both myeloid and lymphoid cells with hematopoietic origin and is critical for the functioning of a healthy immune response to infection [83, 97, 112, 113].

This was further corroborated when mice deficient in miR-155 displayed impaired dendritic cell (DC) and T- and B-cell responses, characterized by faulty antigen presentation, reduced pro-inflammatory cytokines, reduced serum antibody titers, and inappropriate class-switched immunoglobulins, when challenged with infection [97, 113, 121]. Whereas its transgenic overexpression in hematopoietic cells resulted in a myeloproliferative disorder characterized by the gross expansion of myeloid cells, while overexpression in B cells promoted the development of leukemia and lymphoma in mice [15, 82]. Overall, this early data clearly indicates that miR-155 is essential for mounting an appropriate immune response to infection, yet its overexpression can contribute to immune-related disorders [84, 103, 120]. The purpose of this review is to consolidate the current literature with the aim to decode the impact of miR-155 dysregulation in the pathogenesis of multiple sclerosis (MS).

5.2 Multiple Sclerosis as an Inflammatory Disorder

Multiple sclerosis (MS) is the most common inflammatory disease to affect the central nervous system (CNS). It is a demyelinating disease in which the insulating myelin sheaths that surround nerve cells in the brain and spinal cord are damaged. Clinical manifestation is characterized by disturbances in sensory, motor, and cognitive function, with symptoms of pain and fatigue, while the pathology is characterized by lesions detected by magnetic resonance imaging (MRI) within the CNS. MS typically affects young adults where the average age of onset is 30 years old and affects two to three times more females than males [63, 92]. Its prevalence and incidence rate is increasing globally, especially in the northern hemisphere (140 per 100,000) compared to the global prevalence (30 per 100,000) [63, 92]. As well as causing a major personal burden to young adults, diagnosis and treatment requires a highly integrated and complex multidisciplinary approach resulting in significant economic burdens.

Several subtypes of MS have been described and are important for understanding the prognosis and type of treatment decisions. Eighty-five percent of new diagnoses are relapsing-remitting MS (RRMS) which produces attacks followed by periods of remission that can last months or years. However, it usually tends to get worse over time and often progresses to a secondary progressive MS (SPMS) subtype which begins to decline without periods of remission. In primary progressive MS (PPMS), there is an initial attack with a steady decline in disability without any periods of remission [63].

Despite extensive research conducted worldwide, a cause for all subtypes of MS remains unknown but may possibly arise due to genetic predisposition and/or environmental factors. With the advent of genome-wide association studies, 110 distinct genetic regions have been associated with MS [39]. However, only a handful of these gene variants such as human leukocyte antigen (HLA) class II genes (HLA-DQ, HLA-DR), IL-2RA, and IL-7RA have shown functional and correlative association with MS [32, 35, 73]. Environmental factors include a lack of vitamin D, human cytomegalovirus, and Epstein-Barr virus infection, and geographical latitude may have a role [5].

While the cause is not clear, the underlying mechanism is thought to be mediated by the immune system. Particularly in RRMS, it is extraordinarily conclusive that the symptoms and pathology of MS are due to an influx of immune cells crossing the blood-brain barrier (BBB) into the CNS. This immune cell infiltration results in chronic inflammation and the release of inflammatory cytokines and toxins that cause demyelination and neuroaxonal degeneration [21]. Considering the obvious immunopathology of MS, it was a very natural progression that the impact of miR-155 would be explored in MS.

5.3 miR-155 in Human MS Samples

The first clear indication that miR-155 was dysregulated in MS came about when Meinl and colleagues isolated white matter lesions from paraffin and frozen multiple sclerosis tissue samples. miR-155 was highly upregulated (11.9-fold) in active white matter lesions compared to healthy control white matter [45]. While another study demonstrated that miR-155 was increased in cerebral white matter juxtaposed to active lesions collected from a mixture of relapsing-remitting, primary progressive, and secondary progressive patients [77]. Using laser capture microdissection, miR-155 expression was isolated from individual cell types, namely, myeloid-derived macrophages, microglia, T/B lymphocytes, and astrocytes, suggesting that infiltrating immune cells as well as resident brain cells have the capacity to generate miR-155 ([45, [72]). Interestingly, miR-155 expression was also significantly increased in the neurovascular unit of active lesions from MS brain samples [59]. The neurovascular unit is a sub-anatomical region typically representative of blood-brain barrier comprised of endothelial cells, astrocytes, and neurons and suggests that miR-155 expression is not solely restricted to hematopoietic cells.

Elevated miR-155 in peripheral blood mononuclear cells (PBMCs) isolated from MS patient blood samples has been confirmed in multiple studies [64, 86, 124]. The first of these demonstrated that compared to other miRNAs investigated, miR-155 was remarkably upregulated (two- to threefold) in a cohort of patients with RRMS. Interestingly, the increase of miR-155, combined with miR-146a and miR-142-3p, had an 88.0% specificity in predicting MS disease [124], while another study identified that a subset of (10/24) patients with RRMS had elevated miR-155 expression [64]. Increased expression also correlated with increased IL-17, IFNγ, TNF, and IL-6 and suggests that miR-155 elevation may only occur when cells are in an inflammatory state. Enquiring further, certain studies have elucidated that miR-155 expression was specifically elevated in $CD14^+$ monocytes when purified from PBMCs in a cohort of RRMS patients, while others have shown increased miR-155 in sera alone [72, 135].

From these studies, it is clearly evident that miR-155 is overexpressed in a range of human MS samples, yet many more questions remain. For example, what is the source of miR-155 upregulation in MS? Is its overexpression restricted to hematopoietic cells or do other cells play a role? Is it a consequence of inflammation or is it a trigger for the progression and pathology of MS? Can we realistically generate a feasible therapeutic for MS based on miR-155 targeting? Studies from animal models and in vitro studies have greatly contributed to our understanding and are discussed in the following sections.

5.4 miR-155 in MS Mouse Models

The experimental autoimmune encephalomyelitis (EAE) model is the most widely used experimental animal model for MS. Although it is often criticized, it does resemble a model whereby peripheral activation of immune cells by a CNS-originating peptide has the capacity to closely resemble the human MS disease. In this model, an emulsified CNS antigen, typically myelin basic protein (MBP) or myelin oligodendrocyte glycoprotein (MOG), is injected together with Freud's adjuvant and pertussis toxin. The result is the activation and presentation of MBP or MOG by dendritic cells to $CD4^+$ T cells in the lymph nodes, which together result in a massive infiltration of differentiated $CD4^+$ T helper (Th1) and Th17 cells, B cells, $CD8^+$ T cells, and innate immune cells infiltrating the CNS, leading to inflammation and tissue damage.

The impact of miR-155 was first implicated when mice deficient in miR-155 were shown to be highly resistant to MOG_{35-55} peptide-induced EAE. miR-155−/− mice displayed a delayed onset, decreased disease severity and paralysis compared to wild-type mice [74, 79]. Brain histological analysis revealed less inflammation and less demyelination consistent with their reduced disease severity [74, 79]. The beneficial prognosis in miR-155−/− mice was primarily associated with reduced numbers of peripheral Th1 and Th17 in the spleen and lymph nodes, as well as reduced numbers in the CNS [76, 79]. When cultured ex vivo, Th1 and Th17 cells had decreased proliferative responses and reduced capabilities to produce IFNγ and IL-17 upon stimulation, suggesting that they are functionally defective. The progression of EAE was intrinsically linked to $CD4^+$ T cells, because the adoptive transfer of miR-155+/+ $CD4^+$ T cells into RAG1−/− recipients (which lack mature T/B lymphocytes) had a substantially more severe and accelerated disease course compared to mice receiving miR-155−/− $CD4^+$ T cells [79]. Moreover, CNS-isolated $CD4^+$ T cells from EAE-induced rats could be reactivated to produce more miR-155 upon reexposure to MBP in vitro, suggesting that $CD4^+$ T cells have the capacity to further increase miR-155 upon contact with reactive antigens [44]. These studies predict that overexpression of miR-155, typically observed in EAE models and human MS samples, acts to promote a Th1/Th17 phenotype.

The contribution of miR-155 in driving Th1 and Th17 responses was consolidated when wild-type mice were treated with a locked nucleic acid (LNA)-miR-155 oligonucleotide (herein called "antagomir") which reduced the clinical manifestation of disease when administered before or during EAE induction [74, 135]. The antagomir resulted in diminished proliferation and reduced IFNγ and IL-17 secretion in $CD4^+$ T cells isolated from the CNS [74]. In contrast, the intravenous injection of a miR-155 mimic aggravated EAE disease severity caused by the prominent inflammatory infiltration and demyelination in the spinal cord [135]. Additionally, increased frequencies of Th1 and Th17 cells, along with increased IL-17 and IFNγ cytokine production, were observed in the spleen, lymph nodes, and CNS of miR-155 mimic administered mice [135].

Further studies have illustrated how miR-155 mechanistically regulates Th1 and Th17 differentiation and function. For example, O'Connell and colleagues illustrated that miR-155 targets the transcription factor Ets1, a well-established negative regulator of Th17 differentiation [37]. Consistently, miR-155−/− Th17 cells had elevated Ets1 expression and lacked the expression of cytokines essential for Th17 differentiation. Namely, the Ets1/miR-155 axis was necessary for IL-23 responsiveness and was critical for normal expansion of Th17 cells in vivo and during induction of EAE [37]. Escobar and colleagues elegantly demonstrated that miR-155 can regulate the chromatin structure and epigenetic changes in Th17 cells [23]. By performing transcriptome analysis, they identified Jarid2, an RNA-binding protein, was upregulated in miR-155−/− differentiated Th17 cells. An increase in Jarid2 reprograms the epigenome of Th17 cells via H3K27 methylation and results in the silencing of specific genes (such as IL-22) that are required for Th17 differentiation [23]. Defects in Th17 differentiation and cytokine expression in the absence of miR-155 could be partially restored by Jarid2 deletion [23]. Another study demonstrated that miR-155 could control Th1 and Th17 proliferation and tissue migration by directly repressing the enzyme heme oxygenase-1 (HO-1) [136]. HO-1 catalyzes the oxidation of heme to generate carbon monoxide, biliverdin, and iron products and is essential for mediating important anti-inflammatory and antioxidant effects. Thus, when miR-155−/− animals, typically resistant to EAE, were injected with the HO-1 inhibitor ZnPP, disease severity increased [136]. In another study, miR-155-3p was found to be more highly expressed in CNS-isolated $CD4^+$ T cells at the peak of EAE, rather than the typical miR-155-5p variant [75]. In fact, miR-155-3p was shown to drive the upregulation of Th17 marker genes *Rora* and *IL17* compared to miR-155-5p and could specifically promote Th17 differentiation [75].

Altogether, it is particularly evident that miR-155 is critical for driving pathogenic Th1 and Th17 responses, whereas its deletion and/or inhibition results in reduced proliferative, functional responsiveness, and migration into the CNS during EAE. This is extremely important considering Th1 and Th17 cells are key drivers of the human disease [31]. Yet, it must also be emphasized that the contribution of miR-155 in myeloid cells, B cells, or brain-resident cells has not been thoroughly investigated in the context of EAE or other types of MS models. This is intriguing considering miR-155 expression has shown to be specifically elevated in PBMCs and tissue-resident brain cells (microglia and astrocytes) from human MS patient samples. It is critical that we delve deeper into understanding cellular origins of miR-155 in a more thorough manner.

5.5 miR-155 in Peripheral Immune Cells

Intensive research has focused on pathogenic CD4+ T cells as the key participants in the pathogenesis of MS. However from as early as 1990, it was demonstrated that up to 50% of the immune cells that infiltrate the CNS in the EAE model are in fact peripheral myeloid-derived monocytes and macrophages [38]. Importantly, the depletion of myeloid cells has been shown to completely prevent EAE disease progression [115]. During the early phase of MS, infiltrated monocytes/macrophages are immediately activated to become M1 macrophages, releasing pro-inflammatory cytokines, reactive oxygen species, and toxic metabolites that cause irreversible damage to neurons within the CNS. In fact, M1 polarized macrophages show prolonged periods of apposition within MS lesions, releasing reactive oxygen and nitrogen species that were shown to be particularly toxic to neurons and their axons [38, 44, 76]. During the later phase of disease and during periods of remission, macrophages are less activated and present as M2 or alternatively activated, releasing anti-inflammatory cytokines accompanied by inflammation resolution and tissue repair [96, 131]. For example, selective depletion of M2 macrophages inhibits experimental remyelination, whereas the transfer or enhancement of M2-polarized macrophages suppresses EAE [6, 71, 117]. M2 macrophages have huge capacity in therapeutics; however the molecular mechanisms which drive M2 polarization remain largely unknown. Evidence from the literature suggests that the levels of miR-155 expression in monocytes/macrophages are intimately associated with M1/M2 polarization states.

For example, miR-155 is potently induced by M1 agonists, namely, LPS, IFNγ, TNF, and GM-CSF [83, 84]. The transcription factors required for M1 polarization, such as NF-κB, AP-1, HIF1α, and most recently by us ETS2, required for the sustenance of an M1 phenotype are also absolutely essential for the transcriptional induction of miR-155 [7, 94]. HIF1α has been shown to bind to the miR-155 promoter and enhance its expression [7]. Interestingly, HIF1α plays a central role in the metabolic reprogramming of macrophages, acting to promote glycolysis essential for the maintenance and functional responses required by M1 macrophages [50]. Overexpression of miR-155 in monocytes and macrophages has been shown to increase reactive oxygen species, pro-inflammatory cytokines, and cell surface markers CD80 and CD86 [72, 78, 123, 139]. Furthermore, miR-155 transfected monocytes can enhance T-cell proliferation and IFNγ production [72].

Mechanistically, miR-155 contributes to pro-inflammatory signaling cascades and effector functions in macrophages by inhibiting numerous targets (SHIP1, FADD, SOCS1, IKK, IL13R1, CEBPβ, and SMAD2), which collectively result in the upregulation of the pro-inflammatory cytokines and release of reactive oxygen species [58, 60, 67, 91, 98]. Perhaps the most striking study illustrated that miR-155 control is not restricted to the above cellular targets, when approximately 650 genes required for M1 polarization were shown to be dependent on miR-155 in a whole genome transcriptome array [41].

On the other hand, macrophages treated with M2-polarizing agonists IL4/IL-13 fail to induce miR-155, while the anti-inflammatory cytokine IL-10 can potently inhibit LPS-induced miR-155 expression [68]. Moreover, miR-155−/− display elevated M2-polarizing cytokines such as IL-10 and IL4 and their serum [97]. This suggests that the inhibition of miR-155 can promote an M2 phenotype. Indeed, IL-10 inhibits both the primary miR-155 transcript and mature form in a STAT3-dependent manner [68]. Additionally, IL-10 reduced both ETS2 protein expression and its ability to bind to the miR-155 promoter, required for the transcriptional induction of miR-155 [94]. Mechanistically, low miR-155 expression can promote the M2 phenotype, by reversing its repression on M2-associated genes, including IL-13R, SMAD2, and CEBPβ. SMAD2 is a transcription factor essential for mediating the anti-inflammatory effects of TGFβ, while CEBPβ is important for the induction of the M2-associated genes Arg1, IL-10, IL-13R, and CD206 [60, 67, 98].

However, the direct implication of miR-155 on monocytes and macrophages during EAE and MS is less established. In the earliest studies, dendritic cells isolated from mice undergoing EAE had reduced pro-inflammatory cytokines, namely, IL-12, IL-1β, IL-6, IL-23, and TNF, that are required for Th1 and Th17 polarization [74, 79]. In human samples, miR-155 expression is elevated in monocyte-derived macrophages within active MS lesions, and others have shown that miR-155 is specifically increased in CD14+ isolated monocytes from MS blood samples [72]. Notably, mice deficient in the M2 agonist, IL-10, have been shown to develop accelerated disease progression following active immunization with CNS autoantigens, whereas IL-10 transgenic mice are resistant to EAE [18, 100]. Gene transfer methods delivering sustained IL-10 expression ameliorated the disease [19, 87]. It would be interesting to determine if the beneficial effects of IL-10 are mediated by its capacity to downregulate miR-155 or whether the conditional deletion of miR-155 can promote M2 macrophage accumulation in the CNS during EAE.

5.6 miR-155 in Brain-Resident Cells

Although the role of miR-155 in microglia and astrocytes has not been directly assessed in EAE, there is overwhelming evidence from other disease models that miR-155 upregulation contributes to neuroinflammation and subsequent neurodegeneration, whereas its deletion has neuroprotective effects. In primary cultured microglia, miR-155 is the most significantly upregulated miRNA under inflammatory M1-skewing conditions, similar to what has previously been shown in periphery myeloid cells [10, 25, 72]. Inhibition of miR-155 decreased the release of pro-inflammatory cytokines and nitric oxide, while the conditioned medium from these microglia could decrease neuronal cell death [10]. One study has demonstrated that IL-1/IFNγ can significantly upregulate miR-155-5p and miR-155-3p in astrocytes. Antagomirs to both isoforms dramatically reduced the production of TNF, IL-6, and IL-8, suggesting that miR-155 is required for polarization of astrocytes into an activated A1 phenotype [111]. This is interesting considering A1 astrocytes can induce the death of neurons and oligodendrocytes and are highly abundant in brain samples from MS, as well as other neurodegenerative disorders including Alzheimer's, Huntington's, Parkinson's, and amyotrophic lateral sclerosis (ALS) [56].

In animal models, sustained transgenic overexpression of miR-155 resulted in aberrations in the proliferation, migration, and differentiation of neural stem cells in the hippocampus. Whereas genetic deletion of miR-155 could restore the neuroinflammation-induced damage [128]. A strong upregulation of miR-155 was observed within the brain of the Alzheimer's disease model, which occurred simultaneously with increased microglia and astrocyte activation and the appearance of β-amyloid aggregates [33]. miR-155 elevation was also observed in spinal cord microglia from mice with amyotrophic lateral sclerosis (ALS), whereby genetic deletion of miR-155 or treatment with miR-155 antagomirs dramatically increased the survival by 38% in these mice [9]. miR-155 deletion and antagomir treatment were similarly shown to promote the recovery of ischemic stroke as a result of decreased neuroinflammation [89, 125]. Primary miR-155−/− microglia cultures displayed reduced inflammatory responses when treated with α-synuclein, a widespread aggregate found in Parkinson's disease, while a striking neuroprotective effect was observed in miR-155−/− mice with α -synuclein-induced Parkinson's disease [114].

5.7 miR-155 and the Blood-Brain Barrier

The blood-brain barrier (BBB) is a highly selective barrier made up of endothelial cells connected by tight junction proteins. It acts to separate circulating blood from the brain architecture and plays an important role in restricting the diffusion of pathogens, leukocytes, and large molecular weight molecules into the CNS. However, dysregulation of the BBB and the trafficking of peripheral activated leukocytes are among the earliest features observed in MS brains [70]. Although the mechanisms are not fully understood, the release of pro-inflammatory cytokines such as IL-1 and TNF from activated leukocytes and/or activated brain-resident cells can alter the physiology of the endothelial cells that make up the BBB, causing them to increase their permeability and change the dynamics of their tight junctions.

Various studies have confirmed that human endothelial cells can produce miR-155 under inflammatory conditions [51, 59, 93, 109, 130, 140]. Inhibition or overexpression of miR-155 in endothelial cells could either decrease or increase vascular endothelial permeability, respectively [109, 138]. Perhaps the most conclusive data for miR-155 in relation to BBB permeability and its impact on MS was demonstrated in Biozzi mice induced with EAE, an animal model with a predictable relapsing-remitting paralysis course associated with the loss BBB integrity at the spinal cord [59]. When FITC-dextran was injected into wild-type mice undergoing EAE, a high abundance of the marker was located in the spinal cord parenchyma. In contrast, there was a 50% reduction of the marker located in these tissues in miR-155−/− animals [59]. Overexpression of miR-155 increased the leakage of fluorescent dextrans across cultured human brain endothelial cells when challenged with cytokines TNF and IFNγ, whereas miR-155 inhibition reversed this effect. Overexpression of miR-155 could increase the permeability of endothelial cells by targeting tight junction proteins annexin-2 and claudin-1, but also focal adhesion molecules such as DOCK-1 and syntenin-1. In fact, endothelial cells accepted exosome-delivered miR-155, an effect that was shown to destroy tight junctions and the integrity of the endothelial barrier [138]. Furthermore, miR-155 could increase the adhesion of monocytes and T cells to endothelial cells under shear forces [11].

LPS activation of choroid plexus epithelium (CPE), a unique layer of epithelial cells that form a blood-brain barrier with cerebral spinal fluid (CSF), was shown to release miR-155-containing exosomes [3]. These exosomes were released into the CSF and taken up by astrocytes and microglia but not by neurons in an LPS-induced neuroinflammation model. Primary mixed cortical cultures incubated with miR-155-containing exosomes could potently increase the secretion of pro-inflammatory cytokines IL-6, IL-1, and TNF [3]. This effect was also mimicked in the brain of LPS-injected mice and was blocked when an exosome inhibitor was injected intracerebroventricularly. It is fascinating to consider that the peripheral activation of barrier cells such as endothelium and epithelium could be secreting miR-155-containing exosomes as a form communication that can alter the behavior of cells in the brain.

5.8 miR-155 and Environmental Risk Factors

Numerous large-scale epidemiology studies have been performed to search for MS environmental risk factors. Consistently, infection with EBV, infectious mononucleosis (caused by EBV), smoking, lack of vitamin D, and genetic risk alleles show the strongest correlation with MS susceptibility [5, 39, 102]. In particular, EBV infection has the strongest epidemiological credibility, and there is a large body of evidence to suggest that it plays a major role in the pathogenesis of MS [5, 90]. Although the exact mechanisms are incompletely understood, EBV has been shown to affect multiple immune cell parameters including increased EBV-transformed peripheral B cells with a concomitant increase in EBV viral shedding and production of anti-EBV antibodies, increased autoreactive CD4+ T cells in the CNS, impaired EBV-specific CD8+ T-cell immunity, and activation of innate immune cells in MS patients [90]. Intriguingly, early studies showed that BIC and mature miR-155 were strongly elevated in EBV-infected B lymphocytes [52, 95, 133]. The EBV latency membrane-associated protein (LMP1) is an important activator of NF-kB and the immortalization of B cells. LMP1 could induce BIC transcription and mature miR-155 primarily through the activation of NF-κB, p38, and AP1 transcription factors [30, 95, 134]. Importantly, miR-155 antagomirs could reduce EBV nuclear antigen (EBNA) mRNA expression and EBV copy number in infected cells, as well as inhibit the growth of proliferating lymphoblastoid cell lines [57, 62]. Children with infectious mononucleosis caused by primary EBV infection also displayed elevated miR-155 expression in blood-isolated B cells [29]. Large-scale transcriptome analysis of both viral- and cellular-induced miRNA in EBV transformed cells highlighted that BACH1 is a likely target for miR-155 [106]. In fact, BACH1 has been suggested to play a key regulatory role in EAE and MS [29, 107].

The lack of vitamin D, especially in countries located in latitudes correlated with poor sunlight, has been associated with MS susceptibility. Intriguingly, dendritic cells treated with 1 alpha,25-dihydroxyvitamin D(3) (vitamin D) gave rise to an immature phenotype, characterized by low levels of miR-155 and IL-23 [88]. Subsequent studies showed vitamin D could strongly prevent miR-155 induction in human adipocytes and disrupt the formation of miR-155-containing exosomes in chronic lymphocytic leukemia monocytes [8, 46]. Vitamin D could attenuate LPS-induced signaling through a mechanism that involved inhibition of miR-155 and an increase in the target, SOCS1 [13]. Moreover, the suppressive effect of vitamin D on miR-155 could not be achieved when Ago2, a key protein required for RNA- and miRNA-induced silencing complex, was deleted [40].

Of the 110 genetic risk variants identified from a cohort of 14,496 subjects with MS, 97 of these were associated with immunological function [39]. Although a SNP for miR-155 was not reported, five SNPs (TNFSF14, IL2RA, TNFSF1A, IL12A, and STAT4) accounted for more than 50% of the association. Interestingly, enforced expression of miR-155 in PBMCs could enhance IL-2 expression, while stimulated T cells from miR-155−/− mice have deficient IL-2 production [54, 97]. Similarly, miR-155 inhibition or overexpression could inhibit or promote IL-12 production in DCs [61]. Moreover we have shown that genetic deletion of Ets2 in myeloid cells, a critical transcription factor required for miR-155 induction, failed to produce IL-12 cytokine [94]. It is plausible to consider that hyper-expression of miR-155 in MS patients contributes to the dysregulation of signaling pathways controlled by these risk variants. In fact, in an Italian cohort of 360 MS patients, 4 SNPs were located in close proximity to the BIC gene. Three of these formed a unique haplotype (rs2829803, rs2282471, rs2829806) that was overrepresented in MS patients (13.5%) compared to controls (10.3%) and conferred a 1.36-fold increased genetic risk of developing MS [86]. It will be interesting to see if larger-scale studies can recapitulate this result and find an association of the BIC/miR-155 haplotype with MS.

5.9 miR-155 and Disease-Modifying Treatments

There is no cure for MS. However, there are currently >10 FDA-approved disease-modifying treatments (DMF) for RRMS [20]. Early intervention and treatment with these DMFs significantly slows the progression of the disease, as well as lowering the relapse rate and the formation of new lesions. These medications predominantly act on the immune system as immunosuppressants and can be broadly divided based on their ability to block immune cell infiltration into the CNS (natalizumab, fingolimod, mitoxantrone), to reduce immune cell activity (interferon-β, glatiramer acetate, dimethyl fumarate), or to inhibit immune cell proliferation (teriflunomide, alemtuzumab, ocrelizumab) [17]. Considering miR-155 is elevated in MS patients and in animal models undergoing EAE, it is worth understanding the impact of these treatments on miR-155 expression.

IFNβ and glatiramer acetate were the earliest drugs to be approved for MS and are often used as first-line treatment options [20]. IFNβ helps regulate the immune system, decreasing the amount of immune cells infiltrating the CNS, particularly Th1/Th17 subsets and their respective cytokines. Glatiramer acetate is a synthetic amino acid polymer that mimics myelin basic protein and has been shown to divert the generation of Th1 cells to Th2 cells which can suppress the inflammatory response. In two separate studies, whole blood and PBMC samples isolated from MS patients treated with IFNβ and glatiramer acetate failed to demonstrate any effect on miR-155 expression, whereas other miRNAs were found to be reduced significantly [49, 124]. However, glatiramer acetate did reduce miR-155 expression in urine-isolated exosomes from EAE-induced mice at peak disease [105]. In some respects, the lack of repression on miR-155 in IFNβ-treated patients is not surprising considering IFNβ is an established agonist for miR-155 induction in macrophages [83].

Dimethyl fumarate (DMF) is a methyl ester of fumaric acid and was approved by the FDA as an effective oral treatment for RRMS [20]. Although we still do not fully understand its mechanism of action, studies have shown in MS patients treated with DMF that there is a reduction in Th1/Th17 subsets, an increase in Th2 subsets, and a shift from M1 to M2 macrophages [110, 129], while others have shown that DMF can inhibit microglia and astrocyte inflammation and has neuroprotective effects in vitro and in EAE animal models [1, 126]. Promisingly, miR-155 expression was found to be significantly reduced in monocytes from MS patients receiving DMF [69].

Natalizumab was the first humanized monoclonal antibody approved for MS in 2007 and is classed as a highly effective treatment. It blocks the cellular adhesion molecule α4 integrin on immune cells, inhibiting their ability to bind and migrate through the endothelial BBB. Natalizumab has been shown to reduce miR-155 expression in PBMCs and monocytes isolated from MS patients which also correlated with a decrease in IL-17, IFNγ, and TNF gene expression [64, 69]. Furthermore, patients with the highest expression of miR-155 expression pre-natalizumab therapy had higher levels of anti-EBV nuclear antigen titers in their serum [64]. Fingolimod, although not a monoclonal antibody, works in a similar manner to natalizumab by trapping immune cells in lymph nodes and preventing their migration to the CNS. Fingolimod also significantly reduced miR-155 expression in human monocytes [69].

Alemtuzumab is a humanized monoclonal antibody against CD52, a cell surface receptor expressed on mature T and B lymphocytes, while ocrelizumab, the first approved drug for PPMS, is a humanized monoclonal antibody that binds to CD20, a cell surface marker specifically expressed on B cells. Both treatments result in tagging their respective cells for destruction, and patients have shown very promising improvements in disability and disease progression. Although miR-155 expression has not been explored with either drug, ibrutinib, a B-cell-depleting therapy for chronic lymphocytic leukemia has been shown to significantly decrease miR-155 expression [34].

Overall, this data strongly suggests that miR-155 could be a very effective biomarker for monitoring responsiveness to treatment. Moreover, it indicates that inhibiting miR-155 itself could be a very legitimate and realistic target for the treatment of MS. This is even more pertinent when we consider the astounding efficacy miR-155 antagomirs have had in EAE models, as well as consider the overall impact that miR-155 inhibition has in skewing cells of the immune system to an anti-inflammatory phenotype both in vitro and in vivo [74, 135]. Indeed, judging by the extraordinary number of submitted patent applications for miR-155 antagonists, it strongly suggests that generation of miR-155 antagomirs as a therapeutic treatment is already underway.

5.10 Future Directions

One of the biggest challenges in MS is the lack of a biomarker to effectively stratify different MS subtypes, which is critical when considering the most effective treatment strategy. In addition, a biomarker that aids our understanding of disease severity, responsiveness to treatment, and disease progression is still required. The attractiveness of miRNAs as biomarkers for disease cannot be underestimated [27]. Their presence, not only in cells and tissues but also in readily available body fluids and extracellular vesicles, suggests that they represent a gold mine of noninvasive biomarkers for disease. miRNAs are extremely stable; those that are found in blood are highly resistant to ribonuclease due to their packaging in lipid vesicles, binding to RNA-binding molecules, or their association with high-density lipoproteins [14]. Once expressed, they are long-lived. For example, from a range of immunologically-related miRNAs tested in macrophages, we showed that miRNAs have an average half-life of 5 days [28]. Once isolated from relevant samples, they are resistant to extended storage, freeze-thaw, and extreme pH. The development of sensitive platforms for detection and quantification using methods such as miRNA assays, bead-based assays, NanoString techniques, and deep sequencing has ensured that results are quantifiable, reproducible, and accurate.

To date, miRNA expression profiles have been conducted in MS patients from a variety of tissues including whole blood, PBMCs, serum, CSF, MS lesions, as well as sorted lymphocyte and myeloid-derived immune cell populations [27, 42]. Interestingly, Jagot and Davoust consolidated 19 miRNA profiling metadata studies conducted in plasma, CNS, and immune cells in an attempt to find the most commonly dysregulated miRNAs in MS patients. miR-155 was identified to be elevated across all MS patient samples tested, along with altered expression of miR-23a, miR-223, miR-22, miR-326, and miR-21 suggesting that these miRNAs may form an important signature in MS [43]. However, it is also important to note that other studies have found no changes in miR-155 expression in PBMCs from RRMS, SPMS, and/or PPMS [16, 36, 47, 48, 66, 85, 108]. In fact, miR-155 was found to be significantly downregulated in $CD4^+$ T cells from a cohort of secondary progressive MS patients [101]. Moving forward, it will be critical to assess if miR-155 elevation is specific for a particular cell type or subtype of MS, while isolating miR-155 from urine or CSF could provide an alternative avenue of exploration. Overall, generating panels of miRNA signatures to include miR-155, rather than looking at miRNAs individually, could be a more beneficial approach for developing diagnostic and prognostic indicators for MS.

The delivery of stem cells, particularly hematopoietic stem cells (HSC), mesenchymal stem cells (MSC), induced pluripotent stem cells (iPSC), and neural stem cells (MSC), is under intense investigation in many neurodegenerative diseases including MS. In particular, early clinical trials have demonstrated that RRMS patients receiving HSC transplants have had the greatest improvements in disease progression, with a dramatic decrease in relapse frequency and MRI activity [65]. Although the contribution of miRNAs in therapeutic stem cell delivery has not been explored in MS, it has been widely examined in cancer, cardiovascular diseases, arthritis, and neurological disease, where they have been shown to play a fundamental role in stem cell differentiation and therapeutic modulation [24, 99, 104, 119, 132]. It is clearly worth pursuing whether these emerging therapies for MS have a direct impact on miR-155.

In MS, the CNS also has the capacity to remyelinate and repair any damage caused to neuronal axons. Clinicians often observe this as a “shadow plaque” in MRI scans when old lesions have been repaired and remyelinated during the remitting phases of the disease. Remyelination is mediated by a population of oligodendrocyte progenitor cells (OPCs) that can proliferate and migrate to areas of damage, where they differentiate into myelin-producing mature oligodendrocytes. Understanding the molecular mechanisms while identifying novel therapies that can promote oligodendrocyte maturation and remyelination is currently under heavy investigation in the MS field. Recent evidence suggests that microRNAs may govern this process [26]. In particular, miR-219 and miR-338 have been shown to be critical for CNS remyelination after injury, while miR-146a could facilitate remyelination by promoting OPC differentiation in cuprizone demyelinating models [122, 137]. Clearly, the impact of miR-155 in these processes could be an exciting avenue for exploration.

5.11 Conclusion

Although many advances have been made understanding the role of miR-155 in immune cell function and regulation, there are many gaps in our knowledge in understanding its role in non-immune cells. Furthermore, the exact contribution of miR-155 in the majority of relevant cell types has not been directly explored in MS. Generating conditional knockout animals or transgenic overexpressing models will help to answer these questions. Moreover, extending the use of typical MS models to focus on its impact in demyelination and remyelination should be considered. Understanding whether miR-155 is simply a marker for inflammation or whether it plays a predominant role in triggering MS requires further study.

Standardizing miRNA detection methods in various human samples will help consolidate whether miR-155 could be used as a realistic biomarker in MS. Most importantly, therapeutic inhibition of miR-155 and understanding whether it will be most efficacious by delivering it in the periphery or directly to the CNS, or directly to specific cell types or specific MS disease subtypes, remains to be elucidated.

References

1. Albrecht P, Bouchachia I, Goebels N, Henke N, Hofstetter HH et al (2012) Effects of dimethyl fumarate on neuroprotection and immunomodulation. J Neuroinflammation 9:163
2. Ambros V (2004) The functions of animal microRNAs. Nature 431:350–355
3. Balusu S, Van Wonterghem E, De Rycke R, Raemdonck K, Stremersch S et al (2016) Identification of a novel mechanism of blood-brain communication during peripheral inflammation via choroid plexus-derived extracellular vesicles. EMBO Mol Med 8:1162–1183
4. Bartel DP (2004) MicroRNAs: genomics, biogenesis, mechanism, and function. Cell 116:281–297
5. Belbasis L, Bellou V, Evangelou E, Ioannidis JP, Tzoulaki I (2015) Environmental risk factors and multiple sclerosis: an umbrella review of systematic reviews and meta-analyses. Lancet Neurol 14:263–273
6. Benedek G, Meza-Romero R, Jordan K, Keenlyside L, Offner H et al (2015) HLA-DRalpha1-mMOG-35-55 treatment of experimental autoimmune encephalomyelitis reduces CNS inflammation, enhances M2 macrophage frequency, and promotes neuroprotection. J Neuroinflammation 12:123
7. Bruning U, Cerone L, Neufeld Z, Fitzpatrick SF, Cheong A et al (2011) MicroRNA-155 promotes resolution of hypoxia-inducible factor 1alpha activity during prolonged hypoxia. Mol Cell Biol 31:4087–4096
8. Bruns H, Bottcher M, Qorraj M, Fabri M, Jitschin S et al (2017) CLL-cell-mediated MDSC induction by exosomal miR-155 transfer is disrupted by vitamin D. Leukemia 31:985–988
9. Butovsky O, Jedrychowski MP, Cialic R, Krasemann S, Murugaiyan G et al (2015) Targeting miR-155 restores abnormal microglia and attenuates disease in SOD1 mice. Ann Neurol 77:75–99
10. Cardoso AL, Guedes JR, Pereira de Almeida L, Pedroso de Lima MC (2012) miR-155 modulates microglia-mediated immune response by down-regulating SOCS-1 and promoting cytokine and nitric oxide production. Immunology 135:73–88
11. Cerutti C, Soblechero-Martin P, Wu D, Lopez-Ramirez MA, de Vries H et al (2016) MicroRNA-155 contributes to shear-resistant leukocyte adhesion to human brain endothelium in vitro. Fluids Barriers CNS 13:8
12. Chekulaeva M, Filipowicz W (2009) Mechanisms of miRNA-mediated post-transcriptional regulation in animal cells. Curr Opin Cell Biol 21:452–460
13. Chen Y, Liu W, Sun T, Huang Y, Wang Y et al (2013) 1,25-Dihydroxyvitamin D promotes negative feedback regulation of TLR signaling via targeting microRNA-155-SOCS1 in macrophages. J Immunol 190:3687–3695
14. Cortez MA, Bueso-Ramos C, Ferdin J, Lopez-Berestein G, Sood AK et al (2011) MicroRNAs in body fluids--the mix of hormones and biomarkers. Nat Rev Clin Oncol 8:467–477
15. Costinean S, Zanesi N, Pekarsky Y, Tili E, Volinia S et al (2006) Pre-B cell proliferation and lymphoblastic leukemia/high-grade lymphoma in E(mu)-miR155 transgenic mice. Proc Natl Acad Sci U S A 103:7024–7029
16. Cox MB, Cairns MJ, Gandhi KS, Carroll AP, Moscovis S et al (2010) MicroRNAs miR-17 and miR-20a inhibit T cell activation genes and are under-expressed in MS whole blood. PLoS One 5:e12132

17. Cross AH, Naismith RT (2014) Established and novel disease-modifying treatments in multiple sclerosis. J Int Med 275(4):350–363. doi:10.1111/joim.12203. Epub 2014 March 11. Review. PubMed PMID:24444048
18. Cua DJ, Groux H, Hinton DR, Stohlman SA, Coffman RL (1999) Transgenic interleukin 10 prevents induction of experimental autoimmune encephalomyelitis. J Exp Med 189:1005–1010
19. Cua DJ, Hutchins B, LaFace DM, Stohlman SA, Coffman RL (2001) Central nervous system expression of IL-10 inhibits autoimmune encephalomyelitis. J Immunol 166:602–608
20. Dargahi N, Katsara M, Tselios T, Androutsou ME, de Courten M et al (2017) Multiple sclerosis: immunopathology and treatment update. Brain Sci 7
21. Dendrou CA, Fugger L, Friese MA (2015) Immunopathology of multiple sclerosis. Nat Rev Immunol 15:545–558
22. Eis PS, Tam W, Sun L, Chadburn A, Li Z et al (2005) Accumulation of miR-155 and BIC RNA in human B cell lymphomas. Proc Natl Acad Sci U S A 102:3627–3632
23. Escobar TM, Kanellopoulou C, Kugler DG, Kilaru G, Nguyen CK et al (2014) miR-155 activates cytokine gene expression in Th17 cells by regulating the DNA-binding protein Jarid2 to relieve polycomb-mediated repression. Immunity 40:865–879
24. Faravelli I, Corti S (2017) MicroRNA-directed neuronal reprogramming as a therapeutic strategy for neurological diseases. Mol Neurobiol
25. Freilich RW, Woodbury ME, Ikezu T (2013) Integrated expression profiles of mRNA and miRNA in polarized primary murine microglia. PLoS One 8:e79416
26. Galloway DA, Moore CS (2016) miRNAs as emerging regulators of oligodendrocyte development and differentiation. Front cell. Dev Biol 4:59
27. Gandhi R (2015) miRNA in multiple sclerosis: search for novel biomarkers. Mult Scler 21:1095–1103
28. Gantier MP, McCoy CE, Rusinova I, Saulep D, Wang D et al (2011) Analysis of microRNA turnover in mammalian cells following Dicer1 ablation. Nucleic Acids Res 39:5692–5703
29. Gao L, Ai J, Xie Z, Zhou C, Liu C et al (2015) Dynamic expression of viral and cellular microRNAs in infectious mononucleosis caused by primary Epstein-Barr virus infection in children. Virol J 12:208
30. Gatto G, Rossi A, Rossi D, Kroening S, Bonatti S, et al. (2008) Epstein-Barr virus latent membrane protein 1 trans-activates miR-155 transcription through the NF-kappaB
31. Goverman J (2009) Autoimmune T cell responses in the central nervous system. Nat Rev Immunol 9:393–407
32. Gregory SG, Schmidt S, Seth P, Oksenberg JR, Hart J, et al. (2007) Interleukin 7 receptor alpha chain (IL7R) shows allelic and functional association with multiple sclerosis. Nat Genet
33. Guedes JR, Custodia CM, Silva RJ, de Almeida LP, Pedroso de Lima MC et al (2014) Early miR-155 upregulation contributes to neuroinflammation in Alzheimer's disease triple transgenic mouse model. Hum Mol Genet 23:6286–6301
34. Guinn D, Ruppert AS, Maddocks K, Jaglowski S, Gordon A et al (2015) miR-155 expression is associated with chemoimmunotherapy outcome and is modulated by Bruton's tyrosine kinase inhibition with Ibrutinib. Leukemia 29:1210–1213
35. Hartmann FJ, Khademi M, Aram J, Ammann S, Kockum I et al (2014) Multiple sclerosis-associated IL2RA polymorphism controls GM-CSF production in human TH cells. Nat Commun 5:5056
36. Hecker M, Thamilarasan M, Koczan D, Schroder I, Flechtner K et al (2013) MicroRNA expression changes during interferon-beta treatment in the peripheral blood of multiple sclerosis patients. Int J Mol Sci 14:16087–16110
37. Hu R, Huffaker TB, Kagele DA, Runtsch MC, Bake E et al (2013) MicroRNA-155 confers encephalogenic potential to Th17 cells by promoting effector gene expression. J Immunol 190:5972–5980
38. Huitinga I, van Rooijen N, de Groot CJ, Uitdehaag BM, Dijkstra CD (1990) Suppression of experimental allergic encephalomyelitis in Lewis rats after elimination of macrophages. J Exp Med 172:1025–1033

39. International Multiple Sclerosis Genetics C, Beecham AH, Patsopoulos NA, Xifara DK, Davis MF, et al (2013) Analysis of immune-related loci identifies 48 new susceptibility variants for multiple sclerosis. Nat Genet 45:1353–1360
40. Iosue I, Quaranta R, Masciarelli S, Fontemaggi G, Batassa EM et al (2013) Argonaute 2 sustains the gene expression program driving human monocytic differentiation of acute myeloid leukemia cells. Cell Death Dis 4:e926
41. Jablonski KA, Gaudet AD, Amici SA, Popovich PG, Guerau-de-Arellano M (2016) Control of the inflammatory macrophage transcriptional signature by miR-155. PLoS One 11:e0159724
42. Jagot F, Davoust N (2016) Is it worth considering circulating microRNAs in multiple sclerosis? Front Immunol 7:129
43. Jevtic B, Timotijevic G, Stanisavljevic S, Momcilovic M, Mostarica Stojkovic M et al (2015) Micro RNA-155 participates in re-activation of encephalitogenic T cells. Biomed Pharmacother 74:206–210
44. Jiang Z, Jiang JX, Zhang GX (2014) Macrophages: a double-edged sword in experimental autoimmune encephalomyelitis. Immunol Lett 160:17–22
45. Junker A, Krumbholz M, Eisele S, Mohan H, Augstein F et al (2009) MicroRNA profiling of multiple sclerosis lesions identifies modulators of the regulatory protein CD47. Brain 132:3342–3352
46. Karkeni E, Bonnet L, Marcotorchino J, Tourniaire F, Astier J, et al. (2017) Vitamin D limits inflammation-linked microRNA expression in adipocytes in vitro and in vivo: A new mechanism for the regulation of inflammation by vitamin D. Epigenetics:0
47. Keller A, Leidinger P, Lange J, Borries A, Schroers H et al (2009) Multiple sclerosis: microRNA expression profiles accurately differentiate patients with relapsing-remitting disease from healthy controls. PLoS One 4:e7440
48. Keller A, Leidinger P, Steinmeyer F, Stähler C, Franke A, Hemmrich-Stanisak G, Kappel A, Wright I, Dörr J, Paul F, Diem R, Tocariu-Krick B, Meder B, Backes C, Meese E, Ruprecht K (2014) Comprehensive analysis of microRNA profiles in multiple sclerosis including next-generation sequencing. Mult Scler 20(3):295–303. doi:10.1177/1352458513496343. Epub 2013 Jul 8. PubMed PMID:23836875
49. Keller A, Leidinger P, Steinmeyer F, Stahler C, Franke A et al (2014) Comprehensive analysis of microRNA profiles in multiple sclerosis including next-generation sequencing. Mult Scler 20:295–303
50. Kelly B, O'Neill LA (2015) Metabolic reprogramming in macrophages and dendritic cells in innate immunity. Cell Res 25:771–784
51. Kin K, Miyagawa S, Fukushima S, Shirakawa Y, Torikai K et al (2012) Tissue- and plasma-specific MicroRNA signatures for atherosclerotic abdominal aortic aneurysm. J Am Heart Assoc 1:e000745
52. Kluiver J, Poppema S, de Jong D, Blokzijl T, Harms G, Jacobs S, Kroesen BJ, van den Berg A (2005) BIC and miR-155 are highly expressed in Hodgkin, primary mediastinal and diffuse large B cell lymphomas. J Pathol 207(2):243–249. PubMed PMID:16041695
53. Landgraf P, Rusu M, Sheridan R, Sewer A, Iovino N et al (2007) A mammalian microRNA expression atlas based on small RNA library sequencing. Cell 129:1401–1414
54. Lashine YA, Salah S, Aboelenein HR, Abdelaziz AI (2015) Correcting the expression of miRNA-155 represses PP2Ac and enhances the release of IL-2 in PBMCs of juvenile SLE patients. Lupus 24(3):240–247. doi:10.1177/0961203314552117. Epub 2014 September 24. PubMed PMID:25253569
55. Li Z, Rana TM (2014) Therapeutic targeting of microRNAs: current status and future challenges. Nat Rev Drug Discov 13:622–638
56. Liddelow SA, Guttenplan KA, Clarke LE, Bennett FC, Bohlen CJ et al (2017) Neurotoxic reactive astrocytes are induced by activated microglia. Nature 541:481–487
57. Linnstaedt SD, Gottwein E, Skalsky RL, Luftig MA, Cullen BR (2010) Virally induced cellular microRNA miR-155 plays a key role in B-cell immortalization by Epstein-Barr virus. J Virol 84:11670–11678

58. Liu G, Abraham E (2013) MicroRNAs in immune response and macrophage polarization. Arterioscler Thromb Vasc Biol 33:170–177
59. Lopez-Ramirez MA, Wu D, Pryce G, Simpson JE, Reijerkerk A et al (2014) MicroRNA-155 negatively affects blood-brain barrier function during neuroinflammation. FASEB J 28:2551–2565
60. Louafi F, Martinez-Nunez RT, Sanchez-Elsner T (2010) MicroRNA-155 targets SMAD2 and modulates the response of macrophages to transforming growth factor-{beta}. J Biol Chem 285:41328–41336
61. Lu C, Huang X, Zhang X, Roensch K, Cao Q et al (2011) miR-221 and miR-155 regulate human dendritic cell development, apoptosis, and IL-12 production through targeting of p27kip1, KPC1, and SOCS-1. Blood 117:4293–4303
62. Lu F, Weidmer A, Liu CG, Volinia S, Croce CM et al (2008) Epstein-Barr virus-induced miR-155 attenuates NF-kappaB signaling and stabilizes latent virus persistence. J Virol 82:10436–10443
63. Lund C, Nakken KO, Edland A, Celius EG (2014) Multiple sclerosis and seizures: incidence and prevalence over 40 years. Acta Neurol Scand 130:368–373
64. Mameli G, Arru G, Caggiu E, Niegowska M, Leoni S et al (2016) Natalizumab therapy modulates miR-155, miR-26a and Proinflammatory cytokine expression in MS patients. PLoS One 11:e0157153
65. Mancardi G, Saccardi R (2008) Autologous haematopoietic stem-cell transplantation in multiple sclerosis. Lancet Neurol 7:626–636
66. Martinelli-Boneschi F, Fenoglio C, Brambilla P, Sorosina M, Giacalone G et al (2012) MicroRNA and mRNA expression profile screening in multiple sclerosis patients to unravel novel pathogenic steps and identify potential biomarkers. Neurosci Lett 508:4–8
67. Martinez-Nunez RT, Louafi F, Sanchez-Elsner T (2011) The interleukin 13 (IL-13) pathway in human macrophages is modulated by microRNA-155 via direct targeting of interleukin 13 receptor alpha1 (IL13Ralpha1). J Biol Chem 286:1786–1794
68. McCoy CE, Sheedy FJ, Qualls JE, Doyle SL, Quinn SR et al (2010) IL-10 inhibits miR-155 induction by toll-like receptors. J Biol Chem 285:20492–20498
69. Michell-Robinson MA, Moore CS, Healy LM, Osso LA, Zorko N et al (2016) Effects of fumarates on circulating and CNS myeloid cells in multiple sclerosis. Ann Clin Transl Neurol 3:27–41
70. Minagar A, Alexander JS (2003) Blood-brain barrier disruption in multiple sclerosis. Mult Scler 9:540–549
71. Miron VE, Boyd A, Zhao JW, Yuen TJ, Ruckh JM et al (2013) M2 microglia and macrophages drive oligodendrocyte differentiation during CNS remyelination. Nat Neurosci 16:1211–1218
72. Moore CS, Rao VT, Durafourt BA, Bedell BJ, Ludwin SK, Bar-Or A, Antel JP (2013) miR-155 as a multiple sclerosisrelevant regulator of myeloid cell polarization. Ann Neurol 74(5):709–720. doi:10.1002/ana.23967. Epub 2013 September 23. PubMed PMID: 23818336
73. Moutsianas L, Jostins L, Beecham AH, Dilthey AT, Xifara DK et al (2015) Class II HLA interactions modulate genetic risk for multiple sclerosis. Nat Genet 47:1107–1113
74. Murugaiyan G, Beynon V, Mittal A, Joller N, Weiner HL (2011) Silencing microRNA-155 ameliorates experimental autoimmune encephalomyelitis. J Immunol 187:2213–2221
75. Mycko MP, Cichalewska M, Cwiklinska H, Selmaj KW (2015) miR-155-3p drives the development of autoimmune demyelination by regulation of heat shock protein 40. J Neurosci 35:16504–16515
76. Nikic I, Merkler D, Sorbara C, Brinkoetter M, Kreutzfeldt M et al (2011) A reversible form of axon damage in experimental autoimmune encephalomyelitis and multiple sclerosis. Nat Med 17:495–499
77. Noorbakhsh F, Ellestad KK, Maingat F, Warren KG, Han MH et al (2011) Impaired neurosteroid synthesis in multiple sclerosis. Brain 134:2703–2721
78. O'Connell RM, Chaudhuri AA, Rao DS, Baltimore D (2009) Inositol phosphatase SHIP1 is a primary target of miR-155. Proc Natl Acad Sci U S A 106:7113–7118

79. O'Connell RM, Kahn D, Gibson WS, Round JL, Scholz RL et al (2010) MicroRNA-155 promotes autoimmune inflammation by enhancing inflammatory T cell development. Immunity 33:607–619
80. O'Connell RM, Rao DS, Baltimore D (2012) microRNA regulation of inflammatory responses. Annu Rev Immunol 30:295–312
81. O'Connell RM, Rao DS, Chaudhuri AA, Baltimore D (2010) Physiological and pathological roles for microRNAs in the immune system. Nat Rev Immunol 10:111–122
82. O'Connell RM, Rao DS, Chaudhuri AA, Boldin MP, Taganov KD et al (2008) Sustained expression of microRNA-155 in hematopoietic stem cells causes a myeloproliferative disorder. J Exp Med 205:585–594
83. O'Connell RM, Taganov KD, Boldin MP, Cheng G, Baltimore D (2007) MicroRNA-155 is induced during the macrophage inflammatory response. Proc Natl Acad Sci U S A 104:1604–1609
84. O'Neill LA, Sheedy FJ, McCoy CE (2011) MicroRNAs: the fine-tuners of toll-like receptor signalling. Nat Rev Immunol 11:163–175
85. Otaegui D, Baranzini SE, Armananzas R, Calvo B, Munoz-Culla M et al (2009) Differential micro RNA expression in PBMC from multiple sclerosis patients. PLoS One 4:e6309
86. Paraboschi EM, Solda G, Gemmati D, Orioli E, Zeri G et al (2011) Genetic association and altered gene expression of mir-155 in multiple sclerosis patients. Int J Mol Sci 12:8695–8712
87. Payne NL, Sun G, McDonald C, Moussa L, Emerson-Webber A et al (2013) Human adipose-derived mesenchymal stem cells engineered to secrete IL-10 inhibit APC function and limit CNS autoimmunity. Brain Behav Immun 30:103–114
88. Pedersen AW, Holmstrom K, Jensen SS, Fuchs D, Rasmussen S et al (2009) Phenotypic and functional markers for 1alpha,25-dihydroxyvitamin D(3)-modified regulatory dendritic cells. Clin Exp Immunol 157:48–59
89. Pena-Philippides JC, Caballero-Garrido E, Lordkipanidze T, Roitbak T (2016) In vivo inhibition of miR-155 significantly alters post-stroke inflammatory response. J Neuroinflammation 13:287
90. Pender MP, Burrows SR (2014) Epstein-Barr virus and multiple sclerosis: potential opportunities for immunotherapy. Clin Transl Immunol 3:e27
91. Ponomarev ED, Veremeyko T, Weiner HL (2013) MicroRNAs are universal regulators of differentiation, activation, and polarization of microglia and macrophages in normal and diseased CNS. Glia 61:91–103
92. Pugliatti M, Rosati G, Carton H, Riise T, Drulovic J et al (2006) The epidemiology of multiple sclerosis in Europe. Eur J Neurol 13:700–722
93. Pulkkinen KH, Yla-Herttuala S, Levonen AL (2011) Heme oxygenase 1 is induced by miR-155 via reduced BACH1 translation in endothelial cells. Free Radic Biol Med 51:2124–2131
94. Quinn SR, Mangan NE, Caffrey BE, Gantier MP, Williams BR et al (2014) The role of Ets2 transcription factor in the induction of microRNA-155 (miR-155) by lipopolysaccharide and its targeting by interleukin-10. J Biol Chem 289:4316–4325
95. Rahadiani N, Takakuwa T, Tresnasari K, Morii E, Aozasa K (2008) Latent membrane protein-1 of Epstein-Barr virus induces the expression of B-cell integration cluster, a precursor form of microRNA-155, in B lymphoma cell lines. Biochem Biophys Res Commun 377:579–583
96. Rawji KS, Yong VW (2013) The benefits and detriments of macrophages/microglia in models of multiple sclerosis. Clin Dev Immunol 2013:948976
97. Rodriguez A, Vigorito E, Clare S, Warren MV, Couttet P, Soond DR, van Dongen S, Grocock RJ, Das PP, Miska EA, Vetrie D, Okkenhaug K, Enright AJ, Dougan G, Turner M, Bradley A (2007) Requirement of bic/microRNA-155 for normal immune function. Science 316(5824):608–611. PubMed PMID:17463290; PubMed Central PMCID:PMC2610435
98. Ruffell D, Mourkioti F, Gambardella A, Kirstetter P, Lopez RG et al (2009) A CREB-C/EBPbeta cascade induces M2 macrophage-specific gene expression and promotes muscle injury repair. Proc Natl Acad Sci U S A 106:17475–17480
99. Rupaimoole R, Slack FJ (2017) MicroRNA therapeutics: towards a new era for the management of cancer and other diseases. Nat Rev Drug Discov 16:203–222

100. Samoilova EB, Horton JL, Chen Y (1998) Acceleration of experimental autoimmune encephalomyelitis in interleukin-10-deficient mice: roles of interleukin-10 in disease progression and recovery. Cell Immunol 188:118–124
101. Sanders KA, Benton MC, Lea RA, Maltby VE, Agland S et al (2016) Next-generation sequencing reveals broad down-regulation of microRNAs in secondary progressive multiple sclerosis CD4+ T cells. Clin Epigenetics 8:87
102. Schmidt C (2016) Biology: a degenerative affliction. Nature 540:S2–S3
103. Seddiki N, Brezar V, Ruffin N, Levy Y, Swaminathan S (2014) Role of miR-155 in the regulation of lymphocyte immune function and disease. Immunology 142:32–38
104. Seeger FH, Zeiher AM, Dimmeler S (2013) MicroRNAs in stem cell function and regenerative therapy of the heart. Arterioscler Thromb Vasc Biol 33:1739–1746
105. Singh J, Deshpande M, Suhail H, Rattan R, Giri S (2016) Targeted stage-specific inflammatory microRNA profiling in urine during disease progression in experimental autoimmune encephalomyelitis: markers of disease progression and drug response. J Neuroimmune Pharmacol 11:84–97
106. Skalsky RL, Corcoran DL, Gottwein E, Frank CL, Kang D et al (2012) The viral and cellular microRNA targetome in lymphoblastoid cell lines. PLoS Pathog 8:e1002484
107. So AY, Garcia-Flores Y, Minisandram A, Martin A, Taganov K et al (2012) Regulation of APC development, immune response, and autoimmunity by Bach1/HO-1 pathway in mice. Blood 120:2428–2437
108. Sondergaard HB, Hesse D, Krakauer M, Sorensen PS, Sellebjerg F (2013) Differential microRNA expression in blood in multiple sclerosis. Mult Scler 19:1849–1857
109. Sun HX, Zeng DY, Li RT, Pang RP, Yang H et al (2012) Essential role of microRNA-155 in regulating endothelium-dependent vasorelaxation by targeting endothelial nitric oxide synthase. Hypertension 60:1407–1414
110. Tannahill GM, Iraci N, Gaude E, Frezza C, Pluchino S (2015) Metabolic reprograming of mononuclear phagocytes in progressive multiple sclerosis. Front Immunol 6:106
111. Tarassishin L, Loudig O, Bauman A, Shafit-Zagardo B, Suh HS et al (2011) Interferon regulatory factor 3 inhibits astrocyte inflammatory gene expression through suppression of the proinflammatory miR-155 and miR-155*. Glia 59:1911–1922
112. Teng G, Hakimpour P, Landgraf P, Rice A, Tuschl T et al (2008) MicroRNA-155 is a negative regulator of activation-induced cytidine deaminase. Immunity 28:621–629
113. Thai TH, Calado DP, Casola S, Ansel KM, Xiao C et al (2007) Regulation of the germinal center response by microRNA-155. Science 316:604–608
114. Thome AD, Harms AS, Volpicelli-Daley LA, Standaert DG (2016) microRNA-155 regulates alpha-Synuclein-induced inflammatory responses in models of Parkinson disease. J Neurosci 36:2383–2390
115. Tran EH, Hoekstra K, van Rooijen N, Dijkstra CD, Owens T (1998) Immune invasion of the central nervous system parenchyma and experimental allergic encephalomyelitis, but not leukocyte extravasation from blood, are prevented in macrophage-depleted mice. J Immunol 161:3767–3775
116. Tsitsiou E, Lindsay MA (2009) microRNAs and the immune response. Curr Opin Pharmacol 9:514–520
117. Vaknin I, Kunis G, Miller O, Butovsky O, Bukshpan S et al (2011) Excess circulating alternatively activated myeloid (M2) cells accelerate ALS progression while inhibiting experimental autoimmune encephalomyelitis. PLoS One 6:e26921
118. van den Berg A, Kroesen BJ, Kooistra K, de Jong D, Briggs J et al (2003) High expression of B-cell receptor inducible gene BIC in all subtypes of Hodgkin lymphoma. Genes Chromosomes Cancer 37:20–28
119. Vicente R, Noel D, Pers YM, Apparailly F, Jorgensen C (2016) Deregulation and therapeutic potential of microRNAs in arthritic diseases. Nat Rev Rheumatol 12:211–220
120. Vigorito E, Kohlhaas S, Lu D, Leyland R (2013) miR-155: an ancient regulator of the immune system. Immunol Rev 253:146–157

121. Vigorito E, Perks KL, Abreu-Goodger C, Bunting S, Xiang Z et al (2007) microRNA-155 regulates the generation of immunoglobulin class-switched plasma cells. Immunity 27:847–859
122. Wang H, Moyano AL, Ma Z, Deng Y, Lin Y, et al. (2017) miR-219 cooperates with miR-338 in myelination and promotes myelin repair in the CNS. Dev cell 40: 566-582 e565
123. Wang J, Wu M, Wen J, Yang K, Li M et al (2014) MicroRNA-155 induction by Mycobacterium Bovis BCG enhances ROS production through targeting SHIP1. Mol Immunol 62:29–36
124. Waschbisch A, Atiya M, Linker RA, Potapov S, Schwab S et al (2011) Glatiramer acetate treatment normalizes deregulated microRNA expression in relapsing remitting multiple sclerosis. PLoS One 6:e24604
125. Wen Y, Zhang X, Dong L, Zhao J, Zhang C et al (2015) Acetylbritannilactone modulates MicroRNA-155-mediated inflammatory response in ischemic cerebral tissues. Mol Med 21:197–209
126. Wilms H, Sievers J, Rickert U, Rostami-Yazdi M, Mrowietz U et al (2010) Dimethylfumarate inhibits microglial and astrocytic inflammation by suppressing the synthesis of nitric oxide, IL-1beta, TNF-alpha and IL-6 in an in-vitro model of brain inflammation. J Neuroinflammation 7:30
127. Winter J, Jung S, Keller S, Gregory RI, Diederichs S (2009) Many roads to maturity: microRNA biogenesis pathways and their regulation. Nat Cell Biol 11:228–234
128. Woodbury ME, Freilich RW, Cheng CJ, Asai H, Ikezu S et al (2015) miR-155 is essential for inflammation-induced hippocampal neurogenic dysfunction. J Neurosci 35:9764–9781
129. Wu Q, Wang Q, Mao G, Dowling CA, Lundy SK et al (2017) Dimethyl fumarate selectively reduces memory T cells and shifts the balance between Th1/Th17 and Th2 in multiple sclerosis patients. J Immunol 198:3069–3080
130. Wu XY, Fan WD, Fang R, Wu GF (2014) Regulation of microRNA-155 in endothelial inflammation by targeting nuclear factor (NF)-kappaB P65. J Cell Biochem 115:1928–1936
131. Yamasaki R, Lu H, Butovsky O, Ohno N, Rietsch AM et al (2014) Differential roles of microglia and monocytes in the inflamed central nervous system. J Exp Med 211:1533–1549
132. Yang Y, Ye Y, Su X, He J, Bai W et al (2017) MSCs-derived exosomes and Neuroinflammation, neurogenesis and therapy of traumatic brain injury. Front Cell Neurosci 11:55
133. Yin Q, McBride J, Fewell C, Lacey M, Wang X et al (2008) MicroRNA-155 is an Epstein-Barr virus-induced gene that modulates Epstein-Barr virus-regulated gene expression pathways. J Virol 82:5295–5306
134. Yin Q, Wang X, Roberts C, Flemington EK, Lasky JA (2016) Methylation status and AP1 elements are involved in EBV-mediated miR-155 expression in EBV positive lymphoma cells. Virology 494:158–167
135. Zhang J, Cheng Y, Cui W, Li M, Li B et al (2014) MicroRNA-155 modulates Th1 and Th17 cell differentiation and is associated with multiple sclerosis and experimental autoimmune encephalomyelitis. J Neuroimmunol 266:56–63
136. Zhang J, Vandevenne P, Hamdi H, Van Puyvelde M, Zucchi A et al (2015) Micro-RNA-155-mediated control of heme oxygenase 1 (HO-1) is required for restoring adaptively tolerant CD4+ T-cell function in rodents. Eur J Immunol 45:829–842
137. Zhang J, Zhang ZG, Lu M, Wang X, Shang X et al (2017) MiR-146a promotes remyelination in a cuprizone model of demyelinating injury. Neuroscience 348:252–263
138. Zheng B, Yin WN, Suzuki T, Zhang XH, Zhang Y et al (2017) Exosome-mediated miR-155 transfer from smooth muscle cells to endothelial cells induces endothelial injury and promotes atherosclerosis. Mol Ther 25:1279–1294
139. Zhou H, Huang X, Cui H, Luo X, Tang Y et al (2010) miR-155 and its star-form partner miR-155* cooperatively regulate type I interferon production by human plasmacytoid dendritic cells. Blood 116:5885–5894
140. Zhu N, Zhang D, Chen S, Liu X, Lin L et al (2011) Endothelial enriched microRNAs regulate angiotensin II-induced endothelial inflammation and migration. Atherosclerosis 215:286–293

Chapter 6
Inflammasomes in the Gut Mucosal Homeostasis

Xiaomin Yao and Guangxun Meng

Abstract Inflammasomes are critical checkpoints in inflammation. The activation of inflammasome can cause a series of inflammatory responses including maturation of interleukin (IL)-1β and IL-18 and a specialized form of cell death called pyroptosis. Since its identification in the early 2000s, inflammasomes have been implicated to play multifaceted roles in varied pathological and physiological conditions, especially in the mucosal compartments including the gut. Maintaining gut mucosal homeostasis has always been a remarkable challenge for the host due to both the vast mucosal surface that is exposed to the outside and the enormous amount of local microbiota. To accomplish this challenge, the host mounts a constant dynamic low-grade inflammatory response (physiological inflammation) in coping with insults of microbes in the intestine. This book chapter aims to summarize the current knowledge of how inflammasomes contribute to gut mucosal homeostasis.

Keywords Inflammasome • Gut mucosal homeostasis • IL-1β • IL-18 • Intestinal homeostasis • Pyroptosis

---

X. Yao
CAS Key Laboratory of Molecular Virology & Immunology, Institut Pasteur of Shanghai, Chinese Academy of Sciences, 320 Yueyang Road, Life Science Research Building B-205, Shanghai 200031, China

University of Chinese Academy of Sciences, Beijing 100039, China

G. Meng (✉)
CAS Key Laboratory of Molecular Virology & Immunology, Institut Pasteur of Shanghai, Chinese Academy of Sciences, 320 Yueyang Road, Life Science Research Building B-205, Shanghai 200031, China
e-mail: gxmeng@ips.ac.cn

D. Xu (ed.), *Regulation of Inflammatory Signaling in Health and Disease*, Advances in Experimental Medicine and Biology 1024,
DOI 10.1007/978-981-10-5987-2_6

6.1 Introduction

6.1.1 Characterization of Inflammasomes as a Critical Arm in Innate Immunity

A multicellular organism usually utilizes the barrier system and the immune system to cooperatively segregate body constituents from harmful agents in the surrounding environment while simultaneously leaving some communicable space to absorb benefits such as energy from the outside world. In mammals like humans, the immune system comprises two parts, namely, the innate and adaptive immune systems, which are responsible for non-specific and antigen-specific immune responses, respectively. Although the adaptive immune system is evolutionarily superior and more efficient to target varied intruders, the innate immune system is indispensable for the initiation and full-functioning of the adaptive immune responses. Moreover, since many innate immune cells prefer to reside locally right behind the different barrier systems as well as to patrol in the circulation, innate immunity normally acts as the first line to detect and defend pathogens or organ injuries. Mechanistically, the innate immune system harnesses a set of germ line-encoded pattern recognition receptors (PRR) to recognize different pathogen/danger-associated molecular patterns (PAMP or DAMP). These PRRs are armed with invariable recognition motifs, i.e., each PRR can only target a few limited PAMPs or DAMPs which are ubiquitously presented in different pathogens and tissue injuries, such as lipopolysaccharides(LPS) on gram-negative bacteria cell walls and nucleic acid leaked from dying host cells. Currently, the PRRs contain membrane-associated toll-like receptors (TLRs) and C-type lectin receptors (CLRs), cytosolic NOD-like receptors (NLRs), AIM2-like receptors (ALRs), and RIG-I-like receptors (RLRs) [1]. Of interest, members of the NLRs (NLRP1, NLRP3, NLRP6, NLRC4, and NLRP12) and ALRs (AIM2) can orchestrate a specialized protein complex designated as inflammasomes to license the activation of the profoundly inflammatory caspase-1. The canonical inflammasomes are formed upon either exogenous or endogenous insults, which trigger the PRR molecule to oligomerize and act as nucleate to promote recruited adaptor apoptosis-associated speck-like protein containing a CARD (ASC) and pro-caspase-1 to form prion-like macromolecules, leading to the proximity-induced autoactivation of caspase-1 [2]. Of note, recent studies identified a noncanonical inflammasome formation upon direct binding of cytosolic LPS to caspase-11, which can not only activate caspase-11 itself but also trigger caspase-1 activation through yet unknown mechanisms [3–6]. In addition to members of NLRs and ALRs which are able to form inflammasomes, Shao and colleagues identified a previously neglected PRR, pyrin, to recognize abnormally inactivated RHOA (a small GTPase protein of Rho family) by pathogens and then form classical inflammasome to activate caspase-1 [7]. The activation of inflammatory caspases usually leads to proteolytic maturation and secretion of IL-1β and IL-18, as well as cleavage of pore-forming Gasdermin D (GSDMD), which triggers pyroptosis [2].

6.1.2 A Complex Microbiota-Immune Interaction Underlies the Gut Homeostasis

With more and more newly identified inflammasome-forming PRRs, the landscape of the multifaceted physiological significance of inflammasome has been gradually revealed. Among the many body sites, the intestinal mucosal surface stands out as the major location where the immune system performs its guardian function, since about 70% of immune cells of the entire immune system gather along the intestine [8]. The reason why host deploys such massive immune forces at intestinal mucosal surface is because of the residence of a much more massive amount of microbial flora, namely, microbiota at the same place. It was estimated that a number of 10^{13}–10^{14} microbes reside in the human gut, which is more than ten times of the total number of host cells in our body [9]. Although recently a revised estimation of both the microbial companions and ourselves reduced the ratio from ~10:1 to ~1:1, with each numbering at 3.9×10^{13} and 3×10^{13}, respectively [10], this doesn't affect the magnitude of the huge challenge that gut microbiota has posed to the host. Indeed, in germ-free animals, where no any microbes present in the gut, a substantial shrinkage of the gut local immune system was observed [11]. But surely these tiny bugs have become our neighbors for good reasons during evolution. For example, they help us absorb energy from usually indigestible polysaccharides by degrading and fermenting them to short-chain fatty acid; also, microbiota provides us with precious vitamins to support human health by metabolic processes that we don't possess. More importantly, microbiota can provide vital cues to elicit and shape the sound immune system locally and systematically. Much more than these, microbiota can directly help to expel invasive pathogens trying to penetrate the mucosal surface by competitive "colonization resistance" or by priming the host immune system to better combat infections indirectly [12]. On the other direction, human intestine offers the microbiota with invaluable anaerobic living sphere with exceptionally richness of food for them to prosper, which are rare to find in the inorganic world outside [13]. Such significant mutualism between us and microbiota demands careful maintenance, which is a dynamic status now known as gut homeostasis [13]. As the inflammasomes constitute an important arm of innate immunity, and many mutations identified in inflammasome component genes have been associated with intestinal disorders [14], their roles in keeping gut homeostasis need in-depth investigation. So, in this book chapter, we summarize the current understanding of the inflammasomes' contribution to gut mucosal homeostasis.

6.2 Inflammasome Effectors in Maintaining Gut Homeostasis

The discovery of different inflammasome effector molecules was much earlier than identification of the platform nucleating PRRs [15]. In 1977, Dinarello et al. first purified fever-causing factor, called pyrogen, from human blood [16]. Later on this pyrogen was defined as the first interleukin, i.e., IL-1β [15]. Fifteen years later, the enzyme that is responsible for the maturation of IL-1β, i.e., IL-1β converting enzyme (ICE), a cysteine protease, was cloned and now known as caspase-1 [17–19]. Another interleukin substrate of caspase-1, IL-18, was identified soon after the discovery of caspase-1 [20, 21]. In addition to mediating maturation and secretion of the aforementioned cytokines, caspase-1 activation can also lead to a programed but proinflammatory cell death, which is named as pyroptosis [22]. Since the early identification, the roles of these inflammatory molecules in gut homeostasis have been studied extensively.

6.2.1 IL-1β

As the initial context for characterization of IL-1β is in inflammation, early reports in clinical human inflammatory bowel disease (IBD) patients indicated a pathogenic role of IL-1β in promoting IBD [23]. However, these reports were largely based on studies with patients bearing onset IBD, where it's hard to distinguish whether elevated IL-1β is a result of or the causing factor for IBD. Further studies with animal models trying to clarify this point failed to reach a consensus conclusion, as totally opposite results had been obtained from different researches.

With the antibody blocking approaches to either inhibit or enhance the IL-1 signaling by targeting IL-1R and IL-1R antagonist (IL-1RA), respectively, a positive correlation of IL-1 level with colitis severity was revealed, indicating IL-1β is deleterious for maintaining gut homeostasis [24, 25]. However, in an earlier study, with the same model, they found that pretreatment with low dose of recombinant IL-1β significantly reduced the colitis symptoms [26]. Interestingly, a recent study showed that transplantation of IL-1β-primed mesenchymal stem cells (MSC) can alleviate the chemical dextran sulfate sodium (DSS)-induced colitis [27]. Also in a parallel comparison of IL-1α and IL-1β in DSS-induced colitis using *Il1α*$^{-/-}$ and *Il1β*$^{-/-}$ mice, they found that IL-1β is protective while IL-1α is detrimental [28]. But in a separate research, also using *Il1β*$^{-/-}$ mice, IL-1β was found to be selectively activated by certain bacteria such as pathobiont *Proteus mirabilis* through NLRP3 inflammasome, which contributed to DSS-induced colitis, indicating probably the varied commensal microbiota presented in the intestine that affects disease outcomes [29]. Besides the chemical-induced injury model, there were reports showing that IL-1β can also protect mice from intestinal infection of *Clostridium difficile* and *Citrobacter rodentium*, by promoting phagocytosis and eradication of bacteria in mononuclear phagocytes [30–32]. But in the study carried out by Alipour et al, treatment with exogenous IL-1β in wild-type mice resulted in exacerbated disease by *C. rodentium*, emphasizing that the balanced IL-1β in the intestine might be the key to gut homeostasis [30]. Another investigation found IL-1β ignited inflammation by promoting the accumulation of innate lymphoid cells in the Helicobacter hepaticus elicited chronic intestinal inflammation [33]. In the genetically predisposed spontaneous colitis model in *Il10*$^{-/-}$ and *Tlr5*$^{-/-}$ mice, IL-1β has also been shown to play a detrimental role [34, 35]. A notable fact is that *Il1β*$^{-/-}$ mice housed under normal condition do not develop spontaneous intestinal inflammation. Taken together, the rather ambiguous data concerning IL-1β's role in the IBD pathogenesis demands further investigation with careful consideration of, first, whether the microbiota (or the specific pathogen used in infection model) presented in the studied subject's intestine has strong or weak ability to trigger IL-1β release and, second, the timing to measure IL-1β's role in IBD, i.e., the initiation stage of inflammation vs after inflammation onset or vs the inflammation resolution stage.

Despite the abovementioned multiple studies focusing on the role of IL-1β in varied intestinal disorders, little information has been revealed about its physiological role under homeostasis. Until recently, an elegant study for the first time revealed the beneficial role of IL-1β under homeostasis, in which IL-1β was shown to be secreted by lamina propria mononuclear phagocytes (LPMP) to signal through IL-1R-myd88 axis on the local type 3 innate lymphoid cells (ILC3), which in return release an important cytokine GM-CSF. GM-CSF helps LPMPs to maintain their ability to secrete regulatory retinoic acid and IL-10 to promote the proliferation of local Tregs, which is crucial to sustain intestinal tolerance to food/commensal-derived microbial antigens [36]. Of great importance, the secretion of IL-1β is dependent on microbiota-associated cues [36]. Another study indicated that IL-1β can directly boost the intestinal epithelium secretion of certain antimicrobial peptide to restrict commensal translocation and to help maintain gut homeostasis [37], which was at least partially supported by an earlier study using ex vivo system, which showed that the intestinal ILCs (Lin(−)c-Kit(+)Sca-1(−) cells) can readily respond to IL-1β stimulation and secrete IL-22 and several antimicrobial peptides [38]. IL-1β has also been shown as critical effector for intestinal eosinophils to supervise local IgA production, thus contribute to gut microbial symbiosis [39]. So accumulating evidences support a beneficial role of IL-1β in gut mucosal homeostasis under resting conditions, which suggests that it is necessary to reexamine any treatment of IBD targeting IL-1β in a clinic to avoid over-medication.

6.2.2 IL-18

The initial identification of IL-18 has pointed out its structural similarity to IL-1β, although functionally it was found to induce interferon (IFN)-γ and promote Th1 response [20, 21]. IL-18's role in inflammatory bowel disease was firstly reported in 1999, by examination of its expression in Crohn's disease (CD) and ulcerative colitis (UC) patients in comparison with control people. These studies indicated that IL-18 was upregulated in IBD patients (more profoundly in CD), and the major contribution of activated IL-18 was from the epithelium, while lamina propria mononuclear cells had limited contribution [40, 41]. Similar to the case of IL-1β, however, these studies failed to tell whether the IL-18 increase in patients was a consequence or causing factor for IBD. Indeed, in a separate study, researchers claimed that the increase of IL-18 in IBD patients was rather heterogenous; only a minor part of investigated CD patients carried accountable elevation for IL-18 [42]. In later researches, polymorphisms in IL-18 genomic locus were claimed for association with increased risk for IBD [43–45], while other studies that either failed to obtain a positive conclusion [46, 47] or identified the IL-18 associated increased risk for IBD were actually dependent on NOD2 polymorphism, which is one of the most prominent genetic predisposing factors for IBD [48].

Earlier researches using neutralization strategies, in the chemical-induced animal model colitis, had shown that IL-18 was proinflammatory, as blockade of IL-18 could significantly improve disease status [49–52]. In addition, another study using a transgenic IL-18 mouse line supported the detrimental role of IL-18 in DSS colitis [53]. Also in the $Il18^{-/-}$ mice, TNBS was shown to fail in inducing colitis [52]. In the T cell transfer colitis, administration of antisense RNA against IL-18 mRNA reduced colitis severity, again supporting the notion that IL-18 mediated CD-like colitis pathogenesis [54]. The prevalent blame on IL-18 in fueling IBD in the earlier studies encountered few challenges except a study using $Il18^{-/-}$ and $Il18r^{-/-}$ mice demonstrated that actually IL-18 could be beneficial in DSS colitis [55]. The lopsided situation started to change only in recent years when studies were carried out to explore the role of inflammasomes in IBD. A series of studies investigating varied inflammasomes in experimental IBD pointed out that IL-18 functioned downstream of inflammasome to provide protection against colitis and/or colitis-associated cancer [37, 56–67]. Based on these discrepancies, Britta Siegmund proposed that the IL-18 activation in the epithelium might be helpful, but activation in lamina mononuclear cells might be pathogenic for IBD [68]. But this hypothesis was challenged by a recent research with careful reexamination of IL-18 signaling in contribution to intestinal inflammation. This study found that in the epithelium, IL-18 was actually playing detrimental roles by deformation of goblet cells (which is responsible for intact mucus barrier) during DSS colitis induction [69]. The debate that whether IL-18 is a friend or foe in IBD seems unsettled yet, but notable clues have been revealed, including that physiological stress can significantly affect experimental colitis severity in wild-type mice through stress-induced IL-18 [70] and that IL-18 deficiency can lead to outgrowth of colitogenic bacteria and dysbiosis, which could predispose the host to DSS-induced colitis [58, 62]. These intriguing clues enlightened that IL-18's role in pathogenesis of IBD can be an uncertain result of its delicate interaction with local microbiota, neural/hormonal cues, and/or different subsets of cells. Dissecting these relevant pathways may help our understanding, as well as clinical diagnosis of IBD.

Akin to IL-1β, despite the intensive attention being paid to IL-18 in the diseased condition, little effort has been exerted to decipher its physiological role under normal conditions. Except aforementioned studies revealing that IL-18 may help maintain gut microbial symbiosis through supervising local antimicrobial peptide generation [37, 58, 62], currently there is no clear information related to the contribution of IL-18 to gut homeostasis under resting condition, which is in need for future investigation.

6.2.3 Pyroptosis

By studying the *Salmonella typhimurium*-caused caspase-1-dependent cell death, pyroptosis was defined as a new form of programmed inflammatory cell rupture. Since the *S. typhimurium* is a known enteric pathogen, the pyroptosis has been speculated as a countermeasure utilized by infected enterocytes to expose pathogens to be killed by effector cells such as neutrophils [71, 72]. In a spontaneous colitis model caused by TLR2-MDR1A double deficiency, clear activation of inflammasome by commensal bacteria led to myeloid $CD11b^+$ cells to undergo pyroptosis, and similar situation was observed in genetically relevant IBD patients, suggesting that pyroptosis may play a pathogenic role in IBD [73]. Another study using systemic in vivo delivery of flagellin caused NLRC4 inflammasome-dependent but IL-1β- and IL-18-independent rapid pyroptosis, which contributed to the systemic inflammation-induced lethality [74]. Future study is required to clarify the role of pyroptosis in gut homeostasis and IBD.

6.3 Different Inflammasomes in Intestinal Homeostasis and IBD

6.3.1 NLRP1 Inflammasome

NLRP1 inflammasome is the first identified inflammasome ever [75]; it can respond to MDP challenge in human [76] and *Bacillus anthracis* lethal toxin/*Toxoplasma gondii* challenge in mouse [77–80]. However, the role of NLRP1 inflammasome in intestine health and disease has scarcely been reported. The first implication that it plays a role in intestinal homeostasis came from a report showing that a polymorphism in NLRP1 gene is associated with celiac disease, a common intestinal disorder [81], and also another clinical study showing that the NLRP1 Leu155His polymorphism was more frequent in IBD patients who were unresponsive to glucocorticoid treatment [82]. While in the mouse model study, the only elaborated study carried out with *Nlrp1b*$^{-/-}$ mice showed that NLRP1 was protective in DSS-induced colitis and colitis-associated colon cancer through upregulation of IL-1β and IL-18, with the former providing the majority function [83]. The authors also claimed that NLRP1 mainly functioned in the non-hematopoietic cells to provide protection against experimental IBD [83]. More interestingly, the researchers found that the aggravated colitis was transmissible to WT cagemates through cohousing, which indicated that an outgrowth of colitogenic microbiota was induced in the absence of NLRP1 [83]. Thus, the interaction between NLRP1 and local microbiota (especially certain specific NLRP1 inflammasome-activating bacteria species) and the possible overlapping/complementary relationship between NLRP1 and other NLRs in maintaining gut homeostasis could be of great value that awaits future investigation.

6.3.2 NLRP3 Inflammasome

NLRP3 inflammasome is the most intensively studied inflammasome since its identification to date [84]. The incredibly broad range of stimuli that activate the NLRP3 inflammasome made its exact activation mechanism still foggy; its role in the intestinal homeostasis and diseases is unclear either. First, in the clinical studies for NLRP3 polymorphisms in IBD showed discrepant results, with one group demonstrating that SNPs in the 5-prime region of NLRP3 gene leading to impaired NLRP3 expression and inflammasome activity were associated with CD susceptibility [85], while the other group with a large-scale population-based analysis failed to identify such correlation between NLRP3 polymorphisms and IBD prevalence [86]. The contradictory results in these two studies may be explained by the different genetic background of the subjects they approached, as other two studies revealed that variants in NLRP3 in combination with CARD8 polymorphisms were associated with higher risk to CD [87, 88].

In the animal model study, initially, there were two reports showing that NLRP3 played a protective role in the DSS-induced colitis, as *Nlrp3*$^{-/-}$ mice were more susceptible to DSS challenge [56, 65]. However, these two reports also contradicted with each other upon that in which cell compartment NLRP3 was functioning. By using the same bone marrow transfer strategy, one claimed a profound role of NLRP3 in the epithelium [65], while the other emphasized a function for NLRP3 mainly in the hematopoietic cells [56]. On the contrary, another report showed an even opposite role for NLRP3 in DSS colitis, stating that NLRP3 was pathogenic, as abolishment of NLRP3 protected *Nlrp3*$^{-/-}$ mice from DSS challenge [89]. However, in a successive study, these authors found that the protective effect in *Nlrp3*$^{-/-}$ mice can be blunted by cohousing with WT mice, suggesting that the claimed detrimental effects of NLRP3 in reference [89] are microbiota dependent [90]. Indeed, another study arguing the beneficial role of NLRP3 in both TNBS- and DSS-induced colitis models demonstrated the *Nlrp3*$^{-/-}$ mice contained altered microbiota as compared with WT mice, indicating that loss of NLRP3 might lead to dysbiosis, hence increased susceptibility to experimental IBD [91]. Although it was claimed that NLRP3 plays a detrimental role in DSS colitis without causing dysbiosis in a relevant study mainly focusing on NLRP6 inflammasome [58], in a recent study by a careful screening of NLRP3 inflammasome-activating mouse fecal bacteria, researchers identified that a pathobiont *Proteus mirabilis* can specifically activate NLRP3 inflammasome in the lamina propria mononuclear phagocytes, which is responsible for aggravating inflammation after epithelium breakage mediated by DSS exposure [29]. This study thus gives a strong example of NLRP3-microbiota interaction in regulating intestinal homeostasis and disease outcome. In another common enteric mucosal infection model, i.e., the *Citrobacter rodentium* infection of mice intestine, it was also shown that NLRP3 has an important role in the epithelium, but with unexpected data showing that *Nlrp3*$^{-/-}$ mice had intact intestinal caspase-1 activity [92], which indicated that some hidden facts fill the gap of our current understanding in terms of NLRP3's anti-infection function in the gut. Under resting conditions, in addition to the aforementioned report showing that NLRP3 plays a role in maintaining gut microbial symbiosis [91], a recent report showed that through short-chain fatty acid (SCFA)-GPR43/GPR109A signaling axis on the epithelial cells, potassium efflux was triggered and NLRP3 inflammasome was activated, which promoted the integrity of the intestinal epithelium [93].

In summary, the role of NLRP3 in regulating gut homeostasis and diseases has been increasingly appreciated; however, several key matters still require future studies to be addressed, which include where (epithelium vs lamina propria hematopoietic cells) and how NLRP3 regulates microbiota under homeostasis and diseases (injury vs infection) and how these interactions determine the homeostatic status/final disease outcome.

6.3.3 NLRC4 Inflammasome

NLRC4 was initially characterized with caspase-1-activating ability upon *Salmonella typhimurium* challenge. It forms inflammasome upon Naip-dependent detection of either cytosolic flagellin or type 3 secretion system (T3SS) component proteins [2]. Its role in intestinal homeostasis was implicated by the identification of mutation in NLRC4 causing infantile enterocolitis and autoinflammation syndrome [94]. In mouse models, *Nlrc4*$^{-/-}$ mice showed increased susceptibility to colitis-instigated colon tumorigenesis (AOM + DSS-induced colon cancer (CAC) model) [95, 96]. However, these two reports showed discrepant results in explaining NLRC4's role in the acute DSS colitis model, with Hu et al. arguing a dispensable role while Carvalho et al. pointed out a protective role for NLRC4 in this model [95, 96]. Contradictory results have also been shown, as another study showed *Nlrc4*$^{-/-}$ mice had similar susceptibility to CAC as WT mice [56]. In supporting for NLRC4's protective role, another study using *Naip1–6*-deficient mice showed increased colitis-associated colon tumor formation, but unexpectedly, in the acute colitis model, *Naip1–6* seem to be detrimental, which was speculated to be an NLRC4-independent event [97]. In accordance to its primary function in dictating *Salmonella typhimurium* infection, local or systemic administration of either flagellated bacteria or flagellin alone leads to NLRC4-dependent intestinal inflammation, which would either help expel the pathogen or lead to a severe acute inflammation-induced rapid death of mice [74, 96]. Recently, the epithelium-intrinsic NLRC4 inflammasome activity has been proven essential and sufficient for the expulsion of *S. typhimurium*-infected (or cytosolic flagellin-alerted) intestinal epithelial cells, which highlighted the critical role of NLRC4 in the surveillance of flagellated gut microbes [98, 99]. Thus, despite the in-depth understanding of NLRC4 in restricting specific pathogens such as *S. typhimurium*, its role in the intestinal homeostasis and IBD is quite preliminary and controversial. The reasons behind these contradictory conclusions from different laboratories can vary, but surely the gut microbiota needs to be taken into consideration for future studies.

6.3.4 NLRP6 Inflammasome

Unlike other inflammasomes, the molecular characterization of NLRP6 inflammasome is currently lacking. Nonetheless, in an overexpression system, NLRP6 was shown to interact with ASC to promote caspase-1 activation in vitro [100], but the upstream ligand or signals responsible for NLRP6 activation are not clear, as in the primary *Nlrp6*$^{-/-}$ macrophages, inflammasome activation in response to *Salmonella typhimurium*, *Listeria monocytogenes*, or LPS + ATP was unaffected [101]. However, the role of NLRP6 in the intestinal homeostasis has been well studied. The expression of NLRP6 is constantly high in the intestinal epithelium, and deficiency of this gene leads to dysbiosis, as well as aberrant goblet cell-derived mucus layer [58, 102]. In the intestinal monocyte, a beneficial role of NLRP6 was also revealed during DSS colitis induction, which is mediated by IL-18 in an autocrine manner [64]. NLRP6's role in goblet cell function was supported by a delicate study showing that NLRP6 inflammasome was found critical in licensing sentinel goblet cell activation upon bacterial penetration of the epithelium [103]. In a successive study, the NLRP6 inflammasome was shown to promote gut homeostasis through a positive feedback loop, wherein NLRP6 regulates local antimicrobial peptides production via IL-18, thus supervising microbiota to produce more metabolites such as taurine to support NLRP6 inflammasome activation [62]. Another study indicated that NLRP6 has a pivotal role in detecting intestinal virus through DHX15-MAVS-induced type 1 interferon induction [104]. Thus, NLRP6 has a multifaceted role in regulating intestinal homeostasis and gut diseases.

6.3.5 AIM2 Inflammasome

The AIM2 inflammasome was initially identified in detecting abnormal translocation of cytosolic DNA, mainly from invading pathogens or damaged host cells [2]. The intestine is a place where continuous opportunistic bacterial/viral invasion or physical/chemical damages occur, all of which may lead to aberrant DNA presence in the cellular space, suggesting a potential involvement for AIM2 inflammasome in the gut. Indeed, recent reports using AIM2-deficient mice have revealed such important function of AIM2 in the intestine in varied experimental and clinical disease contexts. In clinical analysis of mutations in small bowel carcinogenesis patients, a high prevalence of AIM2 frame shift was detected [105], and the absence of AIM2 expression in colorectal cancer patients was associated with poor prognosis [106], indicating its importance in tumor control. In addition, an increase of PYHIN-200 family members AIM2 and IFI16 was detected in the mucosa of IBD patients [107]. In the animal study, AIM2-deficient mice showed increased susceptibility to both DSS colitis and colitis-associated colon cancer [37, 63, 108]. But in the tumor patients as well as animal models receiving radiation/chemotherapy, AIM2 in the intestine detects the resultant DNA damage and induces inflammasome-dependent cell death and proinflammatory responses, which leads to severe damages to the host. Thus, targeting AIM2 in such circumstances represents a promising countermeasure to relieve suffering of patients [109, 110].

6.3.6 Noncanonical Inflammasomes: Caspase-11 and Pyrin Inflammasome

Recently, the caspase-11 mediated noncanonical inflammasome, and the pyrin inflammasome has been characterized [3–7]. The role of pyrin in the intestinal homeostasis and disease is still an open question. Unlike pyrin, the noncanonical inflammasome-forming caspase-11 has been studied in the context of intestinal inflammatory processes. Three studies all showed that caspase-11 protects against DSS-induced colitis and epithelial and hematopoietic caspase-11 is required [111–113]. Mechanistically, two reports emphasized IL-18 and IL-1β in mediating the protective effects [111, 112], while the third study denied cytokines' major effects, since they observed normal or even increased IL-18/IL-1β production in caspase-11-deficient mice after DSS challenge. Instead, they attributed the protective effects to pyroptosis downstream of caspase-11 inflammasome [113].

6.4 Concluding Remarks

Our current understanding of inflammasomes' role in intestinal mucosal homeostasis and disease has been summarized in Fig. 6.1, which is complex and severely unequilibrated. On the one hand, due to the relatively long history of identifying different inflammasome components and effectors, the effector molecules characterized earlier such as IL-1β and IL-18 have been studied intensively and controversially, while the latter identified inflammasome orchestrator molecules such as NLRs and AIM2 are still to be investigated, with pyrin even left unstudied. On the other hand, these insights were largely obtained from studies under diseased conditions, while the physiological working model of inflammasome under resting

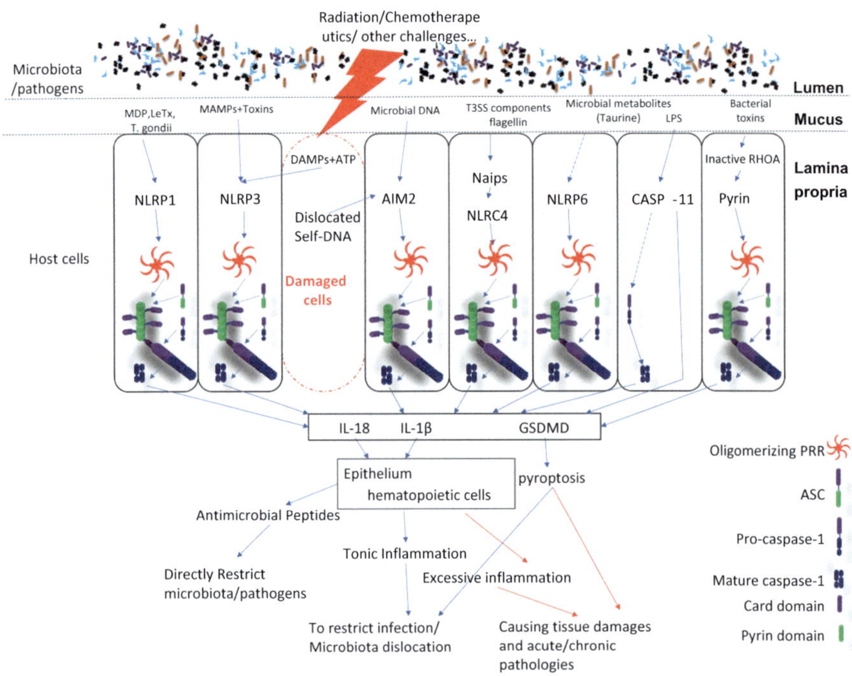

Fig. 6.1 The complex network of different inflammasomes in regulating gut mucosal homeostasis and diseases

The intestinal lumen contains a diverse repertoire of commensal/pathogenic microbes, which often opportunistically impedes the intestinal barrier (mucus layer); the intestine as well will experience injuries caused by ingestion of harmful matters or radio-/chemotherapies, which leads to tissue damages. The varied inflammasome-forming pattern recognition receptors (PRR) in either epithelial or lamina propria cells detect these microbial or danger signals to initiate inflammasome formation, hence activating the downstream effector cascade. The three major effector molecules, IL-1β, IL-18, and GSDMD, would either target the downstream epithelial/hematopoietic cells to regulate inflammatory and antimicrobial responses or form large pores on the cell membrane to mediate pyroptosis, which would both help the expulsion of infected pathogens and boost local inflammation conditions is underinvestigated, probably due to the lack of practical models or inefficient examination tools. For better comprehension of current knowledge and explanation of the contradictories, and for further broadening of our knowledge concerning the physiological role of inflammasomes in intestinal homeostasis, future studies should include more newly developed methods to monitor inflammasome activity. Moreover, full consideration of multiplayers in the gut disease process, including host immune cells, barriers, and intestinal contents (food and microbiota), as well as neural/humoral cues locally and systematically, is also necessary.

Acknowledgments Research in our lab is supported by grants from National Key Basic Research Programs (2015CB554302, 2014CB541905), Natural Science Foundation of China (31570895, 91429307, 31370892, 81761128012), as well as the Strategic Priority Research Program (XDPB0303) and the International Partnership Program of Chinese Academy of Sciences (153831KYSB20160009).

References

1. Brubaker SW, Bonham KS, Zanoni I, Kagan JC (2015) Innate immune pattern recognition: a cell biological perspective. Annu Rev Immunol 33:257–290
2. Broz P, Dixit VM (2016) Inflammasomes: mechanism of assembly, regulation and signalling. Nat Rev Immunol 16:407–420
3. Kayagaki N, Warming S, Lamkanfi M, Vande Walle L, Louie S, Dong J, Newton K, Qu Y, Liu J, Heldens S, Zhang J, Lee WP, Roose-Girma M, Dixit VM (2011) Non-canonical inflammasome activation targets caspase-11. Nature 479:117–121
4. Shi J, Zhao Y, Wang Y, Gao W, Ding J, Li P, Hu L, Shao F (2014) Inflammatory caspases are innate immune receptors for intracellular LPS. Nature 514:187–192
5. Kayagaki N, Wong MT, Stowe IB, Ramani SR, Gonzalez LC, Akashi-Takamura S, Miyake K, Zhang J, Lee WP, Muszynski A, Forsberg LS, Carlson RW, Dixit VM (2013) Noncanonical inflammasome activation by intracellular LPS independent of TLR4. Science 341:1246–1249
6. Hagar JA, Powell DA, Aachoui Y, Ernst RK, Miao EA (2013) Cytoplasmic LPS activates caspase-11: implications in TLR4-independent endotoxic shock. Science 341:1250–1253
7. Xu H, Yang J, Gao W, Li L, Li P, Zhang L, Gong YN, Peng X, Xi JJ, Chen S, Wang F, Shao F (2014) Innate immune sensing of bacterial modifications of Rho GTPases by the Pyrin inflammasome. Nature 513:237–241
8. Lyte M, Cryan JF (2014) Microbial endocrinology. the microbiota-gut-brain axis in health and disease. Springer, New York
9. Savage DC (1977) Microbial ecology of the gastrointestinal tract. Annu Rev Microbiol 31:107–133
10. Sender R, Fuchs S, Milo R (2016) Revised estimates for the number of human and bacteria cells in the body. PLoS Biol 14:e1002533
11. Round JL, Mazmanian SK (2009) The gut microbiota shapes intestinal immune responses during health and disease. Nat Rev Immunol 9:313–323
12. Kamada N, Seo SU, Chen GY, Nunez G (2013) Role of the gut microbiota in immunity and inflammatory disease. Nat Rev Immunol 13:321–335
13. Hooper LV, Macpherson AJ (2010) Immune adaptations that maintain homeostasis with the intestinal microbiota. Nat Rev Immunol 10:159–169

14. Lissner D, Siegmund B (2011) The multifaceted role of the inflammasome in inflammatory bowel diseases. Sci World J 11:1536–1547
15. Dinarello CA (2010) IL-1: discoveries, controversies and future directions. Eur J Immunol 40:599–606
16. Dinarello CA, Renfer L, Wolff SM (1977) Human leukocytic pyrogen: purification and development of a radioimmunoassay. Proc Natl Acad Sci U S A 74:4624–4627
17. Black RA, Kronheim SR, Merriam JE, March CJ, Hopp TP (1989) A pre-aspartate-specific protease from human leukocytes that cleaves pro-interleukin-1 beta. J Biol Chem 264:5323–5326
18. Thornberry NA, Bull HG, Calaycay JR, Chapman KT, Howard AD, Kostura MJ, Miller DK, Molineaux SM, Weidner JR, Aunins J et al (1992) A novel heterodimeric cysteine protease is required for interleukin-1 beta processing in monocytes. Nature 356:768–774
19. Cerretti DP, Kozlosky CJ, Mosley B, Nelson N, Van Ness K, Greenstreet TA, March CJ, Kronheim SR, Druck T, Cannizzaro LA et al (1992) Molecular cloning of the interleukin-1 beta converting enzyme. Science 256:97–100
20. Okamura H, Tsutsi H, Komatsu T, Yutsudo M, Hakura A, Tanimoto T, Torigoe K, Okura T, Nukada Y, Hattori K et al (1995) Cloning of a new cytokine that induces IFN-gamma production by T cells. Nature 378:88–91
21. Bazan JF, Timans JC, Kastelein RA (1996) A newly defined interleukin-1? Nature 379:591
22. Cookson BT, Brennan MA (2001) Pro-inflammatory programmed cell death. Trends Microbiol 9:113–114
23. Cominelli F, Pizarro TT (1996) Interleukin-1 and interleukin-1 receptor antagonist in inflammatory bowel disease. Aliment Pharmacol Ther 10:49–53
24. Cominelli F, Nast CC, Duchini A, Lee M (1992) Recombinant interleukin-1 receptor antagonist blocks the proinflammatory activity of endogenous interleukin-1 in rabbit immune colitis. Gastroenterol 103:65–71
25. Ferretti M, Casini-Raggi V, Pizarro TT, Eisenberg SP, Nast CC, Cominelli F (1994) Neutralization of endogenous IL-1 receptor antagonist exacerbates and prolongs inflammation in rabbit immune colitis. J Clin Invest 94:449–453
26. Cominelli F, Nast CC, Llerena R, Dinarello CA, Zipser RD (1990) Interleukin 1 suppresses inflammation in rabbit colitis Mediation by endogenous prostaglandins. J Clin Invest 85:582–586
27. Fan H, Zhao G, Liu L, Liu F, Gong W, Liu X, Yang L, Wang J, Hou Y (2012) Pre-treatment with IL-1beta enhances the efficacy of MSC transplantation in DSS-induced colitis. Cell Mol Immunol 9:473–481
28. Bersudsky M, Luski L, Fishman D, White RM, Ziv-Sokolovskaya N, Dotan S, Rider P, Kaplanov I, Aychek T, Dinarello CA, Apte RN, Voronov E (2014) Non-redundant properties of IL-1alpha and IL-1beta during acute colon inflammation in mice. Gut 63:598–609
29. Seo SU, Kamada N, Munoz-Planillo R, Kim YG, Kim D, Koizumi Y, Hasegawa M, Himpsl SD, Browne HP, Lawley TD, Mobley HL, Inohara N, Nunez G (2015) Distinct commensals induce interleukin-1beta via NLRP3 inflammasome in inflammatory monocytes to promote intestinal inflammation in response to injury. Immunity 42:744–755
30. Alipour M, Lou Y, Zimmerman D, Bording-Jorgensen MW, Sergi C, Liu JJ, Wine E (2013) A balanced IL-1beta activity is required for host response to Citrobacter rodentium infection. PLoS One 8:e80656
31. Hasegawa M, Kamada N, Jiao Y, Liu MZ, Nunez G, Inohara N (2012) Protective role of commensals against Clostridium difficile infection via an IL-1beta-mediated positive-feedback loop. J Immunol 189:3085–3091
32. Lebeis SL, Powell KR, Merlin D, Sherman MA, Kalman D (2009) Interleukin-1 receptor signaling protects mice from lethal intestinal damage caused by the attaching and effacing pathogen Citrobacter rodentium. Infect Immun 77:604–614

33. Coccia M, Harrison OJ, Schiering C, Asquith MJ, Becher B, Powrie F, Maloy KJ (2012) IL-1beta mediates chronic intestinal inflammation by promoting the accumulation of IL-17A secreting innate lymphoid cells and CD4(+) Th17 cells. J Exp Med 209:1595–1609
34. Li B, Gurung P, Malireddi RK, Vogel P, Kanneganti TD, Geiger TL (2015) IL-10 engages macrophages to shift Th17 cytokine dependency and pathogenicity during T-cell-mediated colitis. Nat Commun 6:6131
35. Carvalho FA, Nalbantoglu I, Ortega-Fernandez S, Aitken JD, Su Y, Koren O, Walters WA, Knight R, Ley RE, Vijay-Kumar M, Gewirtz AT (2012) Interleukin-1beta (IL-1beta) promotes susceptibility of Toll-like receptor 5 (TLR5) deficient mice to colitis. Gut 61:373–384
36. Mortha A, Chudnovskiy A, Hashimoto D, Bogunovic M, Spencer SP, Belkaid Y, Merad M (2014) Microbiota-dependent crosstalk between macrophages and ILC3 promotes intestinal homeostasis. Science 343:1249288
37. Hu S, Peng L, Kwak YT, Tekippe EM, Pasare C, Malter JS, Hooper LV, Zaki MH (2015) The DNA sensor AIM2 maintains intestinal homeostasis via regulation of epithelial antimicrobial host defense. Cell Rep 13:1922–1936
38. Lee Y, Kumagai Y, Jang MS, Kim JH, Yang BG, Lee EJ, Kim YM, Akira S, Jang MH (2013) Intestinal Lin- c-Kit+ NKp46− CD4− population strongly produces IL-22 upon IL-1beta stimulation. J Immunol 190:5296–5305
39. Jung Y, Wen T, Mingler MK, Caldwell JM, Wang YH, Chaplin DD, Lee EH, Jang MH, Woo SY, Seoh JY, Miyasaka M, Rothenberg ME (2015) IL-1beta in eosinophil-mediated small intestinal homeostasis and IgA production. Mucosal Immunol 8:930–942
40. Monteleone G, Trapasso F, Parrello T, Biancone L, Stella A, Iuliano R, Luzza F, Fusco A, Pallone F (1999) Bioactive IL-18 expression is up-regulated in Crohn's disease. J Immunol 163:143–147
41. Pizarro TT, Michie MH, Bentz M, Woraratanadharm J, Smith MF Jr, Foley E, Moskaluk CA, Bickston SJ, Cominelli F (1999) IL-18, a novel immunoregulatory cytokine, is up-regulated in Crohn's disease: expression and localization in intestinal mucosal cells. J Immunol 162:6829–6835
42. Schmidt C, Giese T, Goebel R, Schilling M, Marth T, Ruether A, Schreiber S, Zeuzem S, Meuer SC, Stallmach A (2007) Interleukin-18 is increased only in a minority of patients with active Crohn's disease. Int J Color Dis 22:1013–1020
43. Wang Y, Tong J, Chang B, Wang BF, Zhang D, Wang BY (2014) Genetic polymorphisms in the IL-18 gene and ulcerative colitis risk: a meta-analysis. DNA Cell Biol 33:438–447
44. Aizawa Y, Sutoh S, Matsuoka M, Negishi M, Torii A, Miyakawa Y, Sugisaka H, Nakamura M, Toda G (2005) Association of interleukin-18 gene single-nucleotide polymorphisms with susceptibility to inflammatory bowel disease. Tissue Antigens 65:88–92
45. Tamura K, Fukuda Y, Sashio H, Takeda N, Bamba H, Kosaka T, Fukui S, Sawada K, Tamura K, Satomi M, Yamada T, Yamamura T, Yamamoto Y, Furuyama J, Okamura H, Shimoyama T (2002) IL18 polymorphism is associated with an increased risk of Crohn's disease. J Gastroenterol 37(Suppl 14):111–116
46. Haas SL, Andreas Koch W, Schreiber S, Reinhard I, Koyama N, Singer MV, Bocker U (2005) 137 (G/C) IL-18 promoter polymorphism in patients with inflammatory bowel disease. Scand J Gastroenterol 40:1438–1443
47. Takagawa T, Tamura K, Takeda N, Tomita T, Ohda Y, Fukunaga K, Hida N, Ohnishi K, Hori K, Kosaka T, Fukuda Y, Ikeuchi H, Yamamura T, Miwa H, Matsumoto T (2005) Association between IL-18 gene promoter polymorphisms and inflammatory bowel disease in a Japanese population. Inflamm Bowel Dis 11:1038–1043
48. Glas J, Torok HP, Tonenchi L, Kapser J, Schiemann U, Muller-Myhsok B, Folwaczny M, Folwaczny C (2005) Association of polymorphisms in the interleukin-18 gene in patients with Crohn's disease depending on the CARD15/NOD2 genotype. Inflamm Bowel Dis 11:1031–1037

49. Ten Hove T, Corbaz A, Amitai H, Aloni S, Belzer I, Graber P, Drillenburg P, van Deventer SJ, Chvatchko Y, Te Velde AA (2001) Blockade of endogenous IL-18 ameliorates TNBS-induced colitis by decreasing local TNF-alpha production in mice. Gastroenterol 121:1372–1379
50. Siegmund B, Fantuzzi G, Rieder F, Gamboni-Robertson F, Lehr HA, Hartmann G, Dinarello CA, Endres S, Eigler A (2001) Neutralization of interleukin-18 reduces severity in murine colitis and intestinal IFN-gamma and TNF-alpha production. Am J Physiol Regul Integr Comp Physiol 281:R1264–R1273
51. Sivakumar PV, Westrich GM, Kanaly S, Garka K, Born TL, Derry JM, Viney JL (2002) Interleukin 18 is a primary mediator of the inflammation associated with dextran sulphate sodium induced colitis: blocking interleukin 18 attenuates intestinal damage. Gut 50:812–820
52. Kanai T, Watanabe M, Okazawa A, Sato T, Yamazaki M, Okamoto S, Ishii H, Totsuka T, Iiyama R, Okamoto R, Ikeda M, Kurimoto M, Takeda K, Akira S, Hibi T (2001) Macrophage-derived IL-18-mediated intestinal inflammation in the murine model of Crohn's disease. Gastroenterol 121:875–888
53. Ishikura T, Kanai T, Uraushihara K, Iiyama R, Makita S, Totsuka T, Yamazaki M, Sawada T, Nakamura T, Miyata T, Kitahora T, Hibi T, Hoshino T, Watanabe M (2003) Interleukin-18 overproduction exacerbates the development of colitis with markedly infiltrated macrophages in interleukin-18 transgenic mice. J Gastroenterol Hepatol 18:960–969
54. Wirtz S, Becker C, Blumberg R, Galle PR, Neurath MF (2002) Treatment of T cell-dependent experimental colitis in SCID mice by local administration of an adenovirus expressing IL-18 antisense mRNA. J Immunol 168:411–420
55. Takagi H, Kanai T, Okazawa A, Kishi Y, Sato T, Takaishi H, Inoue N, Ogata H, Iwao Y, Hoshino K, Takeda K, Akira S, Watanabe M, Ishii H, Hibi T (2003) Contrasting action of IL-12 and IL-18 in the development of dextran sodium sulphate colitis in mice. Scand J Gastroenterol 38:837–844
56. Allen IC, TeKippe EM, Woodford R-MTM, Uronis JM, Holl EK, Rogers AB, Herfarth HH, Jobin C, Ting JP (2010) The NLRP3 inflammasome functions as a negative regulator of tumorigenesis during colitis-associated cancer. J Exp Med 207:1045–1056
57. Dupaul-Chicoine J, Yeretssian G, Doiron K, Bergstrom KS, McIntire CR, LeBlanc PM, Meunier C, Turbide C, Gros P, Beauchemin N, Vallance BA, Saleh M (2010) Control of intestinal homeostasis, colitis, and colitis-associated colorectal cancer by the inflammatory caspases. Immunity 32:367–378
58. Elinav E, Strowig T, Kau AL, Henao-Mejia J, Thaiss CA, Booth CJ, Peaper DR, Bertin J, Eisenbarth SC, Gordon JI, Flavell RA (2011) NLRP6 inflammasome regulates colonic microbial ecology and risk for colitis. Cell 145:745–757
59. Engler DB, Leonardi I, Hartung ML, Kyburz A, Spath S, Becher B, Rogler G, Muller A (2015) Helicobacter pylori-specific protection against inflammatory bowel disease requires the NLRP3 inflammasome and IL-18. Inflamm Bowel Dis 21:854–861
60. Hu B, Elinav E, Huber S, Strowig T, Hao L, Hafemann A, Jin C, Wunderlich C, Wunderlich T, Eisenbarth SC, Flavell RA (2013) Microbiota-induced activation of epithelial IL-6 signaling links inflammasome-driven inflammation with transmissible cancer. Proc Natl Acad Sci U S A 110:9862–9867
61. Huber S, Gagliani N, Zenewicz LA, Huber FJ, Bosurgi L, Hu B, Hedl M, Zhang W, O'Connor W Jr, Murphy AJ, Valenzuela DM, Yancopoulos GD, Booth CJ, Cho JH, Ouyang W, Abraham C, Flavell RA (2012) IL-22BP is regulated by the inflammasome and modulates tumorigenesis in the intestine. Nature 491:259–263
62. Levy M, Thaiss CA, Zeevi D, Dohnalova L, Zilberman-Schapira G, Mahdi JA, David E, Savidor A, Korem T, Herzig Y, Pevsner-Fischer M, Shapiro H, Christ A, Harmelin A, Halpern Z, Latz E, Flavell RA, Amit I, Segal E, Elinav E (2015) Microbiota-Modulated metabolites shape the intestinal microenvironment by regulating NLRP6 inflammasome signaling. Cell 163:1428–1443
63. Ratsimandresy RA, Indramohan M, Dorfleutner A, Stehlik C (2017) The AIM2 inflammasome is a central regulator of intestinal homeostasis through the IL-18/IL-22/STAT3 pathway. Cell Mol Immunol 14:127–142

64. Seregin SS, Golovchenko N, Schaf B, Chen J, Eaton KA, Chen GY (2017) NLRP6 function in inflammatory monocytes reduces susceptibility to chemically induced intestinal injury. Mucosal Immunol 10:434–445
65. Zaki MH, Boyd KL, Vogel P, Kastan MB, Lamkanfi M, Kanneganti T-DD (2010) The NLRP3 inflammasome protects against loss of epithelial integrity and mortality during experimental colitis. Immunity 32:379–391
66. Zaki MH, Lamkanfi M, Kanneganti T-DD (2011) The Nlrp3 inflammasome: contributions to intestinal homeostasis. Trends Immunol 32:171–179
67. Salcedo R, Worschech A, Cardone M, Jones Y, Gyulai Z, Dai RM, Wang E, Ma W, Haines D, O'HUigin C, Marincola FM, Trinchieri G (2010) MyD88-mediated signaling prevents development of adenocarcinomas of the colon: role of interleukin 18. J Exp Med 207:1625–1636
68. Siegmund B (2010) Interleukin-18 in intestinal inflammation: friend and foe? Immunity 32:300–302
69. Nowarski R, Jackson R, Gagliani N, de Zoete MR, Palm NW, Bailis W, Low JS, Harman CC, Graham M, Elinav E, Flavell RA (2015) Epithelial IL-18 equilibrium controls barrier function in colitis. Cell 163:1444–1456
70. Matsunaga H, Hokari R, Ueda T, Kurihara C, Hozumi H, Higashiyama M, Okada Y, Watanabe C, Komoto S, Nakamura M, Kawaguchi A, Nagao S, Sekiyama A, Miura S (2011) Physiological stress exacerbates murine colitis by enhancing proinflammatory cytokine expression that is dependent on IL-18. Am J Physiol Gastrointest Liver Physiol 301:G555–G564
71. Chen GY, Nunez G (2011) Inflammasomes in intestinal inflammation and cancer. Gastroenterology 141:1986–1999
72. Martins RP, Aguilar C, Graham JE, Carvajal A, Bautista R, Claros MG, Garrido JJ (2013) Pyroptosis and adaptive immunity mechanisms are promptly engendered in mesenteric lymph-nodes during pig infections with Salmonella enterica serovar Typhimurium. Vet Res 44:120
73. Ey B, Eyking A, Klepak M, Salzman NH, Gothert JR, Runzi M, Schmid KW, Gerken G, Podolsky DK, Cario E (2013) Loss of TLR2 worsens spontaneous colitis in MDR1A deficiency through commensally induced pyroptosis. J Immunol 190:5676–5688
74. von Moltke J, Trinidad NJ, Moayeri M, Kintzer AF, Wang SB, van Rooijen N, Brown CR, Krantz BA, Leppla SH, Gronert K, Vance RE (2012) Rapid induction of inflammatory lipid mediators by the inflammasome in vivo. Nature 490:107–111
75. Martinon F, Burns K, Tschopp J (2002) The inflammasome: a molecular platform triggering activation of inflammatory caspases and processing of proIL-beta. Mol Cell 10:417–426
76. Faustin B, Lartigue L, Bruey JM, Luciano F, Sergienko E, Bailly-Maitre B, Volkmann N, Hanein D, Rouiller I, Reed JC (2007) Reconstituted NALP1 inflammasome reveals two-step mechanism of caspase-1 activation. Mol Cell 25:713–724
77. Boyden ED, Dietrich WF (2006) Nalp1b controls mouse macrophage susceptibility to anthrax lethal toxin. Nat Genet 38:240–244
78. Witola WH, Mui E, Hargrave A, Liu S, Hypolite M, Montpetit A, Cavailles P, Bisanz C, Cesbron-Delauw MF, Fournie GJ, McLeod R (2011) NALP1 influences susceptibility to human congenital toxoplasmosis, proinflammatory cytokine response, and fate of Toxoplasma gondii-infected monocytic cells. Infect Immun 79:756–766
79. Ewald SE, Chavarria-Smith J, Boothroyd JC (2014) NLRP1 is an inflammasome sensor for Toxoplasma gondii. Infect Immun 82:460–468
80. Cirelli KM, Gorfu G, Hassan MA, Printz M, Crown D, Leppla SH, Grigg ME, Saeij JP, Moayeri M (2014) Inflammasome sensor NLRP1 controls rat macrophage susceptibility to Toxoplasma gondii. PLoS Pathog 10:e1003927
81. Pontillo A, Vendramin A, Catamo E, Fabris A, Crovella S (2011) The missense variation Q705K in CIAS1/NALP3/NLRP3 gene and an NLRP1 haplotype are associated with celiac disease. Am J Gastroenterol 106:539–544

82. De Iudicibus S, Stocco G, Martelossi S, Londero M, Ebner E, Pontillo A, Lionetti P, Barabino A, Bartoli F, Ventura A, Decorti G (2011) Genetic predictors of glucocorticoid response in pediatric patients with inflammatory bowel diseases. J Clin Gastroenterol 45:e1–e7
83. Williams TM, Leeth RA, Rothschild DE, Coutermarsh-Ott SL, McDaniel DK, Simmons AE, Heid B, Cecere TE, Allen IC (2015) The NLRP1 inflammasome attenuates colitis and colitis-associated tumorigenesis. J Immunol 194:3369–3380
84. Agostini L, Martinon F, Burns K, McDermott MF, Hawkins PN, Tschopp J (2004) NALP3 forms an IL-1beta-processing inflammasome with increased activity in Muckle-Wells auto-inflammatory disorder. Immunity 20:319–325
85. Villani AC, Lemire M, Fortin G, Louis E, Silverberg MS, Collette C, Baba N, Libioulle C, Belaiche J, Bitton A, Gaudet D, Cohen A, Langelier D, Fortin PR, Wither JE, Sarfati M, Rutgeerts P, Rioux JD, Vermeire S, Hudson TJ, Franchimont D (2009) Common variants in the NLRP3 region contribute to Crohn's disease susceptibility. Nat Genet 41:71–76
86. Lewis GJ, Massey DC, Zhang H, Bredin F, Tremelling M, Lee JC, Berzuini C, Parkes M (2011) Genetic association between NLRP3 variants and Crohn's disease does not replicate in a large UK panel. Inflamm Bowel Dis 17:1387–1391
87. Roberts RL, Topless RK, Phipps-Green AJ, Gearry RB, Barclay ML, Merriman TR (2010) Evidence of interaction of CARD8 rs2043211 with NALP3 rs35829419 in Crohn's disease. Genes Immun 11:351–356
88. Schoultz I, Verma D, Halfvarsson J, Torkvist L, Fredrikson M, Sjoqvist U, Lordal M, Tysk C, Lerm M, Soderkvist P, Soderholm JD (2009) Combined polymorphisms in genes encoding the inflammasome components NALP3 and CARD8 confer susceptibility to Crohn's disease in Swedish men. Am J Gastroenterol 104:1180–1188
89. Bauer C, Duewell P, Mayer C, Lehr HA, Fitzgerald KA, Dauer M, Tschopp J, Endres S, Latz E, Schnurr M (2010) Colitis induced in mice with dextran sulfate sodium (DSS) is mediated by the NLRP3 inflammasome. Gut 59:1192–1199
90. Bauer C, Duewell P, Lehr HA, Endres S, Schnurr M (2012) Protective and aggravating effects of Nlrp3 inflammasome activation in IBD models: influence of genetic and environmental factors. Dig Dis 30(Suppl 1):82–90
91. Hirota SA, Ng J, Lueng A, Khajah M, Parhar K, Li Y, Lam V, Potentier MS, Ng K, Bawa M, McCafferty D-MM, Rioux KP, Ghosh S, Xavier RJ, Colgan SP, Tschopp J, Muruve D, MacDonald JA, Beck PL (2011) NLRP3 inflammasome plays a key role in the regulation of intestinal homeostasis. Inflamm Bowel Dis 17:1359–1372
92. Song-Zhao GX, Srinivasan N, Pott J, Baban D, Frankel G, Maloy KJ (2014) Nlrp3 activation in the intestinal epithelium protects against a mucosal pathogen. Mucosal Immunol 7:763–774
93. Macia L, Tan J, Vieira AT, Leach K, Stanley D, Luong S, Maruya M, Ian McKenzie C, Hijikata A, Wong C, Binge L, Thorburn AN, Chevalier N, Ang C, Marino E, Robert R, Offermanns S, Teixeira MM, Moore RJ, Flavell RA, Fagarasan S, Mackay CR (2015) Metabolite-sensing receptors GPR43 and GPR109A facilitate dietary fibre-induced gut homeostasis through regulation of the inflammasome. Nat Commun 6:6734
94. Romberg N, Al Moussawi K, Nelson-Williams C, Stiegler AL, Loring E, Choi M, Overton J, Meffre E, Khokha MK, Huttner AJ, West B, Podoltsev NA, Boggon TJ, Kazmierczak BI, Lifton RP (2014) Mutation of NLRC4 causes a syndrome of enterocolitis and autoinflammation. Nat Genet 46:1135–1139
95. Hu B, Elinav E, Huber S, Booth CJ, Strowig T, Jin C, Eisenbarth SC, Flavell RA (2010) Inflammation-induced tumorigenesis in the colon is regulated by caspase-1 and NLRC4. Proc Natl Acad Sci 107(50):21635–21640. doi:10.1073/pnas.1016814108
96. Carvalho FA, Nalbantoglu I, Aitken JD, Uchiyama R, Su Y, Doho GH, Vijay-Kumar M, Gewirtz AT (2012) Cytosolic flagellin receptor NLRC4 protects mice against mucosal and systemic challenges. Mucosal Immunol 5:288–298
97. Allam R, Maillard MH, Tardivel A, Chennupati V, Bega H, Yu CW, Velin D, Schneider P, Maslowski KM (2015) Epithelial NAIPs protect against colonic tumorigenesis. J Exp Med 212:369–383
98. Sellin ME, Muller AA, Felmy B, Dolowschiak T, Diard M, Tardivel A, Maslowski KM, Hardt WD (2014) Epithelium-intrinsic NAIP/NLRC4 inflammasome drives infected entero-

cyte expulsion to restrict Salmonella replication in the intestinal mucosa. Cell Host Microbe 16:237–248
99. Rauch I, Deets KA, Ji DX, von Moltke J, Tenthorey JL, Lee AY, Philip NH, Ayres JS, Brodsky IE, Gronert K, Vance RE (2017) NAIP-NLRC4 inflammasomes coordinate intestinal epithelial cell expulsion with eicosanoid and IL-18 release via activation of caspase-1 and -8. Immunity 46:649–659
100. Grenier JM, Wang L, Manji GA, Huang WJ, Al-Garawi A, Kelly R, Carlson A, Merriam S, Lora JM, Briskin M, DiStefano PS, Bertin J (2002) Functional screening of five PYPAF family members identifies PYPAF5 as a novel regulator of NF-kappaB and caspase-1. FEBS Lett 530:73–78
101. Anand PK, Malireddi RK, Lukens JR, Vogel P, Bertin J, Lamkanfi M, Kanneganti TD (2012) NLRP6 negatively regulates innate immunity and host defence against bacterial pathogens. Nature 488:389–393
102. Wlodarska M, Thaiss CA, Nowarski R, Henao-Mejia J, Zhang JP, Brown EM, Frankel G, Levy M, Katz MN, Philbrick WM, Elinav E, Finlay BB, Flavell RA (2014) NLRP6 inflammasome orchestrates the colonic host-microbial interface by regulating goblet cell mucus secretion. Cell 156:1045–1059
103. Birchenough GM, Nystrom EE, Johansson ME, Hansson GC (2016) A sentinel goblet cell guards the colonic crypt by triggering Nlrp6-dependent Muc2 secretion. Science 352:1535–1542
104. Wang P, Zhu S, Yang L, Cui S, Pan W, Jackson R, Zheng Y, Rongvaux A, Sun Q, Yang G, Gao S, Lin R, You F, Flavell R, Fikrig E (2015) Nlrp6 regulates intestinal antiviral innate immunity. Science 350:826–830
105. Schulmann K, Brasch FE, Kunstmann E, Engel C, Pagenstecher C, Vogelsang H, Kruger S, Vogel T, Knaebel HP, Ruschoff J, Hahn SA, Knebel-Doeberitz MV, Moeslein G, Meltzer SJ, Schackert HK, Tympner C, Mangold E, Schmiegel W, German HC (2005) HNPCC-associated small bowel cancer: clinical and molecular characteristics. Gastroenterol 128:590–599
106. Dihlmann S, Tao S, Echterdiek F, Herpel E, Jansen L, Chang-Claude J, Brenner H, Hoffmeister M, Kloor M (2014) Lack of absent in Melanoma 2 (AIM2) expression in tumor cells is closely associated with poor survival in colorectal cancer patients. Int J Cancer 135:2387–2396
107. Vanhove W, Peeters PM, Staelens D, Schraenen A, Van der Goten J, Cleynen I, De Schepper S, Van Lommel L, Reynaert NL, Schuit F, Van Assche G, Ferrante M, De Hertogh G, Wouters EF, Rutgeerts P, Vermeire S, Nys K, Arijs I (2015) Strong upregulation of AIM2 and IFI16 inflammasomes in the mucosa of patients with active inflammatory bowel disease. Inflamm Bowel Dis 21:2673–2682
108. Man SM, Zhu Q, Zhu L, Liu Z, Karki R, Malik A, Sharma D, Li L, Malireddi RK, Gurung P, Neale G, Olsen SR, Carter RA, McGoldrick DJ, Wu G, Finkelstein D, Vogel P, Gilbertson RJ, Kanneganti TD (2015) Critical role for the DNA sensor AIM2 in stem cell proliferation and cancer. Cell 162:45–58
109. Hu B, Jin C, Li HB, Tong J, Ouyang X, Cetinbas NM, Zhu S, Strowig T, Lam FC, Zhao C, Henao-Mejia J, Yilmaz O, Fitzgerald KA, Eisenbarth SC, Elinav E, Flavell RA (2016) The DNA-sensing AIM2 inflammasome controls radiation-induced cell death and tissue injury. Science 354:765–768
110. Lian Q, Xu J, Yan S, Huang M, Ding H, Sun X, Bi A, Ding J, Sun B, Geng M (2017) Chemotherapy-induced intestinal inflammatory responses are mediated by exosome secretion of double-strand DNA via AIM2 inflammasome activation. Cell Res 27(6):784–800. doi:10.1038/cr.2017.54
111. Oficjalska K, Raverdeau M, Aviello G, Wade SC, Hickey A, Sheehan KM, Corr SC, Kay EW, O'Neill LA, Mills KH, Creagh EM (2015) Protective role for caspase-11 during acute experimental murine colitis. J Immunol 194:1252–1260
112. Williams TM, Leeth RA, Rothschild DE, McDaniel DK, Coutermarsh-Ott SL, Simmons AE, Kable KH, Heid B, Allen IC (2015) Caspase-11 attenuates gastrointestinal inflammation and experimental colitis pathogenesis. Am J Physiol Gastrointest Liver Physiol 308:G139–G150
113. Demon D, Kuchmiy A, Fossoul A, Zhu Q, Kanneganti TD, Lamkanfi M (2014) Caspase-11 is expressed in the colonic mucosa and protects against dextran sodium sulfate-induced colitis. Mucosal Immunol 7:1480–1491

Chapter 7
Microbial Factors in Inflammatory Diseases and Cancers

Hong Sheng Ong and Howard Chi Ho Yim

Abstract The intestinal microbes form a symbiotic relationship with their human host to harvest energy for themselves and their host and to shape the immune system of their host. However, alteration of this relationship, which is named as a dysbiosis, has been associated with the development of different inflammatory diseases and cancers. It is found that metabolites, cellular components, and virulence factors derived from the gut microbiota interact with the host locally or systemically to modulate the dysbiosis and the development of these diseases. In this book chapter, we discuss the role of these microbial factors in regulating the host signaling pathways, the composition and load of the gut microbiota, the co-metabolism of the host and the microbiota, the host immune system, and physiology. In particular, we highlight how each microbial factor can contribute in the manifestation of many diseases such as cancers, Inflammatory Bowel Diseases, obesity, type-2 diabetes, non-alcoholic fatty liver diseases, nonalcoholic steatohepatitis, and cardiovascular diseases.

Keywords Microbiota • Inflammation • Inflammasome • TLR • NLR • Microbial metabolites • Inflammatory diseases • Cancers • Dysbiosis • GPCR • Inflammatory Bowel Diseases (IBD) • Obesity • Type 2 diabetes • Non-alcoholic fatty liver diseases (NAFLD) • Non-alcoholic steatohepatitis (NASH) • Atherosclerosis

---

H.S. Ong
Department of Immunology and Infectious Disease, John Curtin School of Medical Research, Australian National University, Building 131, Garran Road, Acton, Canberra 2601, ACT, Australia

H.C.H. Yim (✉)
Department of Medicine, St George & Sutherland Clinical School, Faculty of Medicine, University of New South Wales, Australia, Level 2 Clinical Sciences (WR Pitney) Building, St George Hospital, Short St, Kogarah, NSW 2217, Australia
e-mail: c.yim@unsw.edu.au

D. Xu (ed.), *Regulation of Inflammatory Signaling in Health and Disease*,
Advances in Experimental Medicine and Biology 1024,
DOI 10.1007/978-981-10-5987-2_7

7.1 Introduction

Millions of microbes inhabit the surfaces of the human body. Most of them reside in the gastrointestinal tract and exist in a microbial community named as microbiota. Amazingly, they do not cause any harmful effect to the host. Rather, these microbes foster a symbiotic relationship with their host. While the host provides them a favorable habitat, they participate in the host's digestive process by providing an extra range of enzymatic systems to harvest energy for themselves and the host. They also interact bidirectionally with the host immune system to shape their own ecology and the development of the host immune system, achieving a symbiosis to cope with environmental challenges. This symbiosis plays an important role in maintaining human health.

Many diseases such as cancers, Inflammatory Bowel Diseases (IBD), and metabolic diseases including obesity, type-2 diabetes, non-alcoholic fatty liver diseases (NAFLD), non-alcoholic steatohepatitis (NASH), and atherosclerosis have been linked to different degree of inflammation. Interestingly, this inflammation is associated with an alteration of the symbiotic relationship between the microbiota and its host. This altered relationship is called a dysbiosis. In this chapter, we focus on how metabolites (Table 7.1), cellular components (Table 7.2), and virulence factors (Table 7.3) of the microbiota interact with the host to influence the dysbiosis and involve in the etiology of these inflammatory diseases and cancers.

7.2 Microbial Metabolites

7.2.1 Short-Chain Fatty Acids (SCFAs)

The human intestine does not have enzymes that are capable of degrading most complex carbohydrates and plant polysaccharides [1]. Thus, humans rely heavily on the microbiota present in the colon to ferment undigested carbohydrates including cellulose, xylans, and resistant starch into short-chain fatty acids (SCFAs) such as acetate, propionate, and butyrate [1]. The SCFAs undergo passive diffusion into the host, acting as an important source of energy for the host [1, 2]. For example, butyrate acts as an energy substrate for colonic epithelial cells, while acetate and propionate provide energy to the peripheral tissues [3].

Apart from its role of supplying energy, SCFAs can regulate the host physiology by acting as signaling molecules that bind G-protein-coupled receptors (GPCRs) in cells [2]. For example, SCFAs induce secretion of glucagon-like peptide 1 in the colon via GPR43 activation, leading to the enhancement of insulin sensitivity [4]. SCFAs activate GPR41 to induce the intestinal expression of Pyy which inhibits gut motility, raises intestinal transit rate, and reduces the harvest of energy from the gut [5]. In the adipocytes, SCFAs, via GPR43, inhibit insulin signaling, leading to the reduction of fat accumulation in adipose tissue and the increase of metabolism of

Table 7.1 Involvements of microbial metabolites in inflammatory diseases and cancers

Microbial metabolites	Signaling receptor involved	Associated malignancy/physiology
SCFAs	GPR45	↓ Experimental colitis
	GPR43, GPR109A	↓ Experimental colitis
	GPR43	↑ Gout-associated inflammation
		↑ Insulin sensitivity
		↓ Fat accumulation in adipose tissue
		↑ Lipid and glucose metabolism in non-adipose tissues
	GPR41	↓ Gut motility
		↑ Intestinal transit rate
		↓ Energy harvest from gut
		↑ Intestinal gluconeogenesis
		↓ Allergic airway inflammation
	GPR109A	↓ Colitis-associated colon cancer
		↑ Colon cancer in a genetic model
AHR ligands	AHR	↑ Epithelial integrity
		↓ Experimental IBD
H_2S		↑ IBD
		↑ Colorectal cancer
Primary bile acids	FXR	↓ NAFLD
Secondary bile acids	TGR5	↑ Insulin sensitivity and glucose tolerance
		↓ Obesity
		↓ Hepatic steatosis
		↓ Intraplaque and adipose tissue inflammation
		↓ Vascular lesion formation
		↓ Atherosclerosis
		↑ Obesity-associated hepatocellular carcinoma
Taurine	NLRP6	↓ Experimental colitis
TMAO		↑ Atherosclerosis
Spermine	NLRP6	↑ Experimental colitis
N^1,N^{12}-diacetylspermine		↑ Colorectal cancer

lipids and glucose in other tissues [6]. Moreover, butyrate and propionate activate intestinal gluconeogenesis via cyclic AMP (cAMP)-dependent mechanism and via a GPR41-dependent gut-brain circuit, respectively [7].

An optimal level of SCFAs is required for maintaining a balance between energy harvest and expenditure. However, obesity is closely linked to a higher level of SCFAs. In genetically obese mice, the level of acetate and butyrate is higher in their cecum as compared to their lean littermates [8]. This is attributed to the presence of microbiota in obese mice that is able to degrade dietary polysaccharides into SCFAs [8]. Similarly, overweight human subjects also have higher concentration of SCFA

Table 7.2 Involvements of microbial cellular components in inflammatory diseases and cancers

Microbial components	Signaling receptor	Associated malignancy/physiology
LPS	TLR4	↓ Experimental colitis
		↑ Epithelial integrity
		↑ Obesity
		↑ Type 2 diabetes
		↑ NAFLD and NASH
		↑ Pancreas tumorigenesis
		↑ Colon tumorigenesis
		↑ Liver tumorigenesis
Flagellin	TLR5	↓ IBD
		↑ Insulin sensitivity
		↓ Hyperlipidemia
		↓ Hypertension
		↓ Adiposity
		↑ Tumor progression in a mouse model of sarcoma
		↓ Tumor progression in a mouse model bearing the breast carcinoma cell line
PSA	TLR2	↓ Experimental colitis
		↓ A mouse model of multiple sclerosis
Bacterial DNA	TLR9	↓ Experimental colitis
		↑ Sensitivity to glucose and insulin
		↓ Obesity
		↑ NASH during inflammasome malfunction
MDP	NOD2	↓ Experimental colitis
		↑ Insulin sensitivity
		↓ Colon cancer tumorigenesis

Table 7.3 Involvements of microbial virulence factors in inflammatory diseases and cancers

Microbial virulence factors	Signaling receptor	Associated malignancy
FadA adhesin	E-cadherin	↑ Colon tumorigenesis
PKS islands		↑ Colon tumorigenesis
BFT		↑ Colon tumorigenesis

than the lean subjects [9]. Consistent with the observations in mice, the obese human individuals have enriched microbiota that is able to process carbohydrate for energy [10] and have a higher level of ethanol in their breath [11]. As SCFAs can act as energy source for the host, these findings suggest that the energy harvest in obese people is elevated by the enrichment of SCFA-producing microbiota, which may eventually contribute to the etiology of obesity.

SCFAs act as a double-edged sword in inflammatory diseases and cancers. For example, SCFAs, via GPR45, inhibit histone deacetylase in regulatory T (Treg) cells thereby increasing the frequency and function of these cells, thus alleviating the disease severity of experimental colitis [12, 13]. Furthermore, SCFAs, via GPR43 and GPR109A, induce membrane hyperpolarization and Ca^{2+} ion mobilization in colonic epithelial cell, to activate the NOD-like receptor protein 3 (NLRP3) inflammasome [14]. This leads to the activation of caspase-1 and interleukin (IL)-18, which maintain the integrity of colon epithelium and thus the protection of mice from experimental colitis [14]. As protein kinase R (PKR) has recently been shown to inhibit NLRP3 inflammasome activity and has suggested to regulate the colitis development [15, 16], it would be interesting to investigate if SCFAs activate NLRP3 inflammasome by suppressing PKR activity as well. SCFAs which help reduce the severity of experimental colitis are produced by 17 strains of *Clostridia* in cluster XIVa, IV, and XVIII [17, 18]. Consistently, these bacteria and SCFA level also reduce the disease severity in IBD patients [19–22]. While not limited to the gut, the anti-inflammatory effect of SCFAs is also observed in the airway. SCFAs, via GPR41, suppress allergic airway inflammation by augmenting the generation of dendritic cell precursors in the lung and subsequently decreasing T helper type 2 (Th2) cell function [23]. On the other hand, SCFAs can also exacerbate inflammation. For example, acetate potentiates the monosodium urate crystal-activation of inflammasome possibly via GPR43, leading to the induction of IL-1β and CXCL1 expression and the subsequent exacerbation of gout-associated inflammation [24].

SCFAs activate GPR109A to induce the differentiation of Treg cells and IL-10-producing T cells by triggering anti-inflammatory responses in macrophage and dendritic cells, leading to the protection against colitis-associated colon cancer [25]. In contrast, the butyrate derived from an altered gut microbiota promotes tumorigenesis in a genetic mouse model of colon cancer by inducing colon epithelial cells hyperproliferation [26]. Hence, the role of SCFAs varies among different types of colon cancer.

7.2.2 Aryl Hydrocarbon Receptor Ligands

Tryptophan is an essential amino acid that is found in red meat, fish, eggs, and cruciferous vegetables [27, 28]. Gut microbes such as *Lactobacillus* can catabolize tryptophan into indole-3-aldehyde which is a ligand for the aryl hydrocarbon receptor (AHR) [29]. Upon binding of its ligand, AHR is transported into the nucleus and binds to the AHR nuclear translocator, leading to the transcription of genes including IL-22 in Th17 cells and group 3 innate lymphoid cells [27]. IL-22, in turn, induces the expression of antimicrobial peptides and production of mucus, thereby maintaining the epithelial integrity of the intestine [27]. Interestingly, absence of AHR or AHR ligands in the diet increases the disease severity in a murine IBD model [30]. This suggests that AHR ligands may play a role in preventing IBDs via maintenance of the epithelial integrity.

7.2.3 Hydrogen Sulfide

Food, such as dried fruits, nuts, fermented beverages, and brassica vegetables, and the host colonic mucus layer provide rich sources of inorganic and organic sulfate for the sulfate-reducing bacteria to extract energy [31, 32]. This energy extraction involves oxidation of organic compounds or hydrogen and reduction of sulfate into hydrogen sulfide (H_2S) in the colon [32]. Of note, these sulfate-reducing bacteria in the human colon mainly belong to the genus of *Desulfovibrio* in the class of *Deltaproteobacteria* [33].

Although previous studies have shown contradictory results on the amount of sulfate-reducing bacteria present in the mucosal layer and feces of IBD patients [34–40], the level of H_2S has been shown to be consistently higher in IBD patients as compared to the healthy individuals [34, 41]. There are two hypotheses on how H_2S induces inflammation. It has first been proposed that H_2S causes inflammation in IBDs by inducing injury in colon epithelium via its genotoxic effect and by suppressing the effect of SCFA [42–44]. Recent studies, however, have held a different view that the elevated level of H_2S in IBD patients may degrade the outer and inner mucus layer of the colon of the patients, leading to the invasion of gut microbes into the epithelium and the subsequent colonic inflammation [32, 45].

H_2S has also been associated with colorectal cancer. This is supported by the fact that the level of fecal H_2S is higher in patients who had surgical removal of sigmoid colon cancer and later developed neoplasia as compared to the healthy individuals [46]. In addition, H_2S activates mitogen-activated protein kinase and induces DNA damage [47, 48] which together induce colonic mucosal hyperproliferation [43, 49], thereby leading to colorectal cancer.

7.2.4 Bile Acids

There are two classifications of bile acids: primary and secondary. Primary bile acids include chenodeoxycholic acids and cholic acids in human and cholic acids and muricholic acids in mice and rats [50]. They are synthesized from cholesterol in liver and then conjugated with glycine or taurine in human, or only with taurine in rodents [50]. They are subsequently released into the duodenum to perform its function of solubilizing cholesterol, dietary fats, and fat-soluble vitamins for absorption [1, 50]. Most of these conjugated bile acids are reabsorbed in the ileum and transported back to the liver via the hepatic portal vein, thereby preserving 95% of the bile acid pool [50]. However, a portion of these preserved bile acids are deconjugated by bacteria including *Lactobacillus*, *Bifidobacteria*, *Clostridium*, and *Bacteroides* [50]. To perform such deconjugation, these bacteria utilize the bile salt hydrolase [50]. Deconjugated bile acids are not absorbed by the small intestine but are metabolized into secondary bile acids via 7α/β-dihydroxylation in the colon by bacterial members in genus of *Clostridium* including *C. scindens*, *C. hiranonis*, *C.*

hylemonae, and *C. sordellii* [27]. These secondary bile acids include lithocholic acid deconjugated from chenodeoxycholic acid and deoxycholic acid deconjugated from cholic acid in human, and murideoxycholic acid deconjugated from muricholic acids in mice [50].

Bile acids can act as ligands which bind to cell surface receptors responsible for regulating different metabolic processes. For example, primary bile acids mainly bind to the farnesoid X receptor (FXR) which is highly expressed in the liver, ileum, and kidney [51, 52]. In healthy individuals, FXR controls the synthesis of bile acids via a negative feedback mechanism [50]. However, the dysregulation of primary bile acid activation may result in metabolic diseases. For example, FXR-deficient mice fed on chow diet develop hyperglycemia and hypercholesterolemia [53, 54]. Similarly, mice with double deficiency of FXR and Apolipoprotein E which were fed on a high-fat and high-cholesterol diet, gained weight and increased the severity of atherosclerosis with elevated expression level of inflammatory genes [55]. In contrast, mice with double deficiency of FXR and low-density lipoprotein receptor on a high-fat diet have improved lipid profile and protect against diet-induced development of obesity and atherosclerosis [56]. Similarly, FXR-deficient mice which are fed on a high-fat diet or have genetically obese background are protected against obesity and have improved glucose homeostasis [57–59]. Hence, the role of FXR in obesity and obesity-associated diseases such as atherosclerosis and type-2 diabetes remains controversial. This controversy may have resulted from the different microbiota present across different animal facilities. Further studies are required to identify the specific microbiota which fine-tune the effect of FXR on these diseases. Moreover, patients with NAFLD have reduced expression of hepatic FXR which is associated with the elevated level of liver X receptor, sterol regulatory element-binding protein 1C (SREBP-1C), and enhanced hepatic triglyceride synthesis [60]. This elevation of SREBP-1C is associated with the severity of hepatic steatosis in the NAFLD patients [60]. Therefore, these findings suggest that the primary bile acids play a role in regulating the development of metabolic diseases. However, the role of gut microbiota in these diseases requires further investigation.

Secondary bile acids bind to TGR5 (GPR113). Upon activation, TGR5 induces the release of GLP-1 in enteroendocrine L cells by promoting mitochondrial oxidative phosphorylation and Ca^{2+} influx [61]. This improves the insulin sensitivity and glucose tolerance in obese mice [61]. The TGR5 activation also enhances the energy expenditure in brown adipose tissue, preventing obesity, insulin resistance, and hepatic steatosis [61, 62]. This is partly mediated by the cAMP-dependent thyroid hormone-activating enzyme type 2 iodothyronine deiodinase [62]. In addition to its effect on metabolism, TGR5 also affects the immune system. TGR5 activation by synthetic bile acid inhibits NFκB-dependent expression of cytokines including IL-1β, TNF-α, and IL-6 via cAMP signaling in macrophages [63]. This reduces intraplaque inflammation and vascular lesion formation, thereby preventing the development of atherosclerosis [63]. This macrophage-specific TGR5 activation also inhibits the adipose tissue inflammation by suppressing chemokine expression and macrophage infiltration, leading to the protection against insulin resistance [64]. This suppression is mediated by AKT-mTOR-dependent induction of differential translation of the dominant-negative C/EBPβ isoform [64]. Hence, the binding of TGR5 with secondary bile acids can prevent metabolic diseases including obesity, atherosclerosis, hepatic steatosis, and type-2 diabetes.

Taurine, derived from the deconjugation of primary bile acid, has been shown to activate the NLRP6 inflammasome to induce IL-18 and antimicrobial peptides, providing protection against colitis [65]. However, the clinical relevance of this finding requires further examination.

Development of cancer can be associated with either the dysregulated bile acid secretion or certain types of bile acid. Colorectal cancer patients have increased level of bile secretion and fecal bile acid concentration [66, 67]. Furthermore, the fecal concentration of secondary bile acids in African Americans with a high risk of colon cancer is higher than that in rural native Africans with a low risk of the disease [68]. On the other hand, the deoxycholic acid has suggested to provoke the senescence-associated secretory phenotype in hepatic stellate cells which in turn induces various inflammatory and tumor-promoting factors such as IL-1β in the liver, thus promoting the obesity-associated hepatocellular carcinoma development in mice [69].

*7.2.5 Trimethylamine-*N*-Oxide (TMAO)*

In our diet, trimethylamine-*N*-oxide (TMAO) is directly obtained from fish, or indirectly from meat and high-fat diets that are typically rich in choline and L-carnitine [70]. TMAO can be directly absorbed and transported to the blood for normal function of the body, while in the precursor form, choline and L-carnitine have to first be converted into trimethylamine (TMA) by the gut microbiota. Once absorbed, TMA is then converted into TMAO in the liver by the host hepatic enzyme flavin-containing monooxygenase 3 [70]. The normal function of TMAO includes counteracting the destabilizing effect of urea on proteins and nucleic acids in kidney, and acting as a chaperone that promote protein folding [70].

Diets play a role in shaping the composition of the gut microbiota which in turn determines the plasma level of TMAO. For example, the plasma level of TMAO increases in human subjects who ingest phosphatidylcholine [71]. This elevation is inhibited by antibiotic treatment but reappears after antibiotic withdrawal [71]. Similarly, people who are omnivores have a stronger ability to convert L-carnitine into plasma TMAO than people who are vegans or vegetarians [72]. Such difference in L-carnitine metabolism depends on the differences in specific bacterial taxa in the human gut between these two groups of people [72]. These findings suggest that the long-term dietary habit shapes the composition of microbiota to determine the metabolism of the ingested food.

The increased plasma level of TMAO has been associated with the elevated risk of adverse cardiovascular events including overall mortality, myocardial infarction, or stroke in human [71–73]. This induction of TMAO, which is suggested to be a result of the altered gut microbiota [72], promotes macrophage cholesterol accumulation and foam cell formation [74]. This leads to the promotion of atherosclerosis [74]. On the other hand, the reduced level of available phosphatidylcholine leads to NAFLD in mice [75]. This is caused by the dysregulated enzymatic systems of the altered gut microbiota and the stronger ability of the host to produce excretory methylamine from phosphatidylcholine [75]. Hence, the levels of TMAO and its precursors play a role in regulating metabolic diseases including cardiovascular diseases and NAFLD.

7.2.6 Polyamine

Polyamines are polycationic molecules that are present in living cells to carry out different cellular functions including gene transcription, translation, cell growth, and cell death [76]. These polyamines including putrescine, spermidine, and spermine can be generated by the host cells or the gut microbes to maintain the integrity of the epithelial barrier [76]. However, the upregulation or downregulation of the polyamine level has been associated with diseases. For example, spermine level is increased during colitis caused by NLRP6 deficiency [65]. The elevated spermine inhibits the NLRP6 inflammasome induction of IL-18 and antimicrobial peptide, promoting the disease severity in an experimental colitis model [65]. Intriguingly, the levels of polyamine metabolites including N^1,N^{12}-diacetylspermine are higher in biofilm-positive colon cancer tissues [77]. When the colon cancer patients are treated with oral antibiotics, the isolated colon cancer tissues have no biofilm or culturable bacteria and a reduced level of N^1,N^{12}-diacetylspermine [77]. As biofilm has been associated with the enhanced crypt epithelial cell proliferation in colon mucosa [78], biofilm may act synergistically with the host cancer to generate polyamines that may promote colorectal cancer development [77].

7.3 Microbial Components

7.3.1 Lipopolysaccharide (LPS)

Lipopolysaccharide (LPS) is a cell wall component of the Gram-negative bacteria. It is a ligand of Toll-like receptor 4 (TLR4) and is present in many gut bacteria. Commensal bacteria maintain an optimal TLR4 activation to ensure the gut epithelial integrity, leading to the protection against the intestinal injury-induced colitis [79]. This protection is mediated by MyD88, a downstream signaling molecule of TLR4 [79], which in turn induces the expression of cyclooxygenase 2 from intestinal epithelial cells and macrophage, leading to the enhanced mucosal production of prostaglandin E2 and the subsequent maintenance of gut epithelial survival [80–83].

However, when there is over-activation of TLR4 possibly by LPS from microbiota, the gut barrier is disrupted, and gut bacteria invade the mucosa, leading to the exacerbation of chemically induced colitis [84].

Apart from its role in IBD development, LPS also regulates metabolic inflammation. Mice fed on a high-fat diet have an increased plasma LPS level, elevated intestinal permeability and a higher proportion of LPS-containing gut microbiota, resulting in metabolic endotoxemia [85]. Subcutaneous infusion of LPS into mice leads to insulin resistance and obesity [85]. Consistently, antibiotic treatment decreases the metabolic endotoxemia and the cecal content of LPS in mice that are genetically obese and fed on a high-fat diet, resulting in the reduction of glucose intolerance, body weight gain, fat mass development, and inflammation [86]. Indeed, the metabolic endotoxemia is mediated by CD14, a co-receptor for LPS binding to TLR4 [86]. In support of these studies, patients with metabolic syndrome and type-2 diabetes have endotoxemia [87, 88]. Furthermore, NLRP3 and NLRP6 inflammasomes suppress the progression of metabolic diseases including NAFLD and NASH via IL-18-dependent changes of the gut microbiota [89]. These changes of gut microbiota correlate with the increased hepatic steatosis and inflammation via the influx of TLR4 and TLR9 ligands into the portal circulation, leading to enhanced hepatic TNF-α expression and the subsequent NASH progression [89]. In agreement with the data from mouse models, NAFLD patients have elevated levels of TNF-α, IFN-γ, and IL-6, disruption of microvilli structure in small intestine, and enhanced gut permeability [90–92]. Hence, it is possible that the endotoxemia, which is resulted from the increased gut permeability by the loss of inflammasome activity, is crucial for the development of these metabolic diseases.

LPS, possibly leaked from the gut microbiota, has been shown to promote the development of pancreatic and liver cancer via a TLR4-dependent signaling pathway [93, 94]. This role of LPS for tumorigenesis can also be observed in colitis-related colon cancer. TLR4-deficient mice are protected against the development of such cancer [95], while constitutive activation of TLR4 enhances it [95, 96]. Similarly, overexpression of TLR4 is observed in tumors of the IBD patients [95]. Whether the LPS promotion of colitis-related colon cancer development originated from the gut microbiota requires further studies. Moreover, the intestinal mucus secretion is impaired in tumor in a genetic mouse model of colorectal cancer, leading to the increased gut permeability and translocation of LPS of the gut microbiota into the portal circulation [97]. This in turn activates the myeloid cells to induce IL-17 and IL-23 via the TLR4-MyD88 pathway, promoting colon cancer development [97]. Similarly, in another genetic mouse model of colon cancer, microbiota has been suggested to activate the MyD88 signaling pathway in intestinal epithelial cells which inhibits the degradation of the c-Myc oncoprotein, leading to the promotion of colon tumorigenesis [98]. As c-Myc has been shown to activate IRAK1 which is a signaling molecule downstream of MyD88 [99], it would be interesting to study whether this c-Myc stabilization acts as a positive feedback mechanism to amplify this microbiota-MyD88-dependent tumor formation. In summary, it is suggested that LPS translocates into circulation from the gut microbiota by a leaky gut during the development of liver, pancreatic, and colon cancers. This, in turn, promotes the development of these cancers. However, the causes of this gut leakiness require further examination.

7.3.2 *Flagellin*

Flagellin is a structural component of flagella which allows the pathogenic bacteria to adhere and invade host tissues [100]. It is recognized by TLR5 [100]. Interestingly, TLR5-deficient mice develop spontaneous colitis with increased bacterial load in the colon as compared to wild-type mice [101]. This inflammation may be mediated by the increased colonic expression of hematopoietic cell-derived pro-inflammatory cytokines [101]. Deletion of TLR4 rescues the colitis of TLR5-deficient mice but not the bacterial load in the colon [101], suggesting that TLR4 ligands are involved in induction of mucosal inflammation when TLR5 is absent. This notion is supported by the finding that the polymorphisms in TLR5 gene are associated with ulcerative colitis and simultaneous polymorphisms in TLR4 and TLR5 are correlated with decreased level of pro-inflammatory cytokines [102]. In addition, TLR5 is associated with metabolic syndrome. TLR5-deficient mice exhibit hyperphagia and develop hyperlipidemia, hypertension, insulin resistance, and increased adiposity [103]. These metabolic changes correlate with changes in the composition of the gut microbiota in these mice [103]. Transfer of this microbiota into wild-type germ-free mice results in similar metabolic changes [103]. Similarly, mice lacking TLR5 in intestinal epithelial cells have a low-grade inflammation in the spleen and colon and a delayed clearance of pathobiants, leading to the development of metabolic syndrome and colitis [104]. This inflammation is correlated with the increased localization of gut microbiota into the colonic mucosa and the increased levels of fecal LPS and flagellin [104]. Loss of TLR5 in dendritic cells does not lead to the development of inflammation or the change in the composition of the microbiota but only dampens the flagellin-induced IL-22 production [104]. Consistently, a study using germ-free mice has shown that the gut commensal microbiota is responsible for the gut inflammation when TLR5 is absent [105]. Hence, it is possible that flagellin from the commensal microbes activates TLR5 to maintain the gut epithelial integrity. When this barrier is impaired in the case of TLR5-deficiency such as TLR5 gene polymorphisms, some gut commensal microbes become pathobiants and induce mucosal inflammation via TLR4, contributing to the development of IBD and metabolic diseases.

Depending on the cancer type, commensal bacteria can either potentiate or suppress the extraintestinal cancer progression via interaction with TLR5. Rutkowski et al. has shown that gut commensal microbiota induces systemic IL-6 serum level via TLR5, hence promoting tumor progression in a mouse sarcoma model [106]. This elevation of IL-6 expands the granulocytic myeloid-derived suppressor cells, leading to the secretion of galectin-1 from $\gamma\delta$ T cells [106]. The secreted galectin-1 suppresses the antitumor immunity and enhances malignant progression [106]. On the contrary, when the tumor is unresponsive to IL-6, gut commensal microbiota activation of TLR5 suppresses tumor growth by inhibiting IL-17 production in mice bearing the breast carcinoma cell line [106]. Consistently, breast cancer patients with TLR5 R392X polymorphisms have poorer survival and higher IL-17 levels than the patients with wild-type TLR5 alleles [106]. Whereas in the case of ovarian cancer, patients with the same polymorphisms have improved survival [106]. As discussed above, TLR5 can maintain the epithelial integrity. Hence, it is possible that the gut microbiota modifies the outcome of cancer malignancy only when there is an impairment of gut lining caused by the loss of function of TLR5.

7.3.3 Polysaccharide A (PSA)

Polysaccharide A (PSA), which is a cell wall component of the symbiont *Bacteroides fragilis*, shapes the maturation of the developing immune system including the development and function of T cells [107]. PSA signals through TLR2 on $CD4^{+}$T cells to activate the Treg cells to secrete IL-10, leading to the protection against inflammation in mouse models of IBD and multiple sclerosis [108–111]. In addition, PSA can be released by *Bacteroides fragilis* in outer membrane vesicles (OMV) [112]. This PSA-containing OMV is sensed by TLR2 present in intestinal dendritic cells to promote Treg cells and anti-inflammatory cytokine production, leading to the protection against experimental colitis [112].

7.3.4 Bacterial DNA

Cytosine-guanine dinucleotide (CpG) motifs of bacterial DNA are present in many commensal bacteria. They are recognized by TLR9. Activation of TLR9 at apical surface of intestinal epithelial cells suppresses the NF-κB activation and expression of TNF-α and IL-8, leading to the protection against experimental colitis [113]. This protection is also mediated by TLR9 activation of hairy enhancer of split 1 and expression of vascular endothelial growth factor that promote the intestinal wound healing and CpG-TLR9 enhancement of Treg cell function [114, 115]. Moreover, TLR9 reduces glucose intolerance and insulin resistance, thus protecting against high-fat diet-induced obesity [116]. This is possibly mediated by the TLR9 suppression of M1 macrophages and Th1 cells and TLR9 inhibition of pro-inflammatory cytokine and chemokine expression in the adipose tissue [116]. In contrast, TLR9 ligands promote NASH progression when the permeability of gut is increased by inflammasome malfunction [89]. Hence, the gut integrity may determine whether TLR9 ligands promote or suppress the inflammation in IBD and metabolic diseases.

7.3.5 Muramyl Dipeptide (MDP)

Muramyl dipeptide (MDP), which is a peptidoglycan motif, is present in a wide range of Gram-positive and Gram-negative bacteria including those present in our gut microbiota. It is recognized by nucleotide-binding oligomerization domain protein 2 (NOD2). Previous studies have shown that polymorphisms in NOD2 are associated with Crohn's disease (CD) which is one of the IBD subtypes [117, 118]. The ileal mucosa of CD patients with homozygosity in NOD2 mutation has higher load of *Bacteroidetes* [119]. An elevated load of *Proteobacteria* and a lower load of *Firmicutes* are also associated with the NOD2 mutation [119]. In mice, however, conflicting results have been reported on the role of NOD2 in altering the composition of the gut microbiota, and the mechanism of NOD2 regulation on these changes [117, 119]. Regardless of this inconsistency, translocation of Gram-negative bacteria, Gram-positive bacteria, and yeast *Saccharomyces cerevisiae* on gut Peyer's patches in the ileum is increased in NOD2-deficient mice [120]. These mice are also more susceptible to the experimental colitis [120]. In addition, more adherent-invasive *Escherichia coli* (*E. coli*), which is associated with CD, are attached to and translocate across the Peyer's patches of the NOD2-deficient mice via their long polar fimbriae [121]. Apart from colitis, NOD2-deficient mice have elevation of inflammation in adipose tissue and liver, and insulin resistance when they are fed on a high-fat diet [122]. This is caused by the bacterial translocation from the gut into the adipose tissue and liver [122]. These findings suggest that NOD2, possibly activated by MDP, maintains the gut lining integrity to prevent the translocation of gut microbes, leading to the protection against IBD and metabolic diseases. This integrity maintenance could be mediated by the MDP-NOD2 inhibition of the gut-microbe-activated TLR2 and TLR4-dependent pro-inflammatory responses and by the NOD2 modulation of TGF-β-producing Treg cells in colonic lamina propria [123–126]. However, the origin of the MDP and the mechanism of its translocation across the mucus layer of the gut require further investigation.

NOD2-deficient mice are more susceptible to colitis-associated colorectal cancer [127]. This is mediated by the increased IL-6 secretion resulted from the changes of the gut microbiota in these mice [127]. In addition, variant R702W in NOD2 has been associated with colorectal cancer [128]. These findings suggest that NOD2 plays a role in ensuring a healthy gut microbiota possibly via the above-proposed mechanism of gut integrity maintenance.

7.4 Virulence Factors

7.4.1 FadA Adhesin

Fusobacterium nucleatum (*F. nucleatum*) is an opportunistic commensal bacterium which is present in the oral cavity [129] and is associated with different periodontal diseases [129]. FadA, a virulence factor of *F. nucleatum*, is a surface adhesin that defines the invasiveness of this bacterium [129]. Interestingly, the level of FadA gene in colon tissue from patients with adenomas and adenocarcinomas is higher than that from the healthy individuals [129]. Furthermore, FadA binds to E-cadherin on the gut epithelial cells, leading to the activation of β-catenin induction of epithelial cell proliferation [129]. This results in the *F. nucleatum* potentiation of the tumor growth in a genetic mouse model of colon cancer [130]. Consistently, higher frequency of *F. nucleatum* has also been found in the samples of patients with colorectal cancers [131, 132].

7.4.2 Polyketide Synthases (PKS) Islands

E. coli strains are one of the Gram-negative bacteria that constitute a healthy gut microbiota. However, there is an increased attachment of *E. coli* to the colon mucosa in patients with adenocarcinoma as compared to healthy controls [133–135]. Some strains of *E. coli* have a 54 kb polyketide synthases (PKS) genotoxicity island in their genome. This island encodes multiple enzymes to synthesize colibactin, which is a genotoxin [136]. *E. coli* that contain PKS island have been found in the mucosal lesions of patients with IBD and colorectal cancer [137–139]. They also enhance tumor development in two mouse models of colitis-associated colon cancer and a genetic mouse model of colon cancer [137, 138, 140]. Several mechanisms have been proposed on how colibactin promote colon cancer development. Firstly, it induces cellular senescence by conjugating p53 with small ubiquitin-like modifier, leading to the enhanced tumor growth [140]. Secondly, it induces the colon epithelial cell proliferation [138]. Lastly, it cross-links DNA by its warhead, leading to the induction of DNA damage and the subsequent cell cycle arrest and genomic instability [141–143].

7.4.3 **Bacteroides fragilis** *Toxin (BFT)*

Bacteroides fragilis (*B. fragilis*) are anaerobic symbionts living in the colon [136]. They are classified into non-toxigenic *B. fragilis* (NTBF) and enterotoxigenic *B. fragilis* (ETBF) [136]. While both NTBF and ETBF can colonize mouse colon, only ETBF can promote colitis and tumor development in a multiple intestinal neoplasia mouse model via the activation of Th17 cell responses [144]. This is possibly mediated by *B. fragilis* toxin (BFT) derived from ETBF that can induce the expression of spermine oxidase to catabolize polyamine in colonic epithelial cells, leading to DNA damage [145]. Consistently, higher percentage of colonoscopic biopsies and fecal samples containing the bft gene is found in colorectal cancer patients than in the controls [146, 147]. Hence, both BFT and the entire ETBF are crucial in potentiating the development of IBD and colon cancer.

7.5 Concluding Remarks

Microbial metabolites, microbial components, and virulence factors interact with the host tissues and cells both locally and systemically, contributing to the pathogenesis of inflammatory diseases and cancers. These interactions are made possible by the ability of microbial factors to cross the gut epithelial barrier. While microbial metabolites or secretory toxins can diffuse passively across this barrier, the gut epithelia contain a mucus layer that blocks the passage of intact gut microbes. Thus, it is reasonable to propose that the cellular components of these microbes can only interact with the host cells and tissues when this barrier has been breached during disease development. However, studies have implicated that these microbial components can activate tolerogenic immune responses to maintain this barrier integrity without compromising it in the first place. The mechanisms taken by the gut microbiota or the host to maintain the gut epithelial barrier require further investigation. In addition, the rapid advancement of meta-omic technologies will facilitate the discovery of more microbial factors that contribute to the development of inflammatory diseases and cancers. Once identified, they will provide insights into the establishment of preventive measures and intervention strategies to tackle these diseases.

References

1. Tremaroli V, Bäckhed F (2012) Functional interactions between the gut microbiota and host metabolism. Nature 489(7415):242–249
2. Tilg H, Moschen AR (2014) Microbiota and diabetes: an evolving relationship. Gut 63(9):1513–1521
3. Bergman EN (1990) Energy contributions of volatile fatty acids from the gastrointestinal tract in various species. Physiol Rev 70(2):567–590
4. Tolhurst G, Heffron H, Lam YS, Parker HE, Habib AM et al (2012) Short-chain fatty acids stimulate glucagon-like peptide-1 secretion via the G-protein-coupled receptor FFAR2. Diabetes 61(2):364–371
5. Samuel BS, Shaito A, Motoike T, Rey FE, Backhed F et al (2008) Effects of the gut microbiota on host adiposity are modulated by the short-chain fatty-acid binding G protein-coupled receptor, Gpr41. Proc Natl Acad Sci U S A 105(43):16767–16772

6. Kimura I, Ozawa K, Inoue D, Imamura T, Kimura K et al (2013) The gut microbiota suppresses insulin-mediated fat accumulation via the short-chain fatty acid receptor GPR43. Nat Commun 4:1829
7. De Vadder F, Kovatcheva-Datchary P, Goncalves D, Vinera J, Zitoun C et al (2014) Microbiota-generated metabolites promote metabolic benefits via gut-brain neural circuits. Cell 156(1–2):84–96
8. Turnbaugh PJ, Ley RE, Mahowald MA, Magrini V, Mardis ER, Gordon JI (2006) An obesity-associated gut microbiome with increased capacity for energy harvest. Nature 444(7122):1027–1031
9. Schwiertz A, Taras D, Schäfer K, Beijer S, Bos NA et al (2010) Microbiota and SCFA in lean and overweight healthy subjects. Obesity (Silver Spring) 18(1):190–195
10. Turnbaugh PJ, Hamady M, Yatsunenko T, Cantarel BL, Duncan A et al (2009) A core gut microbiome in obese and lean twins. Nature 457(7228):480–484
11. Nair S, Cope K, Risby TH, Diehl AM (2001) Obesity and female gender increase breath ethanol concentration: potential implications for the pathogenesis of nonalcoholic steatohepatitis. Am J Gastroenterol 96(4):1200–1204
12. Maslowski KM, Vieira AT, Ng A, Kranich J, Sierro F et al (2009) Regulation of inflammatory responses by gut microbiota and chemoattractant receptor GPR43. Nature 461(7268):1282–1286
13. Smith PM, Howitt MR, Panikov N, Michaud M, Gallini CA et al (2013) The microbial metabolites, short-chain fatty acids, regulate colonic T_{reg} cell homeostasis. Science 341(6145):569
14. Macia L, Tan J, Vieira AT, Leach K, Stanley D et al (2015) Metabolite-sensing receptors GPR43 and GPR109A facilitate dietary fibre-induced gut homeostasis through regulation of the inflammasome. Nat Commun 6:6734
15. Yim HCH, Wang D, Yu L, White CL, Faber PW et al (2016) The kinase activity of PKR represses inflammasome activity. Cell Res 26(3):367–379
16. Yim HCH, Williams BRG (2014) Protein kinase R and the inflammasome. J Interf Cytokine Res 34(6):447–454
17. Atarashi K, Tanoue T, Oshima K, Suda W, Nagano Y et al (2013) Treg induction by a rationally selected mixture of Clostridia strains from the human microbiota. Nature 500(7461):232–236
18. Furusawa Y, Obata Y, Fukuda S, Endo TA, Nakato G et al (2013) Commensal microbe-derived butyrate induces the differentiation of colonic regulatory T cells. Nature 504(7480):446–450
19. Frank DN, St Amand AL, Feldman RA, Boedeker EC, Harpaz N, Pace NR (2007) Molecular-phylogenetic characterization of microbial community imbalances in human inflammatory bowel diseases. Proc Natl Acad Sci U S A 104(34):13780–13785
20. Manichanh C, Rigottier-Gois L, Bonnaud E, Gloux K, Pelletier E et al (2006) Reduced diversity of faecal microbiota in Crohn's disease revealed by a metagenomic approach. Gut 55(2):205
21. Treem WR, Ahsan N, Shoup M, Hyams JS (1994) Fecal short-chain fatty acids in children with inflammatory bowel disease. J Pediatr Gastroenterol Nutr 18(2):159–164
22. Vital M, Penton CR, Wang Q, Young VB, Antonopoulos DA et al (2013) A gene-targeted approach to investigate the intestinal butyrate-producing bacterial community. Microbiome 1(1):8
23. Trompette A, Gollwitzer ES, Yadava K, Sichelstiel AK, Sprenger N et al (2014) Gut microbiota metabolism of dietary fiber influences allergic airway disease and hematopoiesis. Nat Med 20(2):159–166
24. Vieira AT, Macia L, Galvão I, Martins FS, Canesso MCC et al (2015) A role for gut microbiota and the metabolite-sensing receptor GPR43 in a murine model of gout. Arthritis Rheumatol 67(6):1646–1656
25. Singh N, Gurav A, Sivaprakasam S, Brady E, Padia R et al (2014) Activation of Gpr109a, receptor for niacin and the commensal metabolite butyrate, suppresses colonic inflammation and carcinogenesis. Immunity 40(1):128–139

26. Belcheva A, Irrazabal T, Robertson SJ, Streutker C, Maughan H et al (2014) Gut microbial metabolism drives transformation of MSH2-deficient colon epithelial cells. Cell 158(2):288–299
27. Basson A, Trotter A, Rodriguez-Palacios A, Cominelli F (2016) Mucosal interactions between genetics, diet, and microbiome in inflammatory bowel disease. Front Immunol 7:290
28. Thorburn AN, Macia L, Mackay CR (2014) Diet, metabolites, and "western-lifestyle" inflammatory diseases. Immunity 40(6):833–842
29. Zelante T, Iannitti RG, Cunha C, De Luca A, Giovannini G et al (2013) Tryptophan catabolites from microbiota engage aryl hydrocarbon receptor and balance mucosal reactivity via interleukin-22. Immunity 39(2):372–385
30. Li Y, Innocentin S, Withers DR, Roberts NA, Gallagher AR et al (2011) Exogenous stimuli maintain intraepithelial lymphocytes via aryl hydrocarbon receptor activation. Cell 147(3):629–640
31. Carbonero F, Benefiel AC, Alizadeh-Ghamsari AH, Gaskins HR (2012) Microbial pathways in colonic sulfur metabolism and links with health and disease. Front Physiol 3:448
32. Ijssennagger N, van der Meer R, van Mil SWC (2016) Sulfide as a Mucus Barrier-Breaker in inflammatory bowel disease? Trends Mol Med 22(3):190–199
33. Scanlan PD, Shanahan F, Marchesi JR (2009) Culture-independent analysis of desulfovibrios in the human distal colon of healthy, colorectal cancer and polypectomized individuals. FEMS Microbiol Ecol 69(2):213–221
34. Gibson GR, Cummings JH, Macfarlane GT (1991) Growth and activities of sulphate-reducing bacteria in gut contents of healthy subjects and patients with ulcerative colitis. FEMS Microbiol Ecol 9(2):103–111
35. Jia W, Whitehead RN, Griffiths L, Dawson C, Bai H et al (2012) Diversity and distribution of sulphate-reducing bacteria in human faeces from healthy subjects and patients with inflammatory bowel disease. FEMS Immunol Med Microbiol 65(1):55–68
36. Loubinoux J, Bronowicki J-P, Pereira IAC, Mougenel J-L, Faou AE (2002) Sulfate-reducing bacteria in human feces and their association with inflammatory bowel diseases. FEMS Microbiol Ecol 40(2):107–112
37. Mills DJS, Tuohy KM, Booth J, Buck M, Crabbe MJC et al (2008) Dietary glycated protein modulates the colonic microbiota towards a more detrimental composition in ulcerative colitis patients and non-ulcerative colitis subjects. J Appl Microbiol 105(3):706–714
38. Pitcher MC, Beatty ER, Cummings JH (2000) The contribution of sulphate reducing bacteria and 5-aminosalicylic acid to faecal sulphide in patients with ulcerative colitis. Gut 46(1):64–72
39. Verma R, Verma AK, Ahuja V, Paul J (2010) Real-time analysis of mucosal flora in patients with inflammatory bowel disease in India. J Clin Microbiol 48(11):4279–4282
40. Zinkevich V, Beech IB (2000) Screening of sulfate-reducing bacteria in colonoscopy samples from healthy and colitic human gut mucosa. FEMS Microbiol Ecol 34(2):147–155
41. Levine J, Ellis CJ, Furne JK, Springfield J, Levitt MD (1998) Fecal hydrogen sulfide production in ulcerative colitis. Am J Gastroenterol 93(1):83–87
42. Babidge W, Millard S, Roediger W (1998) Sulfides impair short chain fatty acid beta-oxidation at acyl-CoA dehydrogenase level in colonocytes: implications for ulcerative colitis. Mol Cell Biochem 181(1–2):117–124
43. Christl SU, Eisner H-D, Dusel G, Kasper H, Scheppach W (1996) Antagonistic effects of sulfide and butyrate on proliferation of colonic mucosa. Dig Dis Sci 41(12):2477–2481
44. Roediger WE, Duncan A, Kapaniris O, Millard S (1993) Reducing sulfur compounds of the colon impair colonocyte nutrition: implications for ulcerative colitis. Gastroenterology 104(3):802–809
45. Ijssennagger N, Belzer C, Hooiveld GJ, Dekker J, van Mil SWC et al (2015) Gut microbiota facilitates dietary heme-induced epithelial hyperproliferation by opening the mucus barrier in colon. Proc Natl Acad Sci U S A 112(32):10038–10043

46. Kanazawa K, Konishi F, Mitsuoka T, Terada A, Itoh K et al (1996) Factors influencing the development of sigmoid colon cancer: Bacteriologic and biochemical studies. Cancer 77(8):1701
47. Attene-Ramos MS, Wagner ED, Gaskins HR, Plewa MJ (2007) Hydrogen sulfide induces direct radical-associated DNA damage. Mol Cancer Res 5(5):455–459
48. Attene-Ramos MS, Wagner ED, Plewa MJ, Gaskins HR (2006) Evidence that hydrogen sulfide is a genotoxic agent. Mol Cancer Res 4(1):9–14
49. Deplancke B, Gaskins HR (2003) Hydrogen sulfide induces serum-independent cell cycle entry in nontransformed rat intestinal epithelial cells. FASEB J 17(10):1310–1312
50. Wahlström A, Sayin SI, Marschall H-U, Bäckhed F (2016) Intestinal crosstalk between bile acids and microbiota and its impact on host metabolism. Cell Metab 24(1):41–50
51. Lefebvre P, Cariou B, Lien F, Kuipers F, Staels B (2009) Role of bile acids and bile acid receptors in metabolic regulation. Physiol Rev 89(1):147–191
52. Teodoro JS, Rolo AP, Palmeira CM (2011) Hepatic FXR: key regulator of whole-body energy metabolism. Trends Endocrinol Metab 22(11):458–466
53. Lambert G, Amar MJA, Guo G, Brewer HB, Gonzalez FJ, Sinal CJ (2003) The farnesoid X-receptor is an essential regulator of cholesterol homeostasis. J Biol Chem 278(4):2563–2570
54. Ma K, Saha PK, Chan L, Moore DD (2006) Farnesoid X receptor is essential for normal glucose homeostasis. J Clin Invest 116(4):1102–1109
55. Hanniman EA, Lambert G, McCarthy TC, Sinal CJ (2005) Loss of functional farnesoid X receptor increases atherosclerotic lesions in apolipoprotein E-deficient mice. J Lipid Res 46(12):2595–2604
56. Zhang Y, Wang X, Vales C, Lee FY, Lee H et al (2006) FXR deficiency causes reduced atherosclerosis in Ldlr/mice. Arterioscler Thromb Vasc Biol 26(10):2316–2321
57. Parséus A, Sommer N, Sommer F, Caesar R, Molinaro A et al (2017) Microbiota-induced obesity requires farnesoid X receptor. Gut 66(3):429–437
58. Prawitt J, Abdelkarim M, Stroeve JHM, Popescu I, Duez H et al (2011) Farnesoid X receptor deficiency improves glucose homeostasis in mouse models of obesity. Diabetes 60(7):1861–1871
59. Zhang Y, Ge X, Heemstra LA, Chen W-D, Xu J et al (2012) Loss of FXR protects against diet-induced obesity and accelerates liver carcinogenesis in ob/ob mice. Mol Endocrinol 26(2):272–280
60. Yang Z-X, Shen W, Sun H (2010) Effects of nuclear receptor FXR on the regulation of liver lipid metabolism in patients with non-alcoholic fatty liver disease. Hepatol Int 4(4):741–748
61. Thomas C, Gioiello A, Noriega L, Strehle A, Oury J et al (2009) TGR5-mediated bile acid sensing controls glucose homeostasis. Cell Metab 10(3):167–177
62. Watanabe M, Houten SM, Mataki C, Christoffolete MA, Kim BW et al (2006) Bile acids induce energy expenditure by promoting intracellular thyroid hormone activation. Nature 439(7075):484–489
63. Pols TWH, Nomura M, Harach T, Lo Sasso G, Oosterveer MH et al (2011) TGR5 activation inhibits atherosclerosis by reducing macrophage inflammation and lipid loading. Cell Metab 14(6):747–757
64. Perino A, Pols TWH, Nomura M, Stein S, Pellicciari R, Schoonjans K (2014) TGR5 reduces macrophage migration through mTOR-induced C/EBPβ differential translation. J Clin Invest 124(12):5424–5436
65. Levy M, Thaiss CA, Zeevi D, Dohnalová L, Zilberman-Schapira G et al (2015) Microbiota-Modulated Metabolites Shape the Intestinal Microenvironment by Regulating NLRP6 Inflammasome Signaling. Cell 163(6):1428–1443
66. Louis P, Hold GL, Flint HJ (2014) The gut microbiota, bacterial metabolites and colorectal cancer. Nat Rev Micro 12(10):661–672
67. Ou J, DeLany JP, Zhang M, Sharma S, O'Keefe SJD (2012) Association between low colonic short-chain fatty acids and high bile acids in high colon cancer risk populations. Nutr Cancer 64(1):34–40

68. Ou J, Carbonero F, Zoetendal EG, DeLany JP, Wang M et al (2013) Diet, microbiota, and microbial metabolites in colon cancer risk in rural Africans and African Americans. Am J Clin Nutr 98(1):111–120
69. Yoshimoto S, Loo TM, Atarashi K, Kanda H, Sato S et al (2013) Obesity-induced gut microbial metabolite promotes liver cancer through senescence secretome. Nature 499(7456):97–101
70. Cho CE, Caudill MA (2017) Trimethylamine-N-oxide: friend, foe, or simply caught in the cross-fire? Trends Endocrinol Metab 28(2):121–130
71. Tang WHW, Wang Z, Levison BS, Koeth RA, Britt EB et al (2013) Intestinal microbial metabolism of phosphatidylcholine and cardiovascular risk. N Engl J Med 368(17):1575–1584
72. Koeth RA, Wang Z, Levison BS, Buffa JA, Org E et al (2013) Intestinal microbiota metabolism of L-carnitine, a nutrient in red meat, promotes atherosclerosis. Nat Med 19(5):576–585
73. Wang Z, Roberts AB, Buffa JA, Levison BS, Zhu W et al (2015) Non-lethal inhibition of gut microbial trimethylamine production for the treatment of atherosclerosis. Cell 163(7):1585–1595
74. Wang Z, Klipfell E, Bennett BJ, Koeth R, Levison BS et al (2011) Gut flora metabolism of phosphatidylcholine promotes cardiovascular disease. Nature 472(7341):57–63
75. Dumas M-E, Barton RH, Toye A, Cloarec O, Blancher C et al (2006) Metabolic profiling reveals a contribution of gut microbiota to fatty liver phenotype in insulin-resistant mice. Proc Natl Acad Sci U S A 103(33):12511–12516
76. Rooks MG, Garrett WS (2016) Gut microbiota, metabolites and host immunity. Nat Rev Immunol 16(6):341–352
77. Johnson CH, Dejea CM, Edler D, Hoang LT, Santidrian AF et al (2015) Metabolism links bacterial biofilms and colon carcinogenesis. Cell Metab 21(6):891–897
78. Dejea CM, Wick EC, Hechenbleikner EM, White JR, Mark Welch JL et al (2014) Microbiota organization is a distinct feature of proximal colorectal cancers. Proc Natl Acad Sci U S A 111(51):18321–18326
79. Rakoff-Nahoum S, Paglino J, Eslami-Varzaneh F, Edberg S, Medzhitov R (2004) Recognition of commensal microflora by toll-like receptors is required for intestinal homeostasis. Cell 118(2):229–241
80. Brown SL, Riehl TE, Walker MR, Geske MJ, Doherty JM et al (2007) Myd88-dependent positioning of Ptgs2-expressing stromal cells maintains colonic epithelial proliferation during injury. J Clin Invest 117(1):258–269
81. Fukata M, Chen A, Klepper A, Krishnareddy S, Vamadevan AS et al (2006) Cox-2 is regulated by Toll-like receptor-4 (TLR4) signaling: Role in proliferation and apoptosis in the intestine. Gastroenterology 131(3):862–877
82. Hernandez Y, Sotolongo J, Breglio K, Conduah D, Chen A et al (2010) The role of prostaglandin E2 (PGE 2) in toll-like receptor 4 (TLR4)-mediated colitis-associated neoplasia. BMC Gastroenterol 10:82
83. Pull SL, Doherty JM, Mills JC, Gordon JI, Stappenbeck TS (2005) Activated macrophages are an adaptive element of the colonic epithelial progenitor niche necessary for regenerative responses to injury. Proc Natl Acad Sci U S A 102(1):99–104
84. Dheer R, Santaolalla R, Davies JM, Lang JK, Phillips MC et al (2016) Intestinal epithelial toll-like receptor 4 signaling affects epithelial function and colonic microbiota and promotes a risk for transmissible colitis. Infect Immun 84(3):798–810
85. Cani PD, Amar J, Iglesias MA, Poggi M, Knauf C et al (2007) Metabolic endotoxemia initiates obesity and insulin resistance. Diabetes 56(7):1761–1772
86. Cani PD, Bibiloni R, Knauf C, Waget A, Neyrinck AM et al (2008) Changes in gut microbiota control metabolic endotoxemia-induced inflammation in high-fat diet-induced obesity and diabetes in mice. Diabetes 57(6):1470–1481
87. Lassenius MI, Pietiläinen KH, Kaartinen K, Pussinen PJ, Syrjänen J et al (2011) Bacterial endotoxin activity in human serum is associated with dyslipidemia, insulin resistance, obesity, and chronic inflammation. Diabetes Care 34(8):1809–1815

88. Pussinen PJ, Havulinna AS, Lehto M, Sundvall J, Salomaa V (2011) Endotoxemia is associated with an increased risk of incident diabetes. Diabetes Care 34(2):392–397
89. Henao-Mejia J, Elinav E, Jin C, Hao L, Mehal WZ et al (2012) Inflammasome-mediated dysbiosis regulates progression of NAFLD and obesity. Nature 482(7384):179–185
90. Jiang W, Wu N, Wang X, Chi Y, Zhang Y et al (2015) Dysbiosis gut microbiota associated with inflammation and impaired mucosal immune function in intestine of humans with non-alcoholic fatty liver disease. Sci Rep 5:8096
91. Miele L, Valenza V, La Torre G, Montalto M, Cammarota G et al (2009) Increased intestinal permeability and tight junction alterations in nonalcoholic fatty liver disease. Hepatology 49(6):1877–1887
92. Wigg AJ, Roberts-Thomson IC, Dymock RB, McCarthy PJ, Grose RH, Cummins AG (2001) The role of small intestinal bacterial overgrowth, intestinal permeability, endotoxaemia, and tumour necrosis factor alpha in the pathogenesis of non-alcoholic steatohepatitis. Gut 48(2):206–211
93. Dapito DH, Mencin A, Gwak G-Y, Pradere J-P, Jang M-K et al (2012) Promotion of hepatocellular carcinoma by the intestinal microbiota and TLR4. Cancer Cell 21(4):504–516
94. Ochi A, Nguyen AH, Bedrosian AS, Mushlin HM, Zarbakhsh S et al (2012) MyD88 inhibition amplifies dendritic cell capacity to promote pancreatic carcinogenesis via Th2 cells. J Exp Med 209(9):1671–1687
95. Fukata M, Chen A, Vamadevan AS, Cohen J, Breglio K et al (2007) Toll-like receptor-4 promotes the development of colitis-associated colorectal tumors. Gastroenterology 133(6):1869–1881
96. Fukata M, Shang L, Santaolalla R, Sotolongo J, Pastorini C et al (2011) Constitutive activation of epithelial TLR4 augments inflammatory responses to mucosal injury and drives colitis-associated tumorigenesis. Inflamm Bowel Dis 17(7):1464–1473
97. Grivennikov SI, Wang K, Mucida D, Stewart CA, Schnabl B et al (2012) Adenoma-linked barrier defects and microbial products drive IL-23/IL-17-mediated tumour growth. Nature 491(7423):254–258
98. Lee SH, Hu L-L, Gonzalez-Navajas J, Seo GS, Shen C et al (2010) ERK activation drives intestinal tumorigenesis in Apc(min/+) mice. Nat Med 16(6):665–670
99. Yim HCH, Li JCB, Pong JCH, Lau ASY (2011) A role for c-Myc in regulating anti-mycobacterial responses. Proc Natl Acad Sci U S A 108(43):17749–17754
100. Belkaid Y, Hand TW (2014) Role of the microbiota in immunity and inflammation. Cell 157(1):121–141
101. Vijay-Kumar M, Sanders CJ, Taylor RT, Kumar A, Aitken JD et al (2007) Deletion of TLR5 results in spontaneous colitis in mice. J Clin Invest 117(12):3909–3921
102. Meena NK, Ahuja V, Meena K, Paul J (2015) Association of TLR5 gene polymorphisms in ulcerative colitis patients of north India and their role in cytokine homeostasis. PLoS One 10(3):e0120697
103. Vijay-Kumar M, Aitken JD, Carvalho FA, Cullender TC, Mwangi S et al (2010) Metabolic syndrome and altered gut microbiota in mice lacking Toll-like receptor 5. Science 328(5975):228–231
104. Chassaing B, Ley RE, Gewirtz AT (2014) Intestinal epithelial cell toll-like receptor 5 regulates the intestinal microbiota to prevent low-grade inflammation and metabolic syndrome in mice. Gastroenterology 147(6):1363–1377.e17
105. Singh V, Yeoh BS, Carvalho F, Gewirtz AT, Vijay-Kumar M (2015) Proneness of TLR5 deficient mice to develop colitis is microbiota dependent. Gut Microbes 6(4):279–283
106. Rutkowski MR, Stephen TL, Svoronos N, Allegrezza MJ, Tesone AJ et al (2015) Microbially driven TLR5-dependent signaling governs distal malignant progression through tumor-promoting inflammation. Cancer Cell 27(1):27–40
107. Mazmanian SK, Liu CH, Tzianabos AO, Kasper DL (2005) An immunomodulatory molecule of symbiotic bacteria directs maturation of the host immune system. Cell 122(1):107–118

108. Mazmanian SK, Round JL, Kasper DL (2008) A microbial symbiosis factor prevents intestinal inflammatory disease. Nature 453(7195):620–625
109. Ochoa-Repáraz J, Mielcarz DW, Ditrio LE, Burroughs AR, Begum-Haque S et al (2010) Central nervous system demyelinating disease protection by the human commensal Bacteroides fragilis depends on polysaccharide A expression. J Immunol 185(7):4101–4108
110. Round JL, Lee SM, Li J, Tran G, Jabri B et al (2011) The Toll-like receptor 2 pathway establishes colonization by a commensal of the human microbiota. Science 332(6032):974–977
111. Round JL, Mazmanian SK (2010) Inducible Foxp3+ regulatory T-cell development by a commensal bacterium of the intestinal microbiota. Proc Natl Acad Sci U S A 107(27):12204–12209
112. Shen Y, Giardino Torchia ML, Lawson GW, Karp CL, Ashwell JD, Mazmanian SK (2012) Outer membrane vesicles of a human commensal mediate immune regulation and disease protection. Cell Host Microbe 12(4):509–520
113. Lee J, Mo J-H, Katakura K, Alkalay I, Rucker AN et al (2006) Maintenance of colonic homeostasis by distinctive apical TLR9 signalling in intestinal epithelial cells. Nat Cell Biol 8(12):1327–1336
114. Obermeier F, Strauch UG, Dunger N, Grunwald N, Rath HC et al (2005) In vivo CpG DNA/toll-like receptor 9 interaction induces regulatory properties in CD4+CD62L+ T cells which prevent intestinal inflammation in the SCID transfer model of colitis. Gut 54(10):1428–1436
115. Rose WA, Sakamoto K, Leifer CA (2012) TLR9 is important for protection against intestinal damage and for intestinal repair. Sci Rep 2:574
116. Hong C-P, Yun CH, Lee G-W, Park A, Kim Y-M, Jang MH (2015) TLR9 regulates adipose tissue inflammation and obesity-related metabolic disorders. Obesity (Silver Spring) 23(11):2199–2206
117. Caruso R, Warner N, Inohara N, Núñez G (2014) NOD1 and NOD2: signaling, host defense, and inflammatory disease. Immunity 41(6):898–908
118. Hold GL, Smith M, Grange C, Watt ER, El-Omar EM, Mukhopadhya I (2014) Role of the gut microbiota in inflammatory bowel disease pathogenesis: what have we learnt in the past 10 years? World J Gastroenterol 20(5):1192–1210
119. Al Nabhani Z, Dietrich G, Hugot J-P, Barreau F (2017) Nod2: The intestinal gate keeper. PLoS Pathog 13(3):e1006177
120. Barreau F, Meinzer U, Chareyre F, Berrebi D, Niwa-Kawakita M et al (2007) CARD15/NOD2 is required for Peyer's patches homeostasis in mice. PLoS One 2(6):e523
121. Chassaing B, Rolhion N, de Vallée A, Salim SY, Prorok-Hamon M et al (2011) Crohn disease--associated adherent-invasive E. coli bacteria target mouse and human Peyer's patches via long polar fimbriae. J Clin Invest 121(3):966–975
122. Denou E, Lolmède K, Garidou L, Pomie C, Chabo C et al (2015) Defective NOD2 peptidoglycan sensing promotes diet-induced inflammation, dysbiosis, and insulin resistance. EMBO Mol Med 7(3):259–274
123. Amendola A, Butera A, Sanchez M, Strober W, Boirivant M (2014) Nod2 deficiency is associated with an increased mucosal immunoregulatory response to commensal microorganisms. Mucosal Immunol 7(2):391–404
124. Barreau F, Madre C, Meinzer U, Berrebi D, Dussaillant M et al (2010) Nod2 regulates the host response towards microflora by modulating T cell function and epithelial permeability in mouse Peyer's patches. Gut 59(2):207–217
125. Watanabe T, Asano N, Murray PJ, Ozato K, Tailor P et al (2008) Muramyl dipeptide activation of nucleotide-binding oligomerization domain 2 protects mice from experimental colitis. J Clin Invest 118(2):545–559
126. Yang Z, Fuss IJ, Watanabe T, Asano N, Davey MP et al (2007) NOD2 transgenic mice exhibit enhanced MDP-mediated down-regulation of TLR2 responses and resistance to colitis induction. Gastroenterology 133(5):1510–1521
127. Couturier-Maillard A, Secher T, Rehman A, Normand S, De Arcangelis A et al (2013) NOD2-mediated dysbiosis predisposes mice to transmissible colitis and colorectal cancer. J Clin Invest 123(2):700–711

128. Möckelmann N, von Schönfels W, Buch S, von Kampen O, Sipos B et al (2009) Investigation of innate immunity genes CARD4, CARD8 and CARD15 as germline susceptibility factors for colorectal cancer. BMC Gastroenterol 9:79
129. Rubinstein MR, Wang X, Liu W, Hao Y, Cai G, Han YW (2013) Fusobacterium nucleatum promotes colorectal carcinogenesis by modulating E-cadherin/β-catenin signaling via its FadA adhesin. Cell Host Microbe 14(2):195–206
130. Kostic AD, Chun E, Robertson L, Glickman JN, Gallini CA et al (2013) Fusobacterium nucleatum potentiates intestinal tumorigenesis and modulates the tumor-immune microenvironment. Cell Host Microbe 14(2):207–215
131. Castellarin M, Warren RL, Freeman JD, Dreolini L, Krzywinski M et al (2012) Fusobacterium nucleatum infection is prevalent in human colorectal carcinoma. Genome Res 22(2):299–306
132. Kostic AD, Gevers D, Pedamallu CS, Michaud M, Duke F et al (2012) Genomic analysis identifies association of Fusobacterium with colorectal carcinoma. Genome Res 22(2):292–298
133. Maddocks ODK, Short AJ, Donnenberg MS, Bader S, Harrison DJ (2009) Attaching and effacing Escherichia coli downregulate DNA mismatch repair protein in vitro and are associated with colorectal adenocarcinomas in humans. PLoS One 4(5):e5517
134. Martin HM, Campbell BJ, Hart CA, Mpofu C, Nayar M et al (2004) Enhanced Escherichia coli adherence and invasion in Crohn's disease and colon cancer. Gastroenterology 127(1):80–93
135. Swidsinski A, Khilkin M, Kerjaschki D, Schreiber S, Ortner M et al (1998) Association between intraepithelial Escherichia coli and colorectal cancer. Gastroenterology 115(2):281–286
136. Sun J, Kato I (2016) Gut microbiota, inflammation and colorectal cancer. Genes Dis 3(2):130–143
137. Arthur JC, Perez-Chanona E, Mühlbauer M, Tomkovich S, Uronis JM et al (2012) Intestinal inflammation targets cancer-inducing activity of the microbiota. Science 338(6103):120–123
138. Bonnet M, Buc E, Sauvanet P, Darcha C, Dubois D et al (2014) Colonization of the human gut by E. coli and colorectal cancer risk. Clin Cancer Res 20(4):859–867
139. Buc E, Dubois D, Sauvanet P, Raisch J, Delmas J et al (2013) High prevalence of mucosa-associated *E. coli* producing cyclomodulin and genotoxin in colon cancer. PLoS ONE 8(2):e56964
140. Cougnoux A, Dalmasso G, Martinez R, Buc E, Delmas J et al (2014) Bacterial genotoxin colibactin promotes colon tumour growth by inducing a senescence-associated secretory phenotype. Gut 63(12):1932–1942
141. Cuevas-Ramos G, Petit CR, Marcq I, Boury M, Oswald E, Nougayrède J-P (2010) Escherichia coli induces DNA damage in vivo and triggers genomic instability in mammalian cells. Proc Natl Acad Sci U S A 107(25):11537–11542
142. Nougayrède J-P, Homburg S, Taieb F, Boury M, Brzuszkiewicz E et al (2006) Escherichia coli induces DNA double-strand breaks in eukaryotic cells. Science 313(5788):848–851
143. Vizcaino MI, Crawford JM (2015) The colibactin warhead crosslinks DNA. Nat Chem 7(5):411–417
144. Wu S, Rhee K-J, Albesiano E, Rabizadeh S, Wu X et al (2009) A human colonic commensal promotes colon tumorigenesis via activation of T helper type 17 T cell responses. Nat Med 15(9):1016–1022
145. Goodwin AC, Destefano Shields CE, Wu S, Huso DL, Wu X et al (2011) Polyamine catabolism contributes to enterotoxigenic Bacteroides fragilis-induced colon tumorigenesis. Proc Natl Acad Sci U S A 108(37):15354–15359
146. Boleij A, Hechenbleikner EM, Goodwin AC, Badani R, Stein EM et al (2015) The Bacteroides fragilis toxin gene is prevalent in the colon mucosa of colorectal cancer patients. Clin Infect Dis 60(2):208–215
147. Toprak NU, Yagci A, Gulluoglu BM, Akin ML, Demirkalem P et al (2006) A possible role of Bacteroides fragilis enterotoxin in the aetiology of colorectal cancer. Clin Microbiol Infect 12(8):782–786

Chapter 8
cGAS-STING Activation in the Tumor Microenvironment and Its Role in Cancer Immunity

Geneviève Pépin and Michael P. Gantier

Abstract Stimulator of interferon (IFN) genes (STING) is a key mediator in the immune response to cytoplasmic DNA sensed by cyclic GMP-AMP (cGAMP) synthase (cGAS). After synthesis by cGAS, cGAMP acts as a second messenger activating STING in the cell harboring cytoplasmic DNA but also in adjacent cells through gap junction transfer. While the role of the cGAS-STING pathway in pathogen detection is now well established, its importance in cancer immunity has only recently started to emerge. Nonetheless, STING appears to be an essential component in the recruitment of immune cells to the tumor microenvironment, which is paramount to immune clearance of the tumor. This review presents an overview of the growing literature around the role of the cGAS-STING pathway in the tumor microenvironment, with a specific focus on the role that cancer cells may play in the direct activation of this pathway, and its amplification through cell-cell transfer of cGAMP.

Keywords cGAS • cGAMP • STING • Interferon • Connexin • Cancer

---

The original version of this chapter was revised.
An erratum to this chapter can be found at https://doi.org/10.1007/978-981-10-5987-2_12

G. Pépin (✉) • M.P. Gantier
Centre for Innate Immunity and Infectious Diseases, Hudson Institute of Medical Research, Clayton, VIC 3168, Australia

Department of Molecular and Translational Science, Monash University, Clayton, VIC 3168, Australia
e-mail: genevieve.pepin@hudson.org.au; michael.gantier@hudson.org.au

D. Xu (ed.), *Regulation of Inflammatory Signaling in Health and Disease*, Advances in Experimental Medicine and Biology 1024,
DOI 10.1007/978-981-10-5987-2_8

8.1 Introduction

Detection of pathogen-associated molecular patterns (PAMPs) by the immune system is the foundation of innate immunity and is one of the first lines of defense against infections. Viral nucleic acids and bacterial cell wall components (such as lipopolysaccharides or flagellar proteins) are PAMPs selectively detected by different types of innate immune receptors. These include membrane-bound Toll-like receptors (TLRs)—located at the surface of the cells or in the endosomes (e.g., TLR3, TLR4, and TLR7/8/9)—or cytoplasmic sensors such as retinoic acid-inducible gene I (RIG-I)-like helicases, which detect foreign RNAs [1, 2]. Activation of such innate immune sensors results in the production of antiviral and antibacterial proteins, including cytokines such as TNF-α or type I interferons (IFNs). Type I IFNs exert a strong antiviral effect through transcriptional induction of more than 2000 genes [3].

8.1.1 Stimulator of Interferon Genes: STING

Stimulator of interferon genes (STING, also known as MITA, ERIS, MPYS, or TMEM173) was independently discovered by four different groups in 2008 in an attempt to characterize mechanisms of DNA recognition resulting in the production of type I interferon (IFN), independently of TLR9 [4–7]. STING is an ER-localized protein containing N-terminal transmembrane helices and a large C-terminal cytosolic domain. Its activation promotes signal transduction through the TBK1-IRF3 axis and production of type I IFNs [8, 9]. STING is directly involved in intracellular bacterial detection through sensing of cyclic dinucleotides (CDNs). During infection, CDNs are produced by bacteria such as *Listeria monocytogenes* and *Mycobacterium tuberculosis* [10, 11], which activate STING to promote strong immune responses to these pathogens.

8.1.2 Cyclic GMP-AMP Synthase: cGAS

Despite the initial demonstration of STING's involvement in intracellular DNA sensing [12], a lack of evidence for a direct dsDNA-STING interaction prompted the community to suggest the existence of an upstream DNA sensor. Several potential cytoplasmic DNA sensors had been described (e.g., DAI, IFI16, DDX41, DNA-PK, MRE11, Sox2, and PQBP1, reviewed in [13]), before the discovery of cyclic GMP-AMP synthase (cGAS) by the Chen laboratory in 2013 [14]. Following cytoplasmic DNA detection, cGAS produces an endogenous second messenger, 2′3′-cyclic GMP-AMP (2′3′-cGAMP), that binds directly to STING to promote its activation [15–19]. It is now clear that cGAS is an important immune receptor for

DNA viruses and retroviruses. As such, its capacity to sense cytoplasmic DNA is not restricted to standard Watson-Crick DNA, and it can also sense DNA-RNA hybrids [20]. The DNA cGAS senses can originate from various sites; in addition to bacterial and viral DNA, there is now good evidence that it senses self-DNA leaked from the nucleus [21] or mitochondria [2, 22].

By activating STING, DNA detection by cGAS culminates in the production of pro-inflammatory cytokines and the recruitment of the IFN response. Soon after the discovery of this pathway, the role of cGAS-STING in sterile inflammation (i.e., in the absence of any pathogen) was investigated. Mutations promoting a gain-of-function of STING have been shown to result in chilblain lupus [23], a rare type of cutaneous auto-inflammation. In addition, *cGas* and/or *Sting* genetic deletion was found to rescue animals from lupus-like diseases [24–26].

8.1.3 cGAS-STING in Tumor Development

Beyond its critical role in innate immunity, the cGAS-STING pathway is rapidly emerging as a critical player in the control of tumor development. Clearance of tumorigenic cells by the immune system initially relies on type I IFN production by dendritic cells (DCs) and the recruitment of $CD8^+$ T cells, which promote the targeted death of such aberrant cells. Accordingly, chemically induced tumors develop better in mice lacking type I IFN signaling compared to their wild-type (WT) counterparts; similarly, a deficiency in type I IFN signaling results in poorer rejection of transplanted immunogenic tumors [27, 28].

Recruitment of $CD8^+$ T cells in the tumor microenvironment is the key step in antitumor immunity and is directly dependent on type I IFN production by DCs [29, 30]. Critically, STING appears to play a unique, non-redundant role in the recruitment of $CD8^+$ T cells to the tumor microenvironment [31, 32]. Indeed, genetic loss of a number of the most important signaling molecules known to date in innate immunity did not have any impact on $CD8^+$ T cell infiltration in the tumor microenvironment in a syngeneic melanoma model [31, 32]. Conversely, the loss of STING abolished the spontaneous infiltration of $CD8^+$ T cells in this melanoma model [32, 33].

8.1.4 STING Agonists as Antitumoral Adjuvants

The use of pathogen-associated molecular pattern (PAMP)-like molecules to potentiate DC activation upon radiation therapy and help tumor clearance is well established, with several clinical trials currently underway [34]. Specifically, there has been a great deal of enthusiasm around the concept that synthetic STING ligands could be designed to facilitate tumor clearance [35, 36]. One STING ligand that showed early promise was DMXAA, a flavonoid compound comprising two phenyl rings and a heterocyclic ring. DMXAA was found to exert strong antitumoral activities in mice and to act as a direct ligand for mouse STING [37]. However, DMXAA failed in human clinical trials [38] and was found not to be a direct ligand for human STING [39]. Other research has focused on cGAMP, the second messenger that activates STING. Although cGAMP can be synthesized in vitro, its application as an adjuvant is partially limited by the fact that it is not membrane permeable. Nonetheless, cGAMP injection has been found to synergize with radiation therapy to control local and distant tumors in pancreatic cancer [40] and to decrease chemotherapy toxicity in colon cancer [41].

To remedy the limitations of natural cGAMP, various types of synthetic STING ligand based on the structures of CDNs have already been synthesized and tested in various cancer models (reviewed in [39]). Most of these ligands have been designed to target human STING and have the potential to be used in clinical trials, as exemplified by compound MIW815 (ADU-S100). In line with a role for STING in $CD8^+$ T cell recruitment and considering prior evidence with DMXAA, intratumoral injection of these CDN molecules into different types of cancer (melanoma, colon, glioma, and breast carcinomas) caused rapid tumor regression and mediated lasting and systemic antigen-specific T cell immunity [37, 35, 39]. The action of STING agonists may not be limited to antitumor immunity, and they could also promote direct apoptosis of cancer cells, as suggested by observations in malignant B cell leukemia [36].

In addition to designing synthetic ligands, loading of otherwise non-permeable cyclic di-GMP into liposomes was found to stimulate clearance of murine melanoma [42]. This suggests that experimentation with packaging and delivery of STING agonists may help enhance the therapeutic potential of these molecules.

8.1.5 STING and Immunotherapy

The tumor microenvironment plays a critical role in the control of $CD8^+$ T cells to kill tumor cells. The specificity of $CD8^+$ T cells relies on the surface expression of different receptors and ligands, creating an immune checkpoint to limit aberrant killing by $CD8^+$ T cells. Programmed cell death 1 (PD-1) expression on $CD8^+$ T cells inhibits their activation upon binding of its ligand, PD-L1. Expression of PD-L1 in the tumor microenvironment promotes inhibition of $CD8^+$ T cell clearance. Consequently, strategies to block PD-L1 or its binding to $CD8^+$ T, ultimately aimed at reigniting immune clearance of tumor cells, have shown great therapeutic potential in several types of cancers [43]. Critically, the cGAS-STING axis appears to be very important in the capacity of PD-L1 blocking strategies to reactivate $CD8^+$ T cells, as evidenced by cGAS- and STING-deficient animal models [44], and in studies using co-administration of PD-L1 with STING ligands [45, 46].

Conversely, there has been evidence that STING activation can dampen immune activity through induction of the immune checkpoint indoleamine 2,3-dioxygenase

(IDO) [47–49]. Further, in a subtype of head and neck cancer where the tumor was in a repressed immune state, STING activation could not mediate recruitment of $CD8^+$ T cells [45].

With these lines of evidence taken together, the role of STING in immunotherapy is context dependent, and further studies are needed to better understand the signaling pathways initiating the different outcomes of its activity. How STING is activated within tumor cells is not fully understood either and is a subject for further study, although the role of STING in immune cells during tumor immunity and tumor clearance is well documented. In this review, we will focus on the role of cGAS and STING within tumor cells. We will also address the role of cell-cell communication in STING activation within the tumor microenvironment.

8.2 Mechanisms Underlying IFN Production in Immune Cells Within the Tumor Microenvironment

8.2.1 STING Activation by Tumor DNA

The current model of IFN production and $CD8^+$ T cell priming by tumor cells indicates a possible role for the activation of antigen-presenting cells (APCs) by DNA released from the tumor as a result of apoptosis and taken up by APC via phagocytosis. This is supported by the demonstration that DNA from the tumor could be observed in APCs where STING was activated and type I IFN produced [32, 31, 50]. In this model, sensing of apoptotic cell-derived nuclear DNA by DCs recruits the STING-IRF3 axis to promote type I IFN production, thereby enhancing the functionality of DCs in an autocrine loop. Activated DCs further mediate the activation and the clonal expansion of $CD8^+$ T cells favoring tumor clearance [31].

However appealing, this model does not account for how the tumor DNA would be released from the endosome/phagolysosome of APCs to reach the cytoplasm, where cGAS is located. In fact, previous work indicates that phagocytosis of apoptotic cells does not result in IFN-I activation, due to sequestration of DNA in the phagosome [51]. Similarly, despite tumor DNA being found in APC cells *in vivo*, dead tumor cells incubated with APCs did not engage an IFN response in vitro [32]. Another limit of the phagocytosis model is that it restricts the role of tumor cells in initiating an immune response to only dying/apoptotic cells.

There is also emerging evidence that the growth and spread of tumors, such as melanoma tumors, which display very little cell death, are regulated by STING. Indeed, mouse melanoma cells transplanted in mice lacking STING were found to develop significantly more lung metastases than WT mice [31]. Collectively, these examples indicate that the essential role of STING in tumor restriction is probably not limited to the activation of cGAS-STING in APCs upon phagocytosis of dead/apoptotic tumor cells.

8.2.2 Cytosolic DNA Activation of cGAS-STING

STING is indirectly activated by cytosolic DNA, through upstream cGAS engagement. Following cytoplasmic DNA recognition, cGAS produces cGAMP which acts as a second messenger between cGAS and STING. Critically, once synthesized, cGAMP has the capacity to be transferred between cells and mediate STING activation in adjacent cells [52], provided they form gap junctions with the cell making cGAMP. In homeostasis, physical restriction of DNA in the nucleus or phagosome (when the DNA is phagocytosed) ensures that self-DNA is not accessible to cytosolic cGAS. However, it is now apparent that self-DNA can escape from its original localization and activate the cGAS-STING pathway in certain contexts. To prevent aberrant cGAS recruitment, several key nucleases are at play to degrade DNA molecules escaping nuclear retention. Accordingly, defects in such DNases—e.g., TREX1 [53], DNase2a [54], or the nuclease-like SamHD1 [55]—lead to cytoplasmic DNA accumulation and cGAS-STING activation by self-DNA. In addition, certain chemical modifications of DNA, such as 8-OHG, protect DNA from DNase degradation (such as by TREX1) thereby favoring cGAS activation [56, 25]. Decreases in genome stability or faults in DNA repair pathways can also lead to STING activation. Loss of ATM, which results in an impaired DNA damage response, increases genomic instability and causes spontaneous type I IFN production through STING activation [57]. In line with this, a lack of RPA and/or RAD51, which normally bind to damaged DNA and sequester it in the nucleus, results in cGAS detection of leaked cytoplasmic DNA [58]. Therefore, the cGAS-STING pathway is indirectly involved in the response activated upon DNA damage.

Genome stability is directly dependent on the capacity of the cell to stop and repair its DNA. During each cell division, multiple mutations are made but are also constantly being repaired. Oncogenes and tumor suppressors regulate cell cycle and proliferation. In cancer, these genes are often mutated to cause sustained proliferation. This constant pressure to proliferate leads to genomic instability by forcing the cells to divide even in the presence of damaged DNA [59]. Accordingly, cytosolic DNA can be observed in B cell lymphomas, but not in normal B cells [60]. Despite being predominantly produced by infiltrating immune cells, some growing tumors can also produce type I IFNs [50]. Such type I IFN production can directly relate to cGAS-STING activation as recently suggested using a model of breast cancer with genomic instability [61]. Although a detailed understanding about how nuclear DNA can be leaked to the cytoplasm remains elusive, there is data to suggest that such leakage results from an active process. Overexpression of RNaseH1 (that degrades the R-loops of DNA-RNA hybrids) reduces the level of cytosolic DNA and type I IFN production and hampers the rejection of the lymphoma tumors [62]. Conversely, in prostate cancer cells, the MUS81 endonuclease cleaves DNA at stalled replication forks, promoting export of DNA products into the cytoplasm [63]. This results in type I IFN production through STING activation by the cancer cells, which consequently enhances the rejection of the tumor *in vivo* [63]. All together, these data strongly suggest that the cGAS-STING can be activated in select tumor cells, independently of APC phagocytosis.

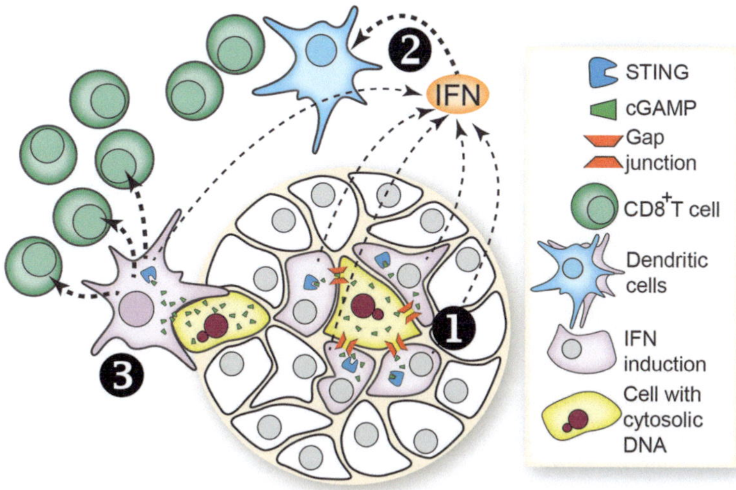

Fig. 8.1 Engagement of cGAS in tumor cells can amplify type I IFN production
Genomic instability, DNA damage, and accelerated cell proliferation can mediate leakage of DNA into the cytoplasm in tumor cells. The released DNA can then be detected by cytosolic cGAS to result in cGAMP production. cGAMP acts as a second messenger, which can directly activate STING in the cGAS engaged cells (if present—in this schematic the cells with cytosolic DNA do not express STING). Independent of intracellular STING activation, cGAMP can transfer to adjacent cells through gap junctions and activate STING in these recipient cells (*1*). Such cells, although not exhibiting cytosolic DNA, respond to cGAMP through STING engagement and produce type I IFN. This amplification of type I IFN production and associated cytokines by adjacent cells promotes the recruitment and activation of dendritic cells (*2*). Recruited dendritic cells scan tumor cells and can be further activated by cGAMP transfer through immune synapse or phagocytosis (*3*). Ultimately, activation of dendritic cells results in $CD8^+$ T activation and tumor immunity

8.2.3 cGAS-STING Activation in Tumor Cells

To date, most reports addressing the *in vivo* role of the cGAS-STING pathway in the tumor environment have relied on the use of $Sting^{-/-}$ tumor-bearing mice compared to WT mice [33, 32, 35]. These models, although supporting a critical role for STING in the tumor microenvironment, are unable to define the source of STING activation in the tumor itself. Given that about 50% of tumor cell lines express cGAS (e.g., 5/11 colorectal cancer cell lines [64] and 7/11 melanoma cell lines [65]), cGAMP production by tumors is likely to be a frequent occurrence, assuming they also have defective DNA repair capacity. Whether such tumor-derived cGAMP activates type I IFN production by cancer cells is dependent on the presence of STING [64, 65], but this cGAMP can also modify the tumor microenvironment through horizontal transfer of the second messenger to adjacent cells [52].

With this in mind, at least two different scenarios involving cGAMP expression in tumor cells can be envisioned as illustrated in Fig. 8.1. First, tumor-derived cGAMP production could lead to type I IFN production by the tumor cells or STING-competent adjacent tumor cells and mediate the subsequent recruitment of immune cells. Alternatively, tumor-derived cGAMP could directly engage STING in immune cells during phagocytosis or during immune synapse formation. These scenarios are not mutually exclusive and could possibly happen simultaneously; they place the focus on tumor-derived cGAMP, rather than immune-derived cGAMP, and illustrate how cGAMP transfer to immune cells better equipped for type I IFN production could be used to amplify the local detection of cytoplasmic DNA, indicative of aberrant cellular replication in this case.

In support of the origin of non-hematopoietic-derived cGAMP, it should be noted that *Trex1*$^{-/-}$ mice initially produce type I IFN in their non-hematopoietic compartment. This primary induction mediates recruitment of inflammatory cells that produce even more type–I IFN [66]. Importantly, cGAMP may also potentiate local type I IFN production through nonimmune adjacent tissues. As such, in an *in vivo* model of melanoma engraftment, cGAMP passage from the tumor to the vasculature may explain the observation that the primary source of IFN was endothelial cells [33].

It is tempting to speculate that cGAS activation in tumor cells could explain why some patients develop spontaneous leukocyte infiltration and antitumor T cell responses. Such spontaneous T cell tumor infiltration has, for instance, been reported in melanoma [67] and ovarian [68], breast [69, 61], and colorectal cancers [70]. Whether such infiltration depends on cGAS activation in the tumor and how such activation would take place are not currently defined. Future studies investigating the correlation between tumor-derived cGAMP and leukocyte infiltration may help refine disease prognosis—leukocyte infiltration is already a powerful prognostic factor in colorectal cancer patients [71]—while being informative about the best therapeutic approach to be selected.

8.2.4 Loss of cGAS-STING Expression in Tumor Cells

Many bacteria and viruses have evolved to block innate immune pathways, thereby facilitating their intracellular survival. Similar selection pressure constantly operates on tumor cells, which attempt to evade clearance by the immune system. Type I IFN is crucial to immune cell recruitment into the tumor microenvironment, and STING is a key factor in such type I IFN production, as discussed previously. One could speculate that selective pressure operates to block the cGAS-STING pathway in tumor cells to facilitate immune evasion. Accordingly, cGAS-STING inhibition has been recently described in human cancers.

The first extensive study on the loss of the cGAS-STING pathway comes from the field of colorectal adenocarcinoma (CA). In that research, cGAS-STING activity was decreased in the vast majority of CA cell lines, which lacked the capacity to produce type-I IFN in response to cytoplasmic DNA [64]. In some cell lines such as HT29 cells, the pathway was altered but still functional [64]. In support of this in vitro data, one third of 48 clinical samples of adenocarcinoma analyzed showed a loss of cGAS expression [64]. Interestingly, upregulation of cGAS expression in the early stages of cancer and disruption of the STING pathway in advanced stages were reported in a similar study [72]. These data indicate that engagement of the pathway during early stages of tumor development, but inhibition of the pathway later in disease, favored tumor growth. Similarly to CA, melanoma cells displayed recurrent loss of cGAS-STING expression, ultimately inhibiting type I IFN production [65]. Epigenetic repression was found to relate to the cGAS and STING inhibition proposed in these studies [64, 65], but other factors are most likely at play. In another study, ovarian cancer cells (serous, clear cell, and endometriosis) had lost responsiveness to DNA transfection via STING-IRF3 activation [73]. These reports collectively suggest that loss of expression or activity of the cGAS-STING pathway may favor the development of the tumor cells.

Therapeutically, such a loss of the cGAS-STING response can be harnessed to obtain clinical benefit. Indeed, cGAS-STING-deficient tumors are more susceptible to viral oncolysis—where modified DNA viruses like HSV-1 have been used to target and kill cancer cells [65, 64].

8.2.5 cGAS-STING in Pathogen-Driven Carcinogenesis

Whether the cGAS-STING pathway is involved in pathogen-driven carcinogenesis is a question that remains elusive. Human papillomavirus (HPV) is the causative agent of cervical cancer and other types of cancer. HPV is a DNA virus, making it a potential target of cGAS sensing [74]. Accordingly, and although not detailed in the current literature, the cGAS-STING pathway may play a role in HPV infection and cancer development. In line with this hypothesis, a single nucleotide polymorphism (SNP) in cGAS (rs311678) has been recently associated with a reduced risk of cervical precancerous lesions. This SNP was found to modulate cGAS expression in vitro, and higher cGAS was associated with a reduced risk of HPV infection [75]. As described for other viruses, HPV can also counteract the activity of the innate immune system. In vitro expression of the HPV E2 protein in human primary keratinocytes downregulates the expression of STING and several other innate immune genes [76]. In addition, STING repression has been shown in HPV^+ low-grade squamous intraepithelial lesions, when compared with HPV^- controls. Furthermore, viral oncogenes such as HPV E7 and E1A of human adenovirus A5 were proposed to bind to the N-terminal region of STING to reduce its downstream signaling [77].

In line with a role for the cGAS-STING pathway in reducing the risk of early stages of pathogen-driven cancer, STING expression is significantly decreased in gastric cancers when compared to non-tumor tissues [78]. *Helicobacter pylori*, the main causative agent of gastric cancer, can activate STING and promote inflammation [78]. These data collectively suggest a role for cGAS and STING in pathogen-mediated carcinogenesis.

8.3 Connexin Expression in Tumor Cells and Its Impact on Tumor Development

Beyond the inhibition of tumor-derived type I IFN production by cGAS, one could argue that loss of cGAS expression by a significant number of tumor cells is also important for stopping the propagation of cGAMP within the tumor microenvironment. As mentioned previously, cGAMP transfers horizontally through gap junctions to activate STING in adjacent cells [52]. With this in mind, it is tempting to revisit previous works on the role of gap junctions in tumorigenesis.

8.3.1 Connexins and Gap Junctions

Gap junctions are formed through the interaction of connexins from both interacting cells to promote the intercellular circulation of ions and small molecules such as cGAMP. Among the family of connexins, connexin (CX) 43 and CX45 were found to be essential in human embryonic kidney 293T (HEK293T) cells for cGAMP horizontal transfer [52]. Critically, transfer of cGAMP to HEK293T CX43/45-deficient cells could be restored through the expression of human CX26, CX31, CX32, CX40, CX43, and CX62 and mouse CX43 and CX45, suggesting that most connexins are able to transfer cGAMP (noting that human CX50 overexpression did not restore transfer) [52].

8.3.2 Loss of Intercellular Cell Communication in Early-Stage Tumors

Loss of intercellular communication by cancer cells was first described over half a century ago [79]. Since their discovery, connexins have been shown to exert both pro- and antitumoral activities, making it a controversial field of research [80]. The overall view is that retention of connexin expression benefits antitumoral activities at early stages but can later favor metastasis. Arguably, loss of connexin expression early in tumor development would be expected to reduce cGAMP horizontal transfer to adjacent cells and inhibit type I IFN induction—thereby favoring tumor initiation and immune evasion. Accordingly, CX43 expression is decreased in prostate cancer patients compared to controls. In line with what is observed with the loss of cGAS, the reduction of CX43 expression was found to correlate with advanced stages of cancer [81]. Similar trends were observed in breast cancers and head and neck squamous cell carcinomas (HNSCC), where low expression of CX43 correlated with a negative prognosis [82]. Given that HPV is also often detected in HNSCC tumors [83], it is of interest to note that HPV-E6 protein expression was associated with a reduction of gap junction formation [84].

Further supporting an antitumoral effect of connexins in early tumor growth, overexpression of CX43 was found to reduce melanoma tumor growth in vivo [85], while its downregulation stimulated the growth of prostate cancer cells [86]. Re-expression of connexins in breast cancer cell lines implanted *in vivo* also reduced tumor growth [87]. Qigesan, a molecule that increases the expression of connexins, reduced cell migration and invasion in esophageal cancer cells [88]. Critically, in a model of chemically induced mammary tumors, enforced expression of connexin decreased the incidence of tumor formation, while no difference was observed on the tumor growth [89]. These findings suggest that connexins may play a greater role in tumor initiation than tumor development [89]. It should, however, be noted that the latter tumor model relied on tumoral expression of the Cre recombinase—which we linked recently to cGAS-STING engagement [90]—opening the possibility for a direct role for the cGAS-STING pathway in the initiation of tumors in this model.

8.3.3 *The Role of Connexin in Metastasis*

Conversely, there are instances where increased expression of connexins was found to enhance metastasis. Indeed, in a mouse melanoma model, the metastatic capacity of the cancer cells was found to be dependent on CX26 expression [91]. Furthermore, CX43 expression is induced in CA cell lines that display greater metastatic potential, while CX43 levels are almost absent in other tumor cell lines [92].

Critically, direct evidence for the role of cGAMP and gap junction in metastasis was recently reported. When breast and lung cancer metastatic cells migrated to the brain, they increased connexin expression, allowing passage of tumor cell-derived cGAMP to adjacent astrocytes [93]. cGAMP transfer to astrocytes promoted astrocyte activation and the subsequent secretion of pro-inflammatory cytokines favoring tumor growth and chemotherapeutic resistance [93].

8.4 cGAS-cGAMP: Connexins-STING in Chemotherapy

The role played by intercellular cell communication during chemotherapy is also controversial: there are reports of both positive and negative roles for connexins on the outcome of chemotherapy. For example, increased expression of connexins enhanced the sensitivity of RKO colon cancer cells to diverse chemotherapeutic agents such as fluorouracil, oxaliplatin, and irinotecan [94]. Critically, increased sensitivity was observed in vitro when the cells were more confluent or when they were treated with retinoic acid, which induces the expression of connexins [94]. In addition, low expression of connexins correlated with reduced sensitivity of hepatocarcinoma cells to oxaliplatin [95]. On the other hand, inhibition of connexins was reported to sensitize glioblastoma cells previously shown to be resistant to chemotherapy [96].

Chemotherapy, which mostly relies on the greater sensitivity of tumor cells to DNA damage, often induces type–I IFN production [97]. We recently discovered that the DNA intercalating agent acriflavine could promote cGAMP synthesis in SV40T immortalized mouse embryonic fibroblasts (MEFs), in association with increased cytoplasmic DNA levels [21]. Despite previous evidence of cGAS and STING activation following DNA damage and cytoplasmic DNA leakage [56, 57, 98], there had been no prior demonstration of the direct engagement of cGAS and cGAMP production in these contexts. Critically, the capacity of the topoisomerase I inhibitor, topotecan (TPT), to restrict breast cancer cell proliferation *in vivo* was abrogated in mice lacking STING [99]. This, however, does not tease out whether cGAMP was generated by the tumor after chemotherapy treatment or by phagocytes. Similar results were found using irradiation of tumors [100]. Interestingly, single-stranded DNA (ssDNA) leaking in the cytoplasm of the tumor cells following chemotherapy was shown to be a by-product of BLOOM syndrome helicase (BLM) and the exonuclease-RNase EXO1 [101]. Along with the discovery that RPA and RAD51 normally work to retain ssDNA generated by BLM and EXO1 during DNA repair [58], these findings suggest that cytoplasmic DNA leakage is likely related to a saturation of the cell's capacity to retain it in the nucleus [101]. Although ssDNA is not supposed to activate cGAS, it can bind to it weakly, and it is likely that its modification upon DNA damage (such as by 8-OHG) somewhat favors cGAS activation [56, 101]. Further work is clearly warranted to better define the modalities of cytoplasmic DNA leakage and cGAS engagement upon DNA damage by chemotherapy.

8.5 DNA Damage Engages cGAS Activity and Horizontal STING Amplification

We have previously demonstrated that DNA damage can mediate cGAMP production [90, 21]. When using inducible Cre recombinase-mediated DNA damage, we observed that only a proportion of cells displayed the hallmarks of DNA damage (through γH2AX staining) [90]. Critically, we demonstrated that the capacity of damaged cells to generate a widespread type I IFN production in the cell monolayer was strongly dependent on cell density—which we attribute to a connexin-dependent transfer of cGAMP [90]. This illustrates the capacity of healthy adjacent cells to amplify the signal of selected damaged cells and suggests that a similar feedback loop could operate in the tumor microenvironment upon induction of DNA damage by chemotherapy. Surprisingly, cGAS-depleted bone marrow-derived DCs can be activated after co-incubation with irradiated tumor cells, albeit modestly, while STING depletion completely thwarted the effect. This residual activation in cGAS-deficient cells suggests that engagement of cGAS in phagocytes is not essential for

STING activation by irradiated tumor cells [100]. From this point of view, the capacity of tumors cells to generate cGAMP and transfer it to adjacent cells may play an important role in the outcome of chemotherapy.

8.6 Conclusion

In summary, the cGAS-STING axis is crucial to cancer immunity. There is accumulating evidence for a role for STING in tumor DCs and cross priming of $CD8^+$ T cells. These findings clearly suggest that STING ligands have strong therapeutic potential. Nonetheless, therapeutic STING activation may also contribute to the tumor expressing interferon-stimulated genes previously linked with chemoresistance [97, 101, 102]. As such, critical questions regarding the modalities of activation of the pathway in the tumor microenvironment and its impact on chemotherapies are still to be answered. For instance, while DCs have a central role in mediating the recruitment of an antitumor immune response through STING, how other cell types like macrophages and neutrophils contribute to this pathway should also be addressed. Critically, the source of cGAMP activating STING in APCs and its putative modalities of transfer to APCs are not defined. In light of the current literature reviewed herein and our own experiments, we propose a model, illustrated in Fig. 8.1, in which cGAMP can be synthesized by tumor cells to play a role in immune cell activation through immune gap junctions. While this is supported by the demonstration that metastatic breast cancer cells could transfer cGAMP to astrocytes in the brain [93], direct evidence of cGAMP production by tumor cells and its transfer to APCs remains to be found. Defining if and how cGAMP can be made by tumors has the potential to help understand why select tumors are devoid of infiltrating immune cells. Given that the efficacy of preferred chemotherapies has been associated with STING signaling *in vivo* (e.g., irinotecan), these may be particularly effective when used in patients with active cGAS and functional connexin tumors.

Acknowledgment We thank Rebecca Smith, Jonathan Ferrand, and Lise Boursinhac for the assistance in the preparation of the manuscript. This work was funded in part by the Australian NHMRC (1062683 and 1081167 to M.P.G.), the Australian Research Council (140100594 Future Fellowship to M.P.G.), and the Victorian Government's Operational Infrastructure Support Program. G.P. is a fellow from Fonds de Recherche du Québec – Santé (*FRQS*), Canada (35071).

References

1. He S, Mao X, Sun H, Shirakawa T, Zhang H, Wang X (2015) Potential therapeutic targets in the process of nucleic acid recognition: opportunities and challenges. Trends Pharmacol Sci 36(1):51–64. doi:10.1016/j.tips.2014.10.013

2. Luecke S, Paludan SR (2016) Molecular requirements for sensing of intracellular microbial nucleic acids by the innate immune system. Cytokine. doi:10.1016/j.cyto.2016.10.003
3. Rusinova I, Forster S, Yu S, Kannan A, Masse M, Cumming H, Chapman R, Hertzog PJ (2013) Interferome v2.0: an updated database of annotated interferon-regulated genes. Nucleic Acids Res 41(Database issue):D1040–D1046. doi:10.1093/nar/gks1215
4. Ishikawa H, Barber GN (2008) STING is an endoplasmic reticulum adaptor that facilitates innate immune signalling. Nature 455(7213):674–678. doi:10.1038/nature07317
5. Sun W, Li Y, Chen L, Chen H, You F, Zhou X, Zhou Y, Zhai Z, Chen D, Jiang Z (2009) ERIS, an endoplasmic reticulum IFN stimulator, activates innate immune signaling through dimerization. Proc Natl Acad Sci U S A 106(21):8653–8658. doi:10.1073/pnas.0900850106
6. Zhong B, Yang Y, Li S, Wang YY, Li Y, Diao F, Lei C, He X, Zhang L, Tien P, Shu HB (2008) The adaptor protein MITA links virus-sensing receptors to IRF3 transcription factor activation. Immunity 29(4):538–550. doi:10.1016/j.immuni.2008.09.003
7. Jin L, Waterman PM, Jonscher KR, Short CM, Reisdorph NA, Cambier JC (2008) MPYS, a novel membrane tetraspanner, is associated with major histocompatibility complex class II and mediates transduction of apoptotic signals. Mol Cell Biol 28(16):5014–5026. doi:10.1128/MCB.00640-08
8. Tanaka Y, Chen ZJ (2012) STING specifies IRF3 phosphorylation by TBK1 in the cytosolic DNA signaling pathway. Sci Signal 5(214):ra20. doi:10.1126/scisignal.2002521
9. Liu S, Cai X, Wu J, Cong Q, Chen X, Li T, Du F, Ren J, Wu YT, Grishin NV, Chen ZJ (2015) Phosphorylation of innate immune adaptor proteins MAVS, STING, and TRIF induces IRF3 activation. Science 347(6227):aaa2630. doi:10.1126/science.aaa2630
10. Dey B, Dey RJ, Cheung LS, Pokkali S, Guo H, Lee JH, Bishai WR (2015) A bacterial cyclic dinucleotide activates the cytosolic surveillance pathway and mediates innate resistance to tuberculosis. Nat Med 21(4):401–406. doi:10.1038/nm.3813
11. Woodward JJ, Iavarone AT, Portnoy DA (2010) c-di-AMP secreted by intracellular Listeria monocytogenes activates a host type I interferon response. Science 328(5986):1703–1705. doi:10.1126/science.1189801
12. Ishikawa H, Ma Z, Barber GN (2009) STING regulates intracellular DNA-mediated, type I interferon-dependent innate immunity. Nature 461(7265):788–792. doi:10.1038/nature08476
13. Bhat N, Fitzgerald KA (2014) Recognition of cytosolic DNA by cGAS and other STING-dependent sensors. Eur J Immunol 44(3):634–640. doi:10.1002/eji.201344127
14. Sun L, Wu J, Du F, Chen X, Chen ZJ (2013) Cyclic GMP-AMP synthase is a cytosolic DNA sensor that activates the type I interferon pathway. Science 339(6121):786–791. doi:10.1126/science.1232458
15. Wu J, Sun L, Chen X, Du F, Shi H, Chen C, Chen ZJ (2013) Cyclic GMP-AMP is an endogenous second messenger in innate immune signaling by cytosolic DNA. Science 339(6121):826–830. doi:10.1126/science.1229963
16. Gao P, Ascano M, Wu Y, Barchet W, Gaffney BL, Zillinger T, Serganov AA, Liu Y, Jones RA, Hartmann G, Tuschl T, Patel DJ (2013) Cyclic [G(2′,5′)pA(3′,5′)p] is the metazoan second messenger produced by DNA-activated cyclic GMP-AMP synthase. Cell 153(5):1094–1107. doi:10.1016/j.cell.2013.04.046
17. Ablasser A, Goldeck M, Cavlar T, Deimling T, Witte G, Rohl I, Hopfner KP, Ludwig J, Hornung V (2013) cGAS produces a 2′-5′-linked cyclic dinucleotide second messenger that activates STING. Nature 498(7454):380–384. doi:10.1038/nature12306
18. Diner EJ, Burdette DL, Wilson SC, Monroe KM, Kellenberger CA, Hyodo M, Hayakawa Y, Hammond MC, Vance RE (2013) The innate immune DNA sensor cGAS produces a noncanonical cyclic dinucleotide that activates human STING. Cell Rep 3(5):1355–1361. doi:10.1016/j.celrep.2013.05.009
19. Zhang X, Shi H, Wu J, Zhang X, Sun L, Chen C, Chen ZJ (2013) Cyclic GMP-AMP containing mixed phosphodiester linkages is an endogenous high-affinity ligand for STING. Mol Cell 51(2):226–235. doi:10.1016/j.molcel.2013.05.022

20. Mankan AK, Schmidt T, Chauhan D, Goldeck M, Honing K, Gaidt M, Kubarenko AV, Andreeva L, Hopfner KP, Hornung V (2014) Cytosolic RNA:DNA hybrids activate the cGAS-STING axis. EMBO J 33(24):2937–2946. doi:10.15252/embj.201488726
21. Pepin G, Nejad C, Thomas BJ, Ferrand J, McArthur K, Bardin PG, Williams BR, Gantier MP (2017) Activation of cGAS-dependent antiviral responses by DNA intercalating agents. Nucleic Acids Res 45(1):198–205. doi:10.1093/nar/gkw878
22. White MJ, McArthur K, Metcalf D, Lane RM, Cambier JC, Herold MJ, van Delft MF, Bedoui S, Lessene G, Ritchie ME, Huang DC, Kile BT (2014) Apoptotic caspases suppress mtDNA-induced STING-mediated type I IFN production. Cell 159(7):1549–1562. doi:10.1016/j.cell.2014.11.036
23. Konig N, Fiehn C, Wolf C, Schuster M, Cura Costa E, Tungler V, Alvarez HA, Chara O, Engel K, Goldbach-Mansky R, Gunther C, Lee-Kirsch MA (2017) Familial chilblain lupus due to a gain-of-function mutation in STING. Ann Rheum Dis 76(2):468–472. doi:10.1136/annrheumdis-2016-209841
24. Mackenzie KJ, Carroll P, Lettice L, Tarnauskaite Z, Reddy K, Dix F, Revuelta A, Abbondati E, Rigby RE, Rabe B, Kilanowski F, Grimes G, Fluteau A, Devenney PS, Hill RE, Reijns MA, Jackson AP (2016) Ribonuclease H2 mutations induce a cGAS/STING-dependent innate immune response. EMBO J 35(8):831–844. doi:10.15252/embj.201593339
25. Gray EE, Treuting PM, Woodward JJ, Stetson DB (2015) Cutting edge: cGAS Is required for lethal autoimmune disease in the trex1-deficient mouse model of aicardi-goutieres syndrome. J Immunol 195(5):1939–1943. doi:10.4049/jimmunol.1500969
26. Ahn J, Ruiz P, Barber GN (2014) Intrinsic self-DNA triggers inflammatory disease dependent on STING. J Immunol 193(9):4634–4642. doi:10.4049/jimmunol.1401337
27. Dunn GP, Bruce AT, Sheehan KC, Shankaran V, Uppaluri R, Bui JD, Diamond MS, Koebel CM, Arthur C, White JM, Schreiber RD (2005) A critical function for type I interferons in cancer immunoediting. Nat Immunol 6(7):722–729. doi:10.1038/ni1213
28. Swann JB, Hayakawa Y, Zerafa N, Sheehan KC, Scott B, Schreiber RD, Hertzog P, Smyth MJ (2007) Type I IFN contributes to NK cell homeostasis, activation, and antitumor function. J Immunol 178(12):7540–7549
29. Diamond MS, Kinder M, Matsushita H, Mashayekhi M, Dunn GP, Archambault JM, Lee H, Arthur CD, White JM, Kalinke U, Murphy KM, Schreiber RD (2011) Type I interferon is selectively required by dendritic cells for immune rejection of tumors. J Exp Med 208(10):1989–2003. doi:10.1084/jem.20101158
30. Fuertes MB, Kacha AK, Kline J, Woo SR, Kranz DM, Murphy KM, Gajewski TF (2011) Host type I IFN signals are required for antitumor CD8+ T cell responses through CD8{alpha}+ dendritic cells. J Exp Med 208(10):2005–2016. doi:10.1084/jem.20101159
31. Klarquist J, Hennies CM, Lehn MA, Reboulet RA, Feau S, Janssen EM (2014) STING-mediated DNA sensing promotes antitumor and autoimmune responses to dying cells. J Immunol 193(12):6124–6134. doi:10.4049/jimmunol.1401869
32. Woo SR, Fuertes MB, Corrales L, Spranger S, Furdyna MJ, Leung MY, Duggan R, Wang Y, Barber GN, Fitzgerald KA, Alegre ML, Gajewski TF (2014) STING-dependent cytosolic DNA sensing mediates innate immune recognition of immunogenic tumors. Immunity 41(5):830–842. doi:10.1016/j.immuni.2014.10.017
33. Demaria O, De Gassart A, Coso S, Gestermann N, Di Domizio J, Flatz L, Gaide O, Michielin O, Hwu P, Petrova TV, Martinon F, Modlin RL, Speiser DE, Gilliet M (2015) STING activation of tumor endothelial cells initiates spontaneous and therapeutic antitumor immunity. Proc Natl Acad Sci U S A 112(50):15408–15413. doi:10.1073/pnas.1512832112
34. Kang J, Demaria S, Formenti S (2016) Current clinical trials testing the combination of immunotherapy with radiotherapy. J Immunother Cancer 4:51. doi:10.1186/s40425-016-0156-7
35. Ohkuri T, Ghosh A, Kosaka A, Zhu J, Ikeura M, David M, Watkins SC, Sarkar SN, Okada H (2014) STING contributes to antiglioma immunity via triggering type I IFN signals in the tumor microenvironment. Cancer Immunol Res 2(12):1199–1208. doi:10.1158/2326-6066.CIR-14-0099

36. Tang CH, Zundell JA, Ranatunga S, Lin C, Nefedova Y, Del Valle JR, Hu CC (2016) Agonist-mediated activation of STING induces apoptosis in malignant B cells. Cancer Res 76(8):2137–2152. doi:10.1158/0008-5472.CAN-15-1885
37. Corrales L, Glickman LH, McWhirter SM, Kanne DB, Sivick KE, Katibah GE, Woo SR, Lemmens E, Banda T, Leong JJ, Metchette K, Dubensky TW Jr, Gajewski TF (2015) Direct activation of STING in the tumor microenvironment leads to potent and systemic tumor regression and immunity. Cell Rep 11(7):1018–1030. doi:10.1016/j.celrep.2015.04.031
38. Lara PN Jr, Douillard JY, Nakagawa K, von Pawel J, McKeage MJ, Albert I, Losonczy G, Reck M, Heo DS, Fan X, Fandi A, Scagliotti G (2011) Randomized phase III placebo-controlled trial of carboplatin and paclitaxel with or without the vascular disrupting agent vadimezan (ASA404) in advanced non-small-cell lung cancer. J Clin Oncol 29(22):2965–2971. doi:10.1200/JCO.2011.35.0660. Epub 2011 Jun 27.
39. Corrales L, Gajewski TF (2016) Endogenous and pharmacologic targeting of the STING pathway in cancer immunotherapy. Cytokine 77:245–247. doi:10.1016/j.cyto.2015.08.258
40. Baird JR, Friedman D, Cottam B, Dubensky TW Jr, Kanne DB, Bambina S, Bahjat K, Crittenden MR, Gough MJ (2016) Radiotherapy combined with novel STING-targeting oligonucleotides results in regression of established tumors. Cancer Res 76(1):50–61 https://doi.org/10.1158/0008-5472.CAN-14-3619
41. Li T, Cheng H, Yuan H, Xu Q, Shu C, Zhang Y, Xu P, Tan J, Rui Y, Li P, Tan X (2016) Antitumor activity of cGAMP via stimulation of cGAS-cGAMP-STING-IRF3 mediated innate immune response. Sci Rep 6:19049 https://doi.org/10.1038/srep19049
42. Nakamura T, Miyabe H, Hyodo M, Sato Y, Hayakawa Y, Harashima H (2015) Liposomes loaded with a STING pathway ligand, cyclic di-GMP, enhance cancer immunotherapy against metastatic melanoma. J Control Release 216:149–157. doi:10.1016/j.jconrel.2015.08.026
43. Topalian SL, Drake CG, Pardoll DM (2015) Immune checkpoint blockade: a common denominator approach to cancer therapy. Cancer Cell 27(4):450–461. doi:10.1016/j.ccell.2015.03.001
44. Wang H, Hu S, Chen X, Shi H, Chen C, Sun L, Chen ZJ (2017) cGAS is essential for the antitumor effect of immune checkpoint blockade. Proc Natl Acad Sci U S A. doi:10.1073/pnas.1621363114
45. Moore E, Clavijo PE, Davis R, Cash H, Van Waes C, Kim Y, Allen C (2016) Established T cell-inflamed tumors rejected after adaptive resistance was reversed by combination STING activation and PD-1 pathway blockade. Cancer Immunol Res 4(12):1061–1071. doi:10.1158/2326-6066.CIR-16-0104
46. Fu J, Kanne DB, Leong M, Glickman LH, McWhirter SM, Lemmens E, Mechette K, Leong JJ, Lauer P, Liu W, Sivick KE, Zeng Q, Soares KC, Zheng L, Portnoy DA, Woodward JJ, Pardoll DM, Dubensky TW Jr, Kim Y (2015) STING agonist formulated cancer vaccines can cure established tumors resistant to PD-1 blockade. Sci Transl Med 7(283):283ra252. doi:10.1126/scitranslmed.aaa4306
47. Lemos H, Mohamed E, Huang L, Ou R, Pacholczyk G, Arbab AS, Munn D, Mellor AL (2016) STING promotes the growth of tumors characterized by low antigenicity via IDO activation. Cancer Res 76(8):2076–2081. doi:10.1158/0008-5472.CAN-15-1456
48. Huang L, Li L, Lemos H, Chandler PR, Pacholczyk G, Baban B, Barber GN, Hayakawa Y, McGaha TL, Ravishankar B, Munn DH, Mellor AL (2013) Cutting edge: DNA sensing via the STING adaptor in myeloid dendritic cells induces potent tolerogenic responses. J Immunol 191(7):3509–3513. doi:10.4049/jimmunol.1301419
49. Lemos H, Huang L, Chandler PR, Mohamed E, Souza GR, Li L, Pacholczyk G, Barber GN, Hayakawa Y, Munn DH, Mellor AL (2014) Activation of the STING adaptor attenuates experimental autoimmune encephalitis. J Immunol 192(12):5571–5578. doi:10.4049/jimmunol.1303258
50. Andzinski L, Spanier J, Kasnitz N, Kroger A, Jin L, Brinkmann MM, Kalinke U, Weiss S, Jablonska J, Lienenklaus S (2016) Growing tumors induce a local STING dependent Type I IFN response in dendritic cells. Int J Cancer 139(6):1350–1357. doi:10.1002/ijc.30159

51. Stetson DB, Medzhitov R (2006) Recognition of cytosolic DNA activates an IRF3-dependent innate immune response. Immunity 24(1):93–103. doi:10.1016/j.immuni.2005.12.003
52. Ablasser A, Schmid-Burgk JL, Hemmerling I, Horvath GL, Schmidt T, Latz E, Hornung V (2013) Cell intrinsic immunity spreads to bystander cells via the intercellular transfer of cGAMP. Nature 503(7477):530–534. doi:10.1038/nature12640
53. Stetson DB, Ko JS, Heidmann T, Medzhitov R (2008) Trex1 prevents cell-intrinsic initiation of autoimmunity. Cell 134(4):587–598. doi:10.1016/j.cell.2008.06.032
54. Lan YY, Londono D, Bouley R, Rooney MS, Hacohen N (2014) Dnase2a deficiency uncovers lysosomal clearance of damaged nuclear DNA via autophagy. Cell Rep 9(1):180–192. doi:10.1016/j.celrep.2014.08.074
55. Kretschmer S, Wolf C, Konig N, Staroske W, Guck J, Hausler M, Luksch H, Nguyen LA, Kim B, Alexopoulou D, Dahl A, Rapp A, Cardoso MC, Shevchenko A, Lee-Kirsch MA (2015) SAMHD1 prevents autoimmunity by maintaining genome stability. Ann Rheum Dis 74(3):e17. doi:10.1136/annrheumdis-2013-204845
56. Gehrke N, Mertens C, Zillinger T, Wenzel J, Bald T, Zahn S, Tuting T, Hartmann G, Barchet W (2013) Oxidative damage of DNA confers resistance to cytosolic nuclease TREX1 degradation and potentiates STING-dependent immune sensing. Immunity 39(3):482–495. doi:10.1016/j.immuni.2013.08.004
57. Hartlova A, Erttmann SF, Raffi FA, Schmalz AM, Resch U, Anugula S, Lienenklaus S, Nilsson LM, Kroger A, Nilsson JA, Ek T, Weiss S, Gekara NO (2015) DNA damage primes the type I interferon system via the cytosolic DNA sensor STING to promote anti-microbial innate immunity. Immunity 42(2):332–343. doi:10.1016/j.immuni.2015.01.012
58. Wolf C, Rapp A, Berndt N, Staroske W, Schuster M, Dobrick-Mattheuer M, Kretschmer S, Konig N, Kurth T, Wieczorek D, Kast K, Cardoso MC, Gunther C, Lee-Kirsch MA (2016) RPA and Rad51 constitute a cell intrinsic mechanism to protect the cytosol from self DNA. Nat Commun 7:11752. doi:10.1038/ncomms11752
59. Hanahan D, Weinberg RA (2011) Hallmarks of cancer: the next generation. Cell 144(5):646–674. doi:10.1016/j.cell.2011.02.013
60. Lam AR, Le Bert N, Ho SS, Shen YJ, Tang ML, Xiong GM, Croxford JL, Koo CX, Ishii KJ, Akira S, Raulet DH, Gasser S (2014) RAE1 ligands for the NKG2D receptor are regulated by STING-dependent DNA sensor pathways in lymphoma. Cancer Res 74(8):2193–2203. doi:10.1158/0008-5472.CAN-13-1703
61. Parkes EE, Walker SM, Taggart LE, McCabe N, Knight LA, Wilkinson R, McCloskey KD, Buckley NE, Savage KI, Salto-Tellez M, McQuaid S, Harte MT, Mullan PB, Harkin DP, Kennedy RD (2017) Activation of STING-dependent innate immune signaling By S-phase-specific DNA damage in breast cancer. J Natl Cancer Inst 109(1). doi:10.1093/jnci/djw199
62. Shen YJ, Le Bert N, Chitre AA, Koo CX, Nga XH, Ho SS, Khatoo M, Tan NY, Ishii KJ, Gasser S (2015) Genome-derived cytosolic DNA mediates type I interferon-dependent rejection of B cell lymphoma cells. Cell Rep 11(3):460–473. doi:10.1016/j.celrep.2015.03.041
63. Ho SS, Zhang WY, Tan NY, Khatoo M, Suter MA, Tripathi S, Cheung FS, Lim WK, Tan PH, Ngeow J, Gasser S (2016) The DNA structure-specific endonuclease MUS81 mediates DNA sensor STING-dependent host rejection of prostate cancer cells. Immunity 44(5):1177–1189. doi:10.1016/j.immuni.2016.04.010
64. Xia T, Konno H, Ahn J, Barber GN (2016) Deregulation of STING signaling in colorectal carcinoma constrains DNA damage responses and correlates with tumorigenesis. Cell Rep 14(2):282–297. doi:10.1016/j.celrep.2015.12.029
65. Xia T, Konno H, Barber GN (2016) Recurrent loss of STING signaling in melanoma correlates with susceptibility to viral oncolysis. Cancer Res. doi:10.1158/0008-5472.CAN-16-1404
66. Gall A, Treuting P, Elkon KB, Loo YM, Gale M Jr, Barber GN, Stetson DB (2012) Autoimmunity initiates in nonhematopoietic cells and progresses via lymphocytes in an interferon-dependent autoimmune disease. Immunity 36(1):120–131. doi:10.1016/j.immuni.2011.11.018

67. Gajewski TF (2006) Identifying and overcoming immune resistance mechanisms in the melanoma tumor microenvironment. Clin Cancer Res 12(7 Pt 2):2326s–2330s. doi:10.1158/1078-0432.CCR-05-2517
68. Hwang WT, Adams SF, Tahirovic E, Hagemann IS, Coukos G (2012) Prognostic significance of tumor-infiltrating T cells in ovarian cancer: a meta-analysis. Gynecol Oncol 124(2):192–198. doi:10.1016/j.ygyno.2011.09.039
69. Mahmoud SM, Paish EC, Powe DG, Macmillan RD, Grainge MJ, Lee AH, Ellis IO, Green AR (2011) Tumor-infiltrating CD8+ lymphocytes predict clinical outcome in breast cancer. J Clin Oncol 29(15):1949–1955. doi:10.1200/JCO.2010.30.5037
70. Galon J, Costes A, Sanchez-Cabo F, Kirilovsky A, Mlecnik B, Lagorce-Pages C, Tosolini M, Camus M, Berger A, Wind P, Zinzindohoue F, Bruneval P, Cugnenc PH, Trajanoski Z, Fridman WH, Pages F (2006) Type, density, and location of immune cells within human colorectal tumors predict clinical outcome. Science 313(5795):1960–1964. doi:10.1126/science.1129139
71. Pages F, Kirilovsky A, Mlecnik B, Asslaber M, Tosolini M, Bindea G, Lagorce C, Wind P, Marliot F, Bruneval P, Zatloukal K, Trajanoski Z, Berger A, Fridman WH, Galon J (2009) In situ cytotoxic and memory T cells predict outcome in patients with early-stage colorectal cancer. J Clin Oncol 27(35):5944–5951. doi:10.1200/JCO.2008.19.6147
72. Yang CA, Huang HY, Chang YS, Lin CL, Lai IL, Chang JG (2017) DNA-sensing and nuclease gene expressions as markers for colorectal cancer progression. Oncology 92(2):115–124. doi:10.1159/000452281
73. Kodigepalli KM, Nanjundan M (2015) Induction of PLSCR1 in a STING/IRF3-dependent manner upon vector transfection in ovarian epithelial cells. PLoS One 10(2):e0117464. doi:10.1371/journal.pone.0117464
74. Small W Jr, Bacon MA, Bajaj A, Chuang LT, Fisher BJ, Harkenrider MM, Jhingran A, Kitchener HC, Mileshkin LR, Viswanathan AN, Gaffney DK (2017) Cervical cancer: a global health crisis. Cancer. doi:10.1002/cncr.30667
75. Xiao D, Huang W, Ou M, Guo C, Ye X, Liu Y, Wang M, Zhang B, Zhang N, Huang S, Zang J, Zhou Z, Wen Z, Zeng C, Wu C, Huang C, Wei X, Yang G, Jing C (2016) Interaction between susceptibility loci in cGAS-STING pathway, MHC gene and HPV infection on the risk of cervical precancerous lesions in Chinese population. Oncotarget. doi:10.18632/oncotarget.12399
76. Sunthamala N, Thierry F, Teissier S, Pientong C, Kongyingyoes B, Tangsiriwatthana T, Sangkomkamhang U, Ekalaksananan T (2014) E2 proteins of high risk human papillomaviruses down-modulate STING and IFN-kappa transcription in keratinocytes. PLoS One 9(3):e91473. doi:10.1371/journal.pone.0091473
77. Lau L, Gray EE, Brunette RL, Stetson DB (2015) DNA tumor virus oncogenes antagonize the cGAS-STING DNA-sensing pathway. Science 350(6260):568–571. doi:10.1126/science.aab3291
78. Song S, Peng P, Tang Z, Zhao J, Wu W, Li H, Shao M, Li L, Yang C, Duan F, Zhang M, Zhang J, Wu H, Li C, Wang X, Wang H, Ruan Y, Gu J (2017) Decreased expression of STING predicts poor prognosis in patients with gastric cancer. Sci Rep 7:39858. doi:10.1038/srep39858
79. Loewenstein WR, Kanno Y (1966) Intercellular communication and the control of tissue growth: lack of communication between cancer cells. Nature 209(5029):1248–1249
80. Aasen T, Mesnil M, Naus CC, Lampe PD, Laird DW (2016) Gap junctions and cancer: communicating for 50 years. Nat Rev Cancer 16(12):775–788. doi:10.1038/nrc.2016.105
81. Xu N, Chen HJ, Chen SH, Xue XY, Chen H, Zheng QS, Wei Y, Li XD, Huang JB, Cai H, Sun XL (2016) Reduced Connexin 43 expression is associated with tumor malignant behaviors and biochemical recurrence-free survival of prostate cancer. Oncotarget 7(41):67476–67484. doi:10.18632/oncotarget.11231
82. Danos K, Brauswetter D, Birtalan E, Pato A, Bencsik G, Krenacs T, Petak I, Tamas L (2016) The potential prognostic value of Connexin 43 expression in head and neck squamous cell carcinomas. Appl Immunohistochem Mol Morphol 24(7):476–481. doi:10.1097/PAI.0000000000000212

83. Ostrow RS, Manias DA, Fong WJ, Zachow KR, Faras AJ (1987) A survey of human cancers for human papillomavirus DNA by filter hybridization. Cancer 59(3):429–434
84. Sun P, Dong L, MacDonald AI, Akbari S, Edward M, Hodgins MB, Johnstone SR, Graham SV (2015) HPV16 E6 controls the gap junction protein Cx43 in cervical tumour cells. Virus 7(10):5243–5256. doi:10.3390/v7102871
85. Tittarelli A, Guerrero I, Tempio F, Gleisner MA, Avalos I, Sabanegh S, Ortiz C, Michea L, Lopez MN, Mendoza-Naranjo A, Salazar-Onfray F (2016) Overexpression of Connexin 43 reduces melanoma proliferative and metastatic capacity. Br J Cancer 115(9):e14. doi:10.1038/bjc.2016.296
86. Li X, Pan JH, Song B, Xiong EQ, Chen ZW, Zhou ZS, Su YP (2012) Suppression of CX43 expression by miR-20a in the progression of human prostate cancer. Cancer Biol Ther 13(10):890–898. doi:10.4161/cbt.20841
87. Qin H, Shao Q, Curtis H, Galipeau J, Belliveau DJ, Wang T, Alaoui-Jamali MA, Laird DW (2002) Retroviral delivery of Connexin genes to human breast tumor cells inhibits in vivo tumor growth by a mechanism that is independent of significant gap junctional intercellular communication. J Biol Chem 277(32):29132–29138. doi:10.1074/jbc.M200797200
88. Shi H, Shi D, Wu Y, Shen Q, Li J (2016) Qigesan inhibits migration and invasion of esophageal cancer cells via inducing connexin expression and enhancing gap junction function. Cancer Lett 380(1):184–190. doi:10.1016/j.canlet.2016.06.015
89. Stewart MK, Bechberger JF, Welch I, Naus CC, Laird DW (2015) Cx26 knockout predisposes the mammary gland to primary mammary tumors in a DMBA-induced mouse model of breast cancer. Oncotarget 6(35):37185–37199. doi:10.18632/oncotarget.5953
90. Pepin G, Ferrand J, Honing K, Jayasekara WS, Cain JE, Behlke MA, Gough DJ, Williams BRG, Hornung V, Gantier MP (2016) Cre-dependent DNA recombination activates a STING-dependent innate immune response. Nucleic Acids Res 44(11):5356–5364. doi:10.1093/nar/gkw405
91. Ito A, Katoh F, Kataoka TR, Okada M, Tsubota N, Asada H, Yoshikawa K, Maeda S, Kitamura Y, Yamasaki H, Nojima H (2000) A role for heterologous gap junctions between melanoma and endothelial cells in metastasis. J Clin Invest 105(9):1189–1197. doi:10.1172/JCI8257
92. Zhang A, Hitomi M, Bar-Shain N, Dalimov Z, Ellis L, Velpula KK, Fraizer GC, Gourdie RG, Lathia JD (2015) Connexin 43 expression is associated with increased malignancy in prostate cancer cell lines and functions to promote migration. Oncotarget 6(13):11640–11651. doi:10.18632/oncotarget.3449
93. Chen Q, Boire A, Jin X, Valiente M, Er EE, Lopez-Soto A, Jacob LS, Patwa R, Shah H, Xu K, Cross JR, Massague J (2016) Carcinoma-astrocyte gap junctions promote brain metastasis by cGAMP transfer. Nature 533(7604):493–498. doi:10.1038/nature18268
94. Zou ZW, Chen HJ, Yu JL, Huang ZH, Fang S, Lin XH (2016) Gap junction composed of connexin43 modulates 5fluorouracil, oxaliplatin and irinotecan resistance on colorectal cancers. Mol Med Rep 14(5):4893–4900. doi:10.3892/mmr.2016.5812
95. Yang Y, Zhu J, Zhang N, Zhao Y, Li WY, Zhao FY, Ou YR, Qin SK, Wu Q (2016) Impaired gap junctions in human hepatocellular carcinoma limit intrinsic oxaliplatin chemosensitivity: a key role of connexin 26. Int J Oncol 48(2):703–713. doi:10.3892/ijo.2015.3266
96. Murphy SF, Varghese RT, Lamouille S, Guo S, Pridham KJ, Kanabur P, Osimani AM, Sharma S, Jourdan J, Rodgers CM, Simonds GR, Gourdie RG, Sheng Z (2016) Connexin 43 inhibition sensitizes chemoresistant glioblastoma cells to temozolomide. Cancer Res 76(1):139–149. doi:10.1158/0008-5472.CAN-15-1286
97. Minn AJ (2015) Interferons and the immunogenic effects of cancer therapy. Trends Immunol 36(11):725–737. doi:10.1016/j.it.2015.09.007
98. Ahn J, Xia T, Konno H, Konno K, Ruiz P, Barber GN (2014) Inflammation-driven carcinogenesis is mediated through STING. Nat Commun 5:5166. doi:10.1038/ncomms6166
99. Kitai Y, Kawasaki T, Sueyoshi T, Kobiyama K, Ishii KJ, Zou J, Akira S, Matsuda T, Kawai T (2017) DNA-containing exosomes derived from cancer cells treated with topotecan activate a STING-dependent pathway and reinforce antitumor immunity. J Immunol 198(4):1649–1659. doi:10.4049/jimmunol.1601694

100. Deng L, Liang H, Xu M, Yang X, Burnette B, Arina A, Li XD, Mauceri H, Beckett M, Darga T, Huang X, Gajewski TF, Chen ZJ, Fu YX, Weichselbaum RR (2014) STING-dependent cytosolic DNA sensing promotes radiation-induced type I interferon-dependent antitumor immunity in immunogenic tumors. Immunity 41(5):843–852. doi:10.1016/j.immuni.2014.10.019
101. Erdal E, Haider S, Rehwinkel J, Harris AL, McHugh PJ (2017) A prosurvival DNA damage-induced cytoplasmic interferon response is mediated by end resection factors and is limited by Trex1. Genes Dev 31(4):353–369. doi:10.1101/gad.289769.116
102. Gaston J, Cheradame L, Yvonnet V, Deas O, Poupon MF, Judde JG, Cairo S, Goffin V (2016) Intracellular STING inactivation sensitizes breast cancer cells to genotoxic agents. Oncotarget. doi:10.18632/oncotarget.12858

Chapter 9
TLR Agonists as Adjuvants for Cancer Vaccines

Ji-Kun Li, Jesse J. Balic, Liang Yu, and Brendan Jenkins

Abstract Toll-like receptors (TLRs) are one of the best characterised families of pattern recognition receptors (PRRs) and play a critical role in the host defence to infection. Accumulating evidence indicates that TLRs also participate in maintaining tissue homeostasis by controlling inflammation and tissue repair, as well as promoting antitumour effects via activation and modulation of adaptive immune responses. TLR agonists have successfully been exploited to ameliorate the efficacy of various cancer therapies. In this chapter, we will discuss the rationales of using TLR agonists as adjuvants to cancer treatments and summarise the recent findings of preclinical and clinical studies of TLR agonist-based cancer therapies.

Keywords Toll-like receptor • Agonist • Adjuvant • Clinical trial

---

J.-K. Li
Department of General Surgery, Shanghai Jiao Tong University Affiliated First People's Hospital, Shanghai Jiao Tong University, Shanghai 200080, China

J.J. Balic • B. Jenkins
Centre for Innate Immunity and Infectious Diseases, Hudson Institute of Medical Research, 27-31 Wright Street, Clayton 3168, VIC, Australia

Department of Molecular Translational Science, Faculty of Medicine, Nursing and Health Sciences, Monash University, Clayton 3800, VIC, Australia

L. Yu (✉)
Department of General Surgery, Shanghai Jiao Tong University Affiliated First People's Hospital, Shanghai Jiao Tong University, Shanghai 200080, China

Centre for Innate Immunity and Infectious Diseases, Hudson Institute of Medical Research, 27-31 Wright Street, Clayton 3168, VIC, Australia

Department of Molecular Translational Science, Faculty of Medicine, Nursing and Health Sciences, Monash University, Clayton 3800, VIC, Australia
e-mail: Liang.yu@hudson.org.au

D. Xu (ed.), *Regulation of Inflammatory Signaling in Health and Disease*,
Advances in Experimental Medicine and Biology 1024,
DOI 10.1007/978-981-10-5987-2_9

9.1 Toll-Like Receptors

TLRs are a class of single membrane-spanning, catalytically inactive receptors. Ten human and 13 mouse TLRs have been classified to date [1, 2]. This protein family is best known for their ability to detect conserved microbial components, so-called PAMPs (pathogen-associated molecular patterns) [3]. The well-characterised TLR microbial ligands include: bacterial lipopolysaccharide (LPS) and its derivatives which activate TLR4; lipoteichoic acid and lipoprotein from bacterial cell wall and fugal zymosan which stimulate TLR1, TLR2 and TLR6; and bacterial flagellin which is sensed by TLR5. Additionally, unmethylated bacterial DNA stimulates TLR9; double-stranded RNA activates TLR3; and single-stranded RNA (ssRNA) is recognised by both TLR7 and TLR8. The cellular localisation of TLRs largely reflects their function and mode of ligand interaction. For example, the TLRs that recognise viral and bacterial nucleic acids such as TLR3, TLR7, TLR8 and TLR9 are mainly localised endosomally, while TLRs on the cell surface such as TLR1, TLR2, TLR4, TLR5, TLR6 and TLR10 are involved in detection of bacterial components in the extracellular space. Several TLRs, particularly TLR2 and TLR4, have been shown to detect not only exogenous PAMPs but also host-derived endogenous "damage-associated molecular patterns" (DAMPs) [4]. Many such ligands are increasingly being identified and include heat-shock proteins, high mobility group box 1 (HMGB1), various metabolic products such as reactive oxygen species (ROS) and uric acid as well as extracellular matrix components such as fibronectin and hyaluronan fragments [4].

Early studies suggested that TLRs are preferentially expressed on innate immune cells where the types of TLRs and level of expression are governed by cell-type specificity and function, which is associated with specific cytokine production [5–8]. More recent data demonstrate that TLRs are also expressed on epithelial cells of the gastrointestinal, urogenital and respiratory tracts where they play important roles in the first-line defence against infection. Additionally, they may also function to preserve epithelial barrier integrity [9–11]. Ligand binding of TLRs induces dimerisation and conformational change which activates two major signalling cascades – the MyD88-dependent or MyD88-independent pathways. Ultimately, this results in the activation and transduction of numerous downstream pathways including nuclear factor kappa B (NF-κB), mitogen-activated protein kinases (MAPKs) and interferon regulatory factors (IRFs) to induce the upregulation of type I interferons (IFNs) and various pro-inflammatory cytokines (e.g. type I IFN, IL-1, IL-6, TNFα, etc.) and chemokines (e.g. MCP-1, CCL and CXCL chemokines). Given the canonical outcomes of TLR activation, the past decade has seen a considerable effort in ways to engage or modulate TLR signalling as potential therapeutic targets capable of enhancing antitumour effect by orchestrating innate responses and activation of the adaptive immune system [12, 13]. The basic immunology of TLR recognition of ligands and mechanisms of signal transduction has been extensively reviewed elsewhere. Rather, this chapter focuses on TLR agonists as adjuvants for cancer immunotherapy in preclinical and clinical studies.

9.2 Cancer Vaccine and Adjuvant

Cancer vaccines, being prophylactic or therapeutic, aim to stimulate or restore the immune system's capacity to protect against persistent infection that initiate and drive oncogenesis. Prophylactic vaccinations against human papillomavirus-induced genital cancers with Gardasil® or Cervarix® vaccines or hepatitis B-induced hepatic cancers with Engerix®-B or Recombivax HB® vaccines have been used worldwide and play an important role in public health [14, 15]. Despite several decades of intense research and clinical evaluation, only one therapeutic vaccine (Provenge®) has been approved by FDA for the treatment cancer – androgen-independent metastatic prostate cancer [16]. In addition to drug safety profiles, there remain a few challenges that impact on vaccine efficacy that must be overcome. For example, the major drawbacks for currently used purified tumour-associated antigens (TAA) such as DNA/RNA, recombinant protein and peptides have poor immunogenicity and may cause inappropriate immune response that elicit no benefit or protection against the targeted infection or malignancy [17–19]. To overcome this and to be more effective, adjuvants are often co-administrated within the vaccine. In most cases, adjuvants were designed to augment the magnitude of an adaptive response to the vaccine administered.

Based on their perceived mechanism of action, adjuvants in the clinical settings can be divided into two main classes: immunostimulators and delivery system adjuvants [18]. Many natural and synthetic agents can be used as adjuvants such as emulsions of liposomes or PAMPs [18]. A carefully designed formulation of an adjuvant-vaccine combination is required for directing appropriate types of responses and for achieving synergism, which is otherwise difficult to achieve with a single adjuvant. For example, Cervarix, the prophylactic cancer vaccine against various types of human papillomavirus (HPV), contains purified virus-like particles (VLPs) of the major capsid (L1) protein of HPV types 6, 11, 16 and 18. Purified VLPs are then absorbed on aluminium hydroxy sulphate particles which act as the first adjuvant, the delivery system as aluminium typically induces Th2 immune profiling. Monophosphoryl lipid A (MPL), the TLR4 agonist, is then added as the second adjuvant which broadens the immune responses [20]. Substantial evidence showed that using natural ligands or synthetic agonists for PRRs as adjuvants can activate multiple elements of innate immunity. A number of TLR agonists are now in clinical or preclinical studies either stand-alone or with many different combinations for improving therapeutic cancer vaccines. So far, three TLR agonists have been licensed by the FDA for use in human cancers, including BCG (the bacillus Calmette-Guerin), a mixed TLR2/TLR4 agonist originally developed as an anti-tuberculosis vaccine but is currently approved for in situ bladder cancer vaccine; MPL (monophosphoryl lipid A), also a mixed TLR2/TLR4 agonist derived from LPS that is used as an immunostimulatory adjuvant as part of Cervarix; and, finally, imiquimod, an imidazoquinoline derivative that can function via TLR7-dependent or TLR7-independent mechanisms [21] that have been tested in many human malignancies (Table 9.1).

Table 9.1 The representative TLR agonists used in clinical trials for cancer therapy

Agent	TLR	Compound source	Indication	Notes	References
852A	TLR7	A synthetic imidazoquinoline	Leukaemia, Lymphoma, Melanoma	Sustained tolerable using prolonged s.c schedule [22]	NCT00276159, NCT00319748, NCT00091689
Ampligen (rintatolimod)	TLR3	poly-IC12U	Breast Ca, Ovarian Ca, Colorectal Ca	Th1 response [23]	NCT01355393, NCT01312389, NCT01545141
BCG	TLR2,4	*Mycobacterium bovis*	Bladder Ca	Intravesical use, FDA approved [24]	NCT02365207, NCT02015104
Cadi-05	polyTLR agonist	*Mycobacterium indicus pranii*	Melanoma	IFNγ-dependent antitumour effect [25]	NCT00675727
CBLB502 (entolimod)	TLR5	*Salmonella enterica flagellin*	Colorectal Ca, SCHNC	Radioprotective and organ-specific immunoadjuvants [26]	NCT02715882, NCT01728480
CpG7909 (Promune)	TLR9	Class B unmethylated CpG oligonucleotides	Renal cell Ca, NSCLC, Breast Ca	Stat1-dependent cell death [25]	NCT00043407, NCT00070629, NCT00043394
DIMS0150 (Kappaproct)	TLR9	ssDNA-based synthetic ODN	UC	Enhance the steroid sensitivity [27]	NCT01493960
Hiltonol	TLR3 MDA5	poly-ICLC	Non-melanoma Skin cancer, Glioma, Lymphoma	Stable and heat tolerant [19]	NCT02423863, NCT01188096, NCT01976585
Imiquimod	TLR7	Imidazoquinoline	Basal cell Ca, CIN	FDA approved [17]	NCT01264731, NCT02917746, NCT02917746

IMM-101	polyTLR agonist	*Mycobacterium obuense*	Melanoma, Unresectable Ca, Pancreas Ca	Restore Th1 and downregulate Th2 [28]	NCT01559818, NCT03009058, NCT01303172
IMO-2055 (EMD1201081)	TLR9	Phosphorothioate oligodeoxynucleotide	Renal cell Ca, NSCLC	Impair EGFR signalling [29]	NCT00729053, NCT00633529
ISS 1018	TLR9	Phosphorothioate oligodeoxynucleotide	Lymphoma	Synergic effects with rituximab [30]	NCT00251394
OM-174 (CRX-527)	TLR2,4	*Escherichia coli*	Solid tumours	Well tolerated in phase I [31]	NCT01800812
Picibanil (OK-432)	TLR2,4	*Streptococcus pyogenes*	Pancreas Ca, AEC	DC maturation [32]	NCT00795977, NCT00291473
Resiquimod	TLR7, TLR8	Imidazoquinoline	Various cancer type	Testing in numerous combination therapy [17]	NCT00821652, NCT00470379, NCT01676831
SD-101	TLR9	Phosphorothioate ODN	Lymphoma, SCCHN	Resensitise to PD-1 blockade [33]	NCT02521870, NCT02927964
TMX-101	TLR7	Imiquimod	Bladder Ca *in situ*	IL-6 and IL-18 induction [34]	NCT01731652
VTX-2337 (motolimod)	TLR8	A small molecule agonist	Ovarian Ca, SCCHN	Strong IL-12 and TNFα production [35]	NCT02431559, NCT02124850

Abbreviation: *pDC* plasmacytoid dendritic cell; *s.c* subcutaneously; *BCG* bacillus Calmette-Guerin; *poly-ICLC* particular formulation of poly-IC that includes carboxymethyl cellulose and poly-L-lysine as stabilising agents; *NSCLC* non-small cell lung cancer; *ODN* oligodeoxyribonucleotide; *SCHNC* squamous cell dead and neck cancer; *UC* ulcerative colitis; *CIN* cervical intraepithelial neoplasia; *AEC* antigen-expressing cancers

9.3 TLR Agonists as Adjuvants

A rationale for adjuvanted vaccines is to induce the synchronous activation of dendritic cell (DC)-presenting antigen and promote Th1 and $CD8^+$ T cell responses with minimal adverse effects. Extensive mouse studies and human trials using synthetic TLR agonists have demonstrated that adding TLR agonists to cancer treatments profoundly influences the extent of adaptive immune responses to tumour antigens. Such enhancing effects include direct activation of DC subsets, type I IFN production, enhanced cross-presentation, augmented $CD8^+$ T cell responses and increased antibody titres; reinvigorated immunosurveillance in patients whose immune system is compromised or in less immunogenic patients such as elderly and children (change the tumour environment); sensitised conventional chemotherapy or radiotherapy; and dose sparing, either minimising the number or amount of antigen introduced or reducing the vaccine schedule for optimal effects [20, 36]. The key effects of TLR agonists on immune system are summarised in Fig. 9.1.

9.3.1 TLR2

Phylogenetic analysis identified TLR2 along with TLR1, 6 and 10 with highly similar primary sequences that cooperate with each other during PAMP recognition. [37, 38]. TLR2 can functionally heterodimerise with TLR1, TLR6 and possibly TLR10 to specifically recognised products from gram-positive bacteria, including triacyl lipopeptides and *Mycoplasma fermentans* macrophage-activating lipopeptide (MALP-2). Pam3Cys, a synthetic triacylated lipoproteins, is widely used ligand to activate TLR2 in the laboratory [39].

SMP-105 is a TLR2 agonist developed from the insoluble fraction of the cell wall skeleton (CWS) of *Mycobacterium bovis*. One study has demonstrated that the activation of TLR2 by SMP-105 significantly increased IFNγ-producing cells and tumour-specific cytotoxic lymphocytes (CTL) in mice inoculated with Lewis lung cancer cells, resulting in a growth suppression of implanted tumours [40]. However, SMP-105 has not yet been trialled, and the use and development of TLR2 agonists as an adjuvant in anticancer immunotherapy has not been investigated. This may be at least partly because of regulatory T cell (Treg) activation by TLR2 agonists and resultant production of the anti-inflammatory IL-10 [41–43]. In contrast, recent studies have shown that TLR2 activation by Pam3Cys can suppress Treg activity and promote a Th17-like phenotype shift in multiple sclerosis patients. Overall, this suggests that the outcomes of TLR2 signalling may be contextually and therefore immunobiologically diverse [44].

CBLB612 is a synthetic lipopeptide that specifically binds to TLR2 and TLR6. However, this molecule was evaluated as only a bone marrow protective agent for cancer patients before and after chemotherapy. A phase II double blind, multicentre study (NCT02778763) was completed on efficacy and safety of this molecule as a

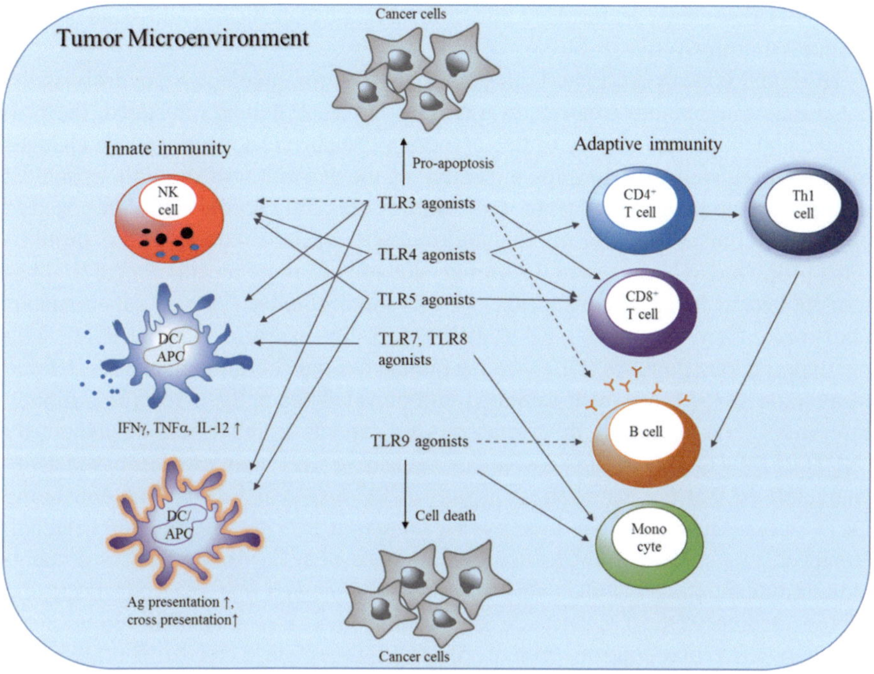

Fig. 9.1 Antitumour responses by TLR ligands through direct and indirect effects

monotherapy for neutropenia prophylaxis in breast cancer patients receiving myelosuppressive chemotherapy.

9.3.2 TLR3

TLR3 is an endosomal receptor that recognise double-stranded RNA molecules. The primary TLR3 ligand is viral dsRNA. Polyinosinic-polycytidylic acid (poly-IC), first developed in 1967, is a synthetic dsRNA molecule used in many studies to activate TLR3 and MDA5 [19, 45]. It has been shown that stimulation with poly-IC induces strong type I interferon production, humoral immunity and Th1 responses. However, the significant toxicity and the degradation by serum nucleases limit its development as a clinical dsRNA adjuvant.

To improve the safety and therapeutic potential, another two poly-IC derivatives, poly-IC12U (Ampligen®) and poly-ICLC (Hiltonol®), were developed and tested in numerous clinical trials as effective adjuvants. The former is a modified poly-IC with shortened half-life and shown to induce lower absolute levels of type I interferon than poly-IC [19]. Early studies showed that Ampligen not only activates NK cell and converts M2 macrophage to M1 counterparts but also elicits direct cytotoxic effects on cancer cells [19, 23]. However, more interest is currently focused on testing this TLR3 agonist against HIV infection, myalgic encephalomyelitis and chronic fatigue syndrome [46].

Hiltonol® is designed to be highly resistant to serum nucleolytic hydrolysis. Its denaturation temperature can be as a high as 40 °C. When administered, the prolonged and enhanced activity of this compound has been shown to induce changes in immune-related gene profiles indicative of the activation of multiple canonical innate immune pathways [47]. In early studies, this drug was initially proposed to be used in clinical trials for treating children with acute leukaemia and neuroblastoma [48]. Due to high dose and systemic side effects, more recent clinical trial has transitioned to local administration such as intramuscular (i.m) or subcutaneous (s.c) route. Current studies have demonstrated that inclusion of poly-ICLC with overlapping long peptides (OLP) from a human tumour self-antigen NY-ESO-1 and Montanide ISA-51 was well tolerated and elevated the antigen-specific antibody titre and T cell responses in ovarian cancer patients immunised subcutaneously (NCT00616941) [49]. A further study of a sensitive CD154 expression-based assay characterised that the major effect of poly-ICLC was achieved through enhancing OLP vaccine-induced $CD4^+$ T cell as an increased IFNγ/IL-4 ratio was detected. However, this study did not assess the influence of using poly-ICLC as a standalone adjuvant [50]. Another very recent study showed that NY-ESO-1-specific IFN-producing $CD8^+$ T cells were significantly increased in patients immunised with poly-ICLC than controls without poly-ICLC treatment [51] (UMIN000007954). Consistent findings were reported in a study of treating low-grade glioma (LGG) patients using a subcutaneous emulsified vaccination of glioma-associated antigen (GAA)-derived peptide with concurrent intramuscular injections of poly-ICLC (NCT00795457). This project was initially to explore an appropriate strategy for treating immunocompetent patients with slow growth rate gliomas as these patients may exhibit more immunogenicity and gain greater benefit from immunisations than those immunocompromised subjects with high-grade gliomas [52]. These results suggest TLR3 agonists as a promising adjuvant to cancer vaccines that target various tumour-associated antigens. Nevertheless, large-scale trials are required to evaluate the safety and efficacy of TLR3 agonist-based cancer therapies.

9.3.3 TLR4

TLR4 was the first TLR identified in mammals and can be activated classically by LPS (also known as endotoxin) from gram-negative bacteria [13]. TLR4 also recognises DAMPS, including heat-shock protein 70 [53] and nonhistone chromatin-binding nuclear constituent HMGB1 [54]. These proteins can be released by cancer cells following heat stress or chemo-/radiotherapy and act as danger signals promoting antitumour immunity by activation of TLR4 in DCs. Patients carrying single nucleotide polymorphisms (SNP) in TLR4 affecting the interaction between TLR4 and HMGB1 relapsed earlier after chemo-/radiotherapy than those with the normal TLR4 allele [54].

Monophosphoryl lipid A (MPL), a derivative of lipid A from gram-negative *Salmonella minnesota* endotoxin with reduced LPS (TLR4 ligand) toxicity in humans, was found to have antitumour activity in vivo early in the 1960s. Since 1984, this molecule as an adjuvant has been extensively investigated for cancer vaccines in the clinical setting. AS04 (Adjuvant System 04), an aluminium salt and MPL-based adjuvant, was endorsed by FDA in 2009 as part of Cervarix®. AS04 was also developed by GSK Biologicals in different formulations and used as proprietary adjuvant in numerous trials of cancer vaccine targeting specific TAAs (MAGE-3, MUC-1, sialyl-Tn and Ras mutant) expressed on multiple cancer types [6, 55].

Bacillus Calmette-Guérin (BCG) is a vaccine primarily developed for the prevention of tuberculosis worldwide [24]. Its anticancer potential began to be evaluated in the clinical settings in the 1970–1980s [24]. Testing BCG as monotherapy in human bladder cancer demonstrated either no clinical benefit or was proven inconclusive due to the small cohort sizes. Since 1977 intravesical instillation of BCG has been the "gold standard" treatment for patients with in situ or non-muscle invasive bladder cancer. Several clinical trials listed were designed to compare BCG as a single agent with combination therapy. Successful trials using BCG for melanoma therapy demonstrated better prognosis overall when BCG was combined with a melanoma cell vaccine or in combination with topical treatment of 5% imiquimod, a cream-formulated TLR7/8 agonist [56]. The non-specific protection by BCG was shown to be mainly through activation of TLR2 and TLR4 in macrophages and DCs, inducing strong cytokine and chemokine production such as IFNγ, IL-2 and TNFα. However, BCG has been found to activate $CD4^+$ $CD25^+$ Treg cells and promotes TGFβ and IL-10 secretion, which could be the reasons of unfavourable results in some trials [56–58].

Picibanil (OK-432) is a lyophilised preparation of *Streptococcus pyogenes* that is approved in Japan for the treatment of cervical cancer, gastric cancer and oral cancer [59]. Studies using this compound in other malignancy appear still active [60–62].

GLA-SE (G100), a synthetic glucopyranosyl lipid A (GLA) which is an oil-in-water emulsion (ES), is a novel TLR4 agonist, and this particular formulation allows to maximise the activation of multiple immune-related signalling pathways [63]. A recent preclinical study showed that intratumoural injection of G100 three times a week significantly suppressed tumour growth and resulted in 60% CR (complete tumour regression) in an A20 lymphoma tumour model via a $CD8^+$ T cell-dependent manner. Gene profiling analysis further demonstrated upregulation of broad immune-related genes, though also including T cell exhaustion marker such as CTL4, LAG3 and PD-L1 in G100-treated tumour. These data provide a rationale for co-administration of this TLR4 agonist with checkpoint blockade therapy to gain a potential synergistic effect [64] and support the current ongoing clinical trial of G100 for patients with non-Hodgkin's lymphoma as a single agent or in combination with pembrolizumab, a humanised antibody against PD-1 (NCT02501473). Furthermore, a single injection of G100 subcutaneously prior to tumour inoculation resulted in reduction of metastatic development of a mammary adenocarcinoma and a colon cancer cell model in both rats and mice with no adverse effects.

The antitumour effect of G100 appeared to be mainly associated with enhanced activation of NK cells [65]. A pilot clinical trial of intratumoural injection of G100 as monotherapy for patients with resectable Merkel cell carcinoma in a neoadjuvant setting exhibited an acceptable tolerability and increased $CD8^+$ T cell antitumour activity [59, 66]. Additionally, in order to trigger a potent tumour antigen-specific antitumour response, G100 was also formulated with recombinant NY-ESO-1 protein and developed to be another immunogenic agent named ID-G305. A novel "priming-boost" combination approach called CMB305 is to sequentially dose LV305 (an in vivo DC-targeting vector expressing the NY-ESO-1 gene) and ID-G305. This recipe allows to synergistically induce multiple level of antitumour immune effects. Of note, CMB305 does not require patient-specific manufacturing or ex vivo manipulation of patient samples. The phase Ib open-label, multicentre trial designed to evaluate the safety, tolerability, immunogenicity and preliminary clinical efficacy of CMB305 in patients with NY-ESO-1-positive tumours is currently recruiting (NCT02387125).

9.3.4 TLR5

Flagellin protein, a constituent protein of bacterial flagella, is the only known natural ligand to activate TLR5. Formulation using liposomal engrafted synthetic peptide containing flagellin fragments can induce DC maturation in vitro and in vivo [25]. In a study employing a mouse xenograft melanoma model, a vaccine formulated with both N-terminal (9Flg and 42Flg) and C-terminal (10Flg and 11Flg) flagellin peptide engrafted ovalbumin (OVA) liposomes suppressed lung metastasis in mice inoculated with B16-OVA cells compared to control mice [67]. More interestingly, TLR5 agonists lack induction of self-amplified driving cytokines, such as TNFα, IL-1β and IL-2, which make its fingerprint as a safe adjuvant for systemic administration [26].

Entolimod (CBLB502), a pharmacological optimised flagellin derivative, has revealed antitumour effects in numerous mouse tumour models [26, 68–71]. Craig et al. also showed that activation of TLR5 signalling by systemic administration of entolimod as a single agent inhibited at least two types of murine tumour metastases to high TLR5 expression organs such as the lung and liver. These antitumour effects were initiated through a CXCR3-dependent NK-DC-$CD8^+$ T cell axis [26]. One phase I clinical trial has been completed to determine the safety and preliminary evidence of efficacy using entolimod (i.m or s.c) in patients with late-stage solid tumours (NCT01527136). Overall, the treatment with entolimod was well tolerated with only common adverse events such as fever, transient hypotension and hyperglycaemia.

Interestingly, entolimod has been shown a protective effect against renal dysfunction in a murine model of acute renal ischemic failure as well as in an ulcerative colitis model [72, 73]. Furthermore, this compound showed radioprotective activity in mouse and primate models without reducing tumour radiosensitivity [74]. Further preclinical study identified the liver and gastrointestinal tract as the major target organs of this molecule [75]. These evidence supported current clinical studies and raise considerable interest to test entolimod in development as a radiation countermeasure in emergency severe condition such as acute radiation syndrome or radiation sickness from radiotherapy against cancer. These findings allow entolimod to be a versatile player in cancer therapy.

Another TLR5 agonist, M-VM3 (Mobilan), is currently in two clinical trials (NCT02654938, NCT02844699) for prostate cancer. Mobilan was designed as a recombinant non-replicating adenovirus encoding human TLR5 and its specific agonistic ligand, flagellin. Delivery of this system into tumour cells would render an autocrine activation of TLR5 and result in subsequent strong adaptive antitumour immune responses. Certain human tumours expressing the coxsackievirus and adenovirus receptor (CAR) such as prostate cancer and several tumours of female reproductive system will be the primary targets using this TLR5 agonist-based system [76].

9.3.5 TLR7/8

TLR7 and 8, both expressed in the endosomes/lysosome, are receptors for ssRNA, especially U or GU-rich oligoribonucleotides [77]. They share high sequence homology and predominately overlap in the ligands they can each respond to. In humans, these TLRs are abundantly expressed in multiple subsets of human DCs. Stimulation of TLR7 and 8 with their agonists significantly augments multiple subset DC maturation, Th1 cellular immunity, cross-presentation and humoral immunity [78, 79]. Of note, conjugation of TLR7/8 agonist rather simply mixing with antigens has been demonstrated more effective to generate $CD8^+$ immunity [17, 80].

Resiquimod, a TLR7/8-bispecific agonist, is a prototypical imidazoquinoline molecule [17]. Early studies showed that soluble molecules like resiquimod distribute quickly from the site of injection throughout the body and fail to induce local immune activation, thereby limiting its clinical utility. To resolve this problem, prototypical imidazoquinolines were formulated as a dermal cream [17]. Aldara®, imiquimod 5% cream, is the one of the only three FDA-approved commercialised small molecule TLR agonists for HPV-mediated external genital warts, superficial basal cell carcinoma and actinic keratosis [24]. In recent clinical trials, imiquimod cream has been further exploited as stand-alone or in combination with various antitumour therapies including chemo-/radio- or laser therapy in various human cancers [17]. A study of topical imiquimod in breast cancer patients with skin metastasis reported that 2 of 10 patients achieved a partial response with histologic evidence of immune-mediated tumour regression. The treatment was well tolerated with exception of frequently local adverse events [81]. In contrast, another randomised controlled trial (NCT00066872) conducted at 12 centres in the UK on 501 participants reported that patients with nodular and superficial basal cell carcinoma treated with imiquimod 5% cream were superior to excision surgery [82].

Considering the diversity of subtype of this skin cancer, to determine surgical or non-surgical modalities alone or combination needs to be optimised by future investigation [83, 84]. Furthermore, in a completed two-part randomised trial on high-risk melanoma patients, NY-ESO-1 antigen emulsified in Montanide was intradermally injected, followed by topical application of 0.2% resiquimod gel or placebo to the vaccine site. Although this formulation with topical resiquimod was safe, the clinical outcome reveals no significant differences between study groups as addition of topical resiquimod was not sufficient to induce consistent specific $CD8^+$ T cells. The reason could be that topical application of the TLR agonist may fail to absorb adequately and activate diverse DC populations in deeper skin layers [85], highlighting the selection of administration route for optimal efficacy using TLR agonists.

3M-052, a novel lipid-modified imidazoquinoline, was developed and evaluated as part of a conventional vaccine formulation. Compared to resiquimod, 3M-052 induced a prolonged response locally with diminished systemic inflammation. It was also evaluated as an adjuvant in many vaccine models such as alum suspensions, liposome formulation and oil-in-water emulsion [17, 86, 87]. In a preclinical study, Singh et al. showed that intratumoural injection 3M-052 in mouse melanoma and prostate tumour models suppressed local and distant tumour growth via not only promoting tumour-specific $CD8^+$ T cells but also shifting M2 macrophages to M1 phenotype. 3M-052 has also exhibited synergic effects with checkpoint inhibitor therapy using CTLA4 and PDL-1 antibodies [88].

9.3.6 TLR9

TLR9 is expressed in the endosome of specific immune cell types – plasmacytoid DC and B cells – where they recognise bacterial or viral DNA containing unmethylated cytosine-guanine (CG) dinucleotides motifs [89]. Activation of TLR9 through the signalling of MyD88 leads to activation of interferon regulatory factor (IRF) 7, resulting in expression of type I IFNs [90].

CpG-7909 (PF-3512676, Promune®), a class B CpG, is the most extensively studied single-stranded CpG ODN. Unfortunately, two phase III clinical studies for advanced non-small cell lung cancer reported that addition of CpG-7909 to chemotherapy gained no improvement to either overall or progression-free survival [91, 92]. Although phase I/II studies examining CpG-7909 in combination therapies were completed in various human cancers such as B-cell lymphoma (NCT00185965), metastatic breast cancer (NCT00824733) and oesophageal cancer (NCT00669292), no trials further assessing this molecule for cancer therapy are active.

MGN1703, a covalently closed natural DNA molecule, is a novel TLR9 agonist, which belongs to a different family called dSLIM (double stem loop immunomodulatory) [89, 93]. While CpG-7909 mainly stimulates B cells and may cause several CG motif-independent immune responses like IL-8 induction, dSLIM elicits significant IFN-α induction and broad activation of human immune cells in vitro [89]. In a phase II study of 59 patients with metastatic colorectal cancer treated with the first-line chemotherapy bevacizumab (anti-VEGF-A), patients who also received MGN1703 showed a superior progression-free survival (PFS) by (NCT01208194) [94–96]. A pivotal phase III trial has been designed to further investigate these data and is currently recruiting patients (NCT02077868). This compound has also been tested in a phase I trial to determine the highest tolerable dose in combination with ipilimumab (anti-CTLA-4) to patients with advanced solid tumours (NCT02668770) and another study in patients with small cell lung cancer (SCLC) (NCT02200081).

SD-101, another synthetic CpG molecule, was tested by intratumoural co-administration with ipilimumab after local radiation therapy in patients with low-grade, recurrent B-cell lymphoma (NCT02254772). The primary objective was to identify the best dose of intratumoural ipilimumab with TLR9 agonist in patients to augment different phases of antitumour immune responses as shown in previous preclinical studies [97, 98]. However, the result has not been released yet. Another trial using this molecule combined with ibrutinib (also known as Imbruvica, a targeted inhibitor for Bruton's tyrosine kinase) and intratumoural radiotherapy for low-grade follicular lymphoma is currently listed as recruiting (NCT02927964) in order to re-evaluate adverse event during this heavy combination therapy.

9.4 Conclusion Remarks

Coupled with chimeric antigen receptor therapy [99] and immune checkpoint blockade [100], engagement of manipulating TLR signalling has drawn considerable interest as a treatment modality in the cancer immunotherapy field. Particularly, synthetic TLR agonists have been actively exploited for their safety and clinical efficacy in various therapeutic settings. Substantial evidence demonstrates that TLR agonists are potent immunostimulators and enhance natural or therapy-triggered antitumour immune responses. Despite the wealth of research in the cancer immunotherapy field, the past few years has witnessed a steady decrease in the number of clinical trials using TLR agonists as cancer treatments, be that unimodal or as adjuvants to cancer vaccines [59]. There may be some reasons for this. Firstly, most TLR ligands initiate complex signalling cascade and may influence various cell types in cell-dependent manners, of which we haven't fully elucidated. Immune cells, cancer cells and tumour stromal cell differ in their specific TLR expression and may all contribute to biological consequences of TLR activation [101]. Noninflammatory roles of TLRs in tumour progression such as regulation of apoptosis and proliferation in context of chronic inflammation and carcinogenic condition have been addressed [9]. Secondly, recent studies reveal that TLR activation can exert not only immunostimulatory effects but also immunosuppressive effects by regulating IL-10, Treg activity and PD-L1 expression [102–104]. In addition, as with many other antitumour agents in general, the underperformance of TLR agonist used in some trials may relate to the fact that patients recruited in most studies are in late-stage disease. Metastatic tumours, which are aberrant in multiple signalling pathways and have mutations in various key genes regulating cellular functions, may all contribute to drug resistance. Additionally, immune system depression as a consequence of late-stage cancer may impede any effective induction of antitumour responses. Last but not least, the delivery system, type of tumour-associated antigen, type of TLR agonist and/or other adjuvants, schedule, route and site of administration need to be carefully considered and further investigated to acquire optimal activation and specificity [19]. Recent convergence of large-scale sequencing, cancer biology and bioinformatics allows researchers to tailor the strategy of priming the immune system against multiple patient-specific neoantigens (tumour mutation-derived antigens) [105, 106]. Additionally, there is scope for these formulations of specific TLR agonists to be personalised for specific patients' tumours. Combination therapies based on stratification of patients by their immune state may lead to better targeted trials using TLR agonists.

Overall, basic research and clinical trials provide a strong rationale for using TLR agonists as adjuvants to cancer treatments.

References

1. Gay NJ, Gangloff M (2007) Structure and function of toll receptors and their ligands. Annu Rev Biochem 76(1):141–165
2. Takeda K, Kaisho T, Akira S (2003) Toll-like receptors. Annu Rev Immunol 21(1):335–376
3. Janeway CA (1989) Approaching the asymptote? Evolution and revolution in immunology. Cold Spring Harb Symp Quant Biol 54:1–13
4. Hernandez C, Huebener P, Schwabe RF (2016) Damage-associated molecular patterns in cancer: a double-edged sword. Oncogene 35(46):5931–5941
5. Kadowaki N et al (2001) Subsets of human dendritic cell precursors express different toll-like receptors and respond to different microbial antigens. J Exp Med 194(6):863–870
6. Adams S (2009) Toll-like receptor agonists in cancer therapy. Immunotherapy 1(6):949–964
7. Hayashi F, Means TK, Luster AD (2003) Toll-like receptors stimulate human neutrophil function. Blood 102(7):2660–2669
8. Peng G et al (2005) Toll-like receptor 8-mediated reversal of $CD4^+$ regulatory T cell function. Science 309(5739):1380–1384
9. Alison CW, Brendan JJ (2015) Inflammatory and non-inflammatory roles for toll-like receptors in gastrointestinal cancer. Curr Pharm Des 21(21):2968–2977
10. Kovach MA, Standiford TJ (2011) Toll like receptors in diseases of the lung. Int Immunopharmacol 11(10):1399–1406
11. LaRue H et al (2013) Toll-like receptors in urothelial cells—targets for cancer immunotherapy. Nat Rev Urol 10(9):537–545
12. Paulos CM et al (2007) Toll-like receptors in tumor immunotherapy. Clin Cancer Res 13(18):5280–5289
13. Dowling JK, Mansell A (2016) Toll-like receptors: the Swiss army knife of immunity and vaccine development. Clin Translat Immunol 5(5):e85
14. Human Papillomavirus (HPV) Vaccines. National Cancer Institute (NCI) (2009). 2009–10-22 (Retrieved 2009–11–11)
15. Hepatitis B vaccines WHO position paper. Weekly epidemiological record (2009) 40(84):405–420
16. Silva A et al (2015) The combination of ISCOMATRIX adjuvant and TLR agonists induces regression of established solid tumors in vivo. J Immunol 194(5):2199–2207

17. Vasilakos JP, Tomai MA (2013) The use of toll-like receptor 7/8 agonists as vaccine adjuvants. Expert Rev Vaccines 12(7):809–819
18. Mutwiri G et al (2011) Combination adjuvants: the next generation of adjuvants? Expert Rev Vaccines 10(1):95–107
19. Martins KAO, Bavari S, Salazar AM (2015) Vaccine adjuvant uses of poly-IC and derivatives. Expert Rev Vaccines 14(3):447–459
20. Banday AH, Jeelani S, Hruby VJ (2015) Cancer vaccine adjuvants – recent clinical progress and future perspectives. Immunopharmacol Immunotoxicol 37(1):1–11
21. Walter A et al (2013) Aldara activates TLR7-independent immune defence. Nat Commun 4:1560
22. Spaner DE et al (2009) A phase I/II trial of TLR-7 agonist immunotherapy in chronic lymphocytic leukemia. Leukemia 24(1):222–226
23. Jasani B, Navabi H, Adams M (2009) Ampligen: a potential toll-like 3 receptor adjuvant for immunotherapy of cancer. Vaccine 27(25–26):3401–3404
24. Vacchelli E et al (2012) Trial watch: FDA-approved toll-like receptor agonists for cancer therapy. OncoImmunology 1(6):894–907
25. Galluzzi L et al (2012) Trial watch: experimental toll-like receptor agonists for cancer therapy. OncoImmunology 1(5):699–739
26. Brackett CM et al (2016) Toll-like receptor-5 agonist, entolimod, suppresses metastasis and induces immunity by stimulating an NK-dendritic-CD8+ T-cell axis. Proc Natl Acad Sci 113(7):E874–E883
27. Musch E et al (2013) Topical treatment with the Toll-like receptor agonist DIMS0150 has potential for lasting relief of symptoms in patients with chronic active ulcerative colitis by restoring glucocorticoid sensitivity. Inflamm Bowel Dis 19(2):283–292
28. Dalgleish AG et al (2016) Randomised, open-label, phase II study of gemcitabine with and without IMM-101 for advanced pancreatic cancer. Br J Cancer 115(7):789–796
29. Smith DA et al (2014) Antitumor activity and safety of combination therapy with the toll-like receptor 9 agonist IMO-2055, erlotinib, and bevacizumab in advanced or metastatic non-small cell lung cancer patients who have progressed following chemotherapy. Cancer Immunol Immunother 63(8):787–796
30. Friedberg JW et al (2009) Phase II study of a TLR-9 agonist (1018 ISS) with rituximab in patients with relapsed or refractory follicular lymphoma. Br J Haematol 146(3):282–291
31. Isambert N et al (2013) Phase I study of OM-174, a lipid A analogue, with assessment of immunological response, in patients with refractory solid tumors. BMC Cancer 13(1):172
32. Vacchelli E et al (2013) Trial watch: toll-like receptor agonists for cancer therapy. OncoImmunology 2(8):e25238
33. Wang S et al (2016) Intratumoral injection of a CpG oligonucleotide reverts resistance to PD-1 blockade by expanding multifunctional CD8+ T cells. Proc Natl Acad Sci 113(46):E7240–E7249
34. Donin NM et al (2017) A phase 2 study of TMX-101, intravesical imiquimod, for the treatment of carcinoma in situ bladder cancer. Urol Oncol Semin Orig Investig 35(2):39.e1–39.e7
35. Lu H et al (2012) VTX-2337 is a novel TLR8 agonist that activates NK cells and augments ADCC. Clin Cancer Res 18(2):499–509
36. Mount A et al (2013) Combination of adjuvants: the future of vaccine design. Expert Rev Vaccines 12(7):733–746
37. Hughes AL, Piontkivska H (2008) Functional diversification of the Toll-like receptor gene family. Immunogenetics 60(5):249–256
38. Roach JC et al (2005) The evolution of vertebrate toll-like receptors. Proc Natl Acad Sci U S A 102(27):9577–9582
39. Dowling JK, Dellacasagrande J (2016) Toll-like receptors: ligands, cell-based models, and readouts for receptor action. In: McCoy CE (ed) Toll-like receptors: practice and methods. Springer, New York, pp 3–27

40. Murata M (2008) Activation of toll-like receptor 2 by a novel preparation of cell wall skeleton from Mycobacterium Bovis BCG Tokyo (SMP-105) sufficiently enhances immune responses against tumors. Cancer Sci 99(7):1435–1440
41. Netea MG et al (2004) Toll-like receptor 2 suppresses immunity against *Candida albicans* through induction of IL-10 and regulatory T cells. J Immunol 172(6):3712–3718
42. Sun X et al (2013) TLR2 mediates helicobacter pylori–induced Tolerogenic immune response in mice. PLoS One 8(9):e74595
43. Yamazaki S et al (2011) TLR2-dependent induction of IL-10 and Foxp3+ CD25+ CD4+ regulatory T cells prevents effective anti-tumor immunity induced by Pam2 Lipopeptides in vivo. PLoS One 6(4):e18833
44. Nyirenda MH et al (2015) TLR2 stimulation regulates the balance between regulatory T cell and Th17 function: a novel mechanism of reduced regulatory T cell function in multiple sclerosis. J Immunol 194(12):5761–5774
45. McCartney S et al (2009) Distinct and complementary functions of MDA5 and TLR3 in poly(I:C)-mediated activation of mouse NK cells. J Exp Med 206(13):2967–2976
46. Mitchell WM (2016) Efficacy of rintatolimod in the treatment of chronic fatigue syndrome/myalgic encephalomyelitis (CFS/ME). Expert Rev Clin Pharmacol 9(6):755–770
47. Caskey M et al (2011) Synthetic double-stranded RNA induces innate immune responses similar to a live viral vaccine in humans. J Exp Med 208(12):2357–2366
48. Lampkin BC et al (1985) Phase II trial of a complex polyriboinosinic-polyribocytidylic acid with poly-L-lysine and carboxymethyl cellulose in the treatment of children with acute leukemia and Neuroblastoma: a report from the Children's cancer study group. Cancer Res 45(11 Part 2):5904–5909
49. Sabbatini P et al (2012) Phase I trial of overlapping long peptides from a tumor self-antigen and poly-ICLC shows rapid induction of integrated immune response in ovarian cancer patients. Clin Cancer Res 18(23):6497–6508
50. Tsuji T et al (2013) Effect of Montanide and poly-ICLC adjuvant on human self/tumor antigen-specific $CD4^+$ T cells in phase I overlapping long peptide vaccine trial. Cancer Immunol Res 1(5):340–350
51. Takeoka T et al (2017) NY-ESO-1 protein cancer vaccine with poly-ICLC and OK-432: rapid and strong induction of NY-ESO-1-specific immune responses by poly-ICLC. J Immunother 40(4):140–147
52. Okada H et al (2015) Induction of robust type-I $CD8^+$ T-cell responses in WHO grade 2 low-grade Glioma patients receiving peptide-based vaccines in combination with poly-ICLC. Clin Cancer Res 21(2):286–294
53. Chen T et al (2009) Heat shock protein 70, released from heat-stressed tumor cells, initiates antitumor immunity by inducing tumor cell chemokine production and activating dendritic cells via TLR4 pathway. J Immunol 182(3):1449–1459
54. Apetoh L et al (2007) Toll-like receptor 4-dependent contribution of the immune system to anticancer chemotherapy and radiotherapy. Nat Med 13(9):1050–1059
55. Cluff CW (2013) Monophosphoryl lipid A (MPL) as an adjuvant for anti-cancer vaccines: clinical results. Austin: Landes Bioscience, p 2000–2013
56. Talat Iqbal N, Hussain R (2014) Non-specific immunity of BCG vaccine: A perspective of BCG immunotherapy. Trials in Vaccinology 3:143–149
57. Coleman MM, Keane J, Mills KHG (2010) Editorial: Tregs and BCG—dangerous liaisons in TB. J Leukoc Biol 88(6):1067–1069
58. Ovchinnikova OA et al (2014) Mycobacterium bovis BCG killed by extended freeze-drying induces an immunoregulatory profile and protects against atherosclerosis. J Intern Med 275(1):49–58
59. Iribarren K et al (2016) Trial watch: Immunostimulation with toll-like receptor agonists in cancer therapy. OncoImmunology 5(3):e1088631
60. Sato T et al (2016) M1 macrophage infiltrations and histological changes in the liver after portal vein embolization using fibrinogen and OK432 in the rat. Cell Immunol 303:66–71

61. Kenjo A et al (2016) Role of intratumoral infiltrating macrophages after transarterial immunoembolization for hepatocellular carcinoma. J Hepatobiliary Pancreat Sci 23(5):298–304
62. Wada H et al (2014) Vaccination with NY-ESO-1 overlapping peptides mixed with Picibanil OK-432 and Montanide ISA-51 in patients with cancers expressing the NY-ESO-1 Antigen. J Immunother 37(2):84–92
63. Lambert SL et al (2012) Molecular and cellular response profiles induced by the TLR4 agonist-based adjuvant Glucopyranosyl lipid A. PLoS One 7(12):e51618
64. Lu H, Jan ter Meulen JH (2016) Intratumoral injection of G100 (TLR4 agonist glucopyranosyl lipid A) modulates tumor microenvironment and induces CD8 T cell-dependent, systemic anti-tumor immunity AACR Annual Meeting, p 4885
65. Matzner P et al (2016) Perioperative treatment with the new synthetic TLR-4 agonist GLA-SE reduces cancer metastasis without adverse effects. Int J Cancer 138(7):1754–1764
66. SID B, Vandeven N et al (2015) Pilot study of intratumoral G100, toll-like receptor-4 (TLR4) agonist, therapy in patients with Merkel cell carcinoma (MCC). ASCO Meeting 33:3083
67. Faham A, Altin JG (2010) Antigen-containing liposomes engrafted with Flagellin-related peptides are effective vaccines that can induce potent antitumor immunity and immunotherapeutic effect. J Immunol 185(3):1744–1754
68. Sfondrini L et al (2006) Antitumor activity of the TLR-5 ligand Flagellin in mouse models of cancer. J Immunol 176(11):6624–6630
69. Garaude J et al (2012) Simultaneous targeting of toll- and nod-like receptors induces effective tumor-specific immune responses. Sci Transl Med 4(120):120ra16–120ra16.
70. Cai Z et al (2011) Activation of Toll-like receptor 5 on breast cancer cells by Flagellin suppresses cell proliferation and tumor growth. Cancer Res 71(7):2466–2475
71. Yang H et al (2016) The toll-like receptor 5 agonist entolimod suppresses hepatic metastases in a murine model of ocular melanoma via an NK cell-dependent mechanism. Oncotarget 7(3):2936–2950
72. Fukuzawa N et al (2011) A TLR5 agonist inhibits acute renal ischemic failure. J Immunol 187(7):3831–3839
73. Xu Y et al (2016) CBLB502 administration protects gut mucosal tissue in ulcerative colitis by inhibiting inflammation. Ann Trans Med 4(16):301
74. Burdelya LG et al (2008) An agonist of toll-like receptor 5 has Radioprotective activity in mouse and primate models. Science 320(5873):226–230
75. Burdelya LG et al (2013) Central role of liver in anticancer and radioprotective activities of toll-like receptor 5 agonist. Proc Natl Acad Sci 110(20):E1857–E1866
76. Reeh M et al (2013) Presence of the Coxsackievirus and adenovirus receptor (CAR) in human neoplasms: a multitumour array analysis. Br J Cancer 109(7):1848–1858
77. Diebold SS et al (2004) Innate antiviral responses by means of TLR7-mediated recognition of single-stranded RNA. Science 303(5663):1529–1531
78. Wille-Reece U et al (2005) HIV Gag protein conjugated to a toll-like receptor 7/8 agonist improves the magnitude and quality of Th1 and CD8+ T cell responses in nonhuman primates. Proc Natl Acad Sci U S A 102(42):15190–15194
79. Coffman RL, Sher A, Seder RA (2010) Vaccine adjuvants: putting innate immunity to work. Immunity 33(4):492–503
80. Tomai MA, Vasilakos JP (2011) TLR-7 and -8 agonists as vaccine adjuvants. Expert Rev Vaccines 10(4):405–407
81. Adams S et al (2012) Topical TLR7 agonist Imiquimod can induce immune-mediated rejection of skin metastases in patients with breast cancer. Clin Cancer Res 18(24):6748–6757
82. Bath-Hextall F et al (2014) Surgical excision versus imiquimod 5% cream for nodular and superficial basal-cell carcinoma (SINS): a multicentre, non-inferiority, randomised controlled trial. Lancet Oncol 15(1):96–105
83. Bassukas ID, Gaitanis G (2014) Basal-cell carcinoma: no response versus relapse. Lancet Oncol 15(3):e104–e105

84. Lear JT (2014) Evidence-based treatment for low-risk basal cell carcinoma. Lancet Oncol 15(1):12–13
85. Sabado RL et al (2015) Resiquimod as an immunologic adjuvant for NY-ESO-1 protein vaccination in patients with high-risk melanoma. Cancer Immunol Res 3(3):278–287
86. Smirnov D et al (2011) Vaccine adjuvant activity of 3M-052: an imidazoquinoline designed for local activity without systemic cytokine induction. Vaccine 29(33):5434–5442
87. Abhyankar MM et al (2017) Nanoformulation of synergistic TLR ligands to enhance vaccination against Entamoeba histolytica. Vaccine 35(6):916–922
88. Singh M et al (2014) Effective innate and adaptive Antimelanoma immunity through localized TLR7/8 activation. J Immunol 193(9):4722–4731
89. Kapp K et al (2016) Distinct immunological activation profiles of dSLIM® and ProMune® depend on their different structural context. Immun Inflammation Dis 4(4):446–462
90. Yang B et al (2014) DNA vaccine for cancer immunotherapy. Hum Vaccin Immunother 10(11):3153–3164
91. Hirsh V et al (2011) Randomized phase III trial of paclitaxel/carboplatin with or without PF-3512676 (toll-like receptor 9 agonist) as first-line treatment for advanced non–small-cell lung cancer. J Clin Oncol 29(19):2667–2674
92. Manegold C et al (2012) A phase III randomized study of gemcitabine and cisplatin with or without PF-3512676 (TLR9 agonist) as first-line treatment of advanced non-small-cell lung cancer. Ann Oncol 23(1):72–77
93. Wittig B et al (2015) MGN1703, an immunomodulator and toll-like receptor 9 (TLR-9) agonist: from bench to bedside. Crit Rev Oncol/Hematol 94(1):31–44
94. Schmoll HJ, Riera-Knorrenschild J, Kopp HG et al (2015) Maintenance therapy with the TLR-9 agonist MGN1703 in the phase II IMPACT study of metastatic colorectal cancer patients: a subgroup with improved overall survival. J Clin Oncol 33(suppl 3):abstr 680
95. Schmoll H-J et al (2014) Maintenance treatment with the immunomodulator MGN1703, a toll-like receptor 9 (TLR9) agonist, in patients with metastatic colorectal carcinoma and disease control after chemotherapy: a randomised, double-blind, placebo-controlled trial. J Cancer Res Clin Oncol 140(9):1615–1624
96. Cunningham D, Zurlo A, Salazar R et al (2015) IMPALA, a randomized phase III study in patients with metastatic colorectal carcinoma: Immunomodulatory maintenance therapy with TLR-9 agonist MGN1703. J Clin Oncol 33(suppl 3):abstr TPS791
97. Houot R, Levy R (2009) T-cell modulation combined with intratumoral CpG cures lymphoma in a mouse model without the need for chemotherapy. Blood 113(15):3546–3552
98. Marabelle A, Kohrt H, Levy R (2013) Intratumoral anti-CTLA-4 therapy: enhancing efficacy while avoiding toxicity. Clin Cancer Res 19(19):5261–5263
99. Barrett DM et al (2014) Chimeric antigen receptor therapy for cancer. Annu Rev Med 65(1):333–347
100. Pardoll DM (2012) The blockade of immune checkpoints in cancer immunotherapy. Nat Rev Cancer 12(4):252–264
101. Dajon M, Iribarren K, Cremer I (2017) Toll-like receptor stimulation in cancer: a pro- and anti-tumor double-edged sword. Immunobiology 222(1):89–100
102. Lu H (2014) TLR agonists for cancer immunotherapy: tipping the balance between the immune stimulatory and inhibitory effects. Front Immunol 5:83
103. Beswick EJ et al (2014) TLR4 activation enhances the PD-L1–mediated Tolerogenic capacity of colonic $CD90^+$ stromal cells. J Immunol 193(5):2218–2229
104. Boes M, Meyer-Wentrup F (2015) TLR3 triggering regulates PD-L1 (CD274) expression in human neuroblastoma cells. Cancer Lett 361(1):49–56
105. Wang R-F, Wang HY (2017) Immune targets and neoantigens for cancer immunotherapy and precision medicine. Cell Res 27(1):11–37
106. Mullard A (2016) The cancer vaccine resurgence. Nat Rev Drug Discov 15(10):663–665

Chapter 10
Telomere Damage Response and Low-Grade Inflammation

Lihui Wang, Xianhua Yu, and Jun-Ping Liu

Abstract Telomeres at the ends of chromosomes safeguard genome integrity and stability in human nucleated cells. However, telomere repeats shed off during cell proliferation and other stress responses. Our recent studies show that telomere attrition induces not only epithelial stem cell senescence but also low-grade inflammation in the lungs. The senescence-associated low-grade inflammation (SALI) is characteristic of alveolar stem cell replicative senescence, increased proinflammatory and anti-inflammatory cytokines, infiltrated immune cells, and spillover effects. To date, the mechanisms underlying SALI remain unclear. Investigations demonstrate that senescent epithelial stem cells with telomere erosion are not the source of secreted cytokines, containing no significant increase in expression of the genes coding for increased cytokines, suggesting an alternative senescence-associated secretory phenotype (A-SASP). Given that telomere loss results in significant alterations in the genomes and accumulations of the cleaved telomeric DNA in the cells and *milieu externe*, we conclude that telomere position effects (TPEs) on gene expression and damage-associated molecular patterns (DAMPs) in antigen presentation are involved in A-SASP and SALI in response to telomere damage in mammals.

Keywords Telomere dysfunction • Cytokines • Immune cells • Inflammation • Tissue senescence

---

L. Wang (✉) • X. Yu
Institute of Ageing Research, Hangzhou Normal University School of Medicine, 1378 Wenyi Road West, Hangzhou, Zhejiang Province 311121, China
e-mail: wanglihui100@163.com

J.-P. Liu
Institute of Ageing Research, Hangzhou Normal University School of Medicine, 1378 Wenyi Road West, Hangzhou, Zhejiang Province 311121, China

Molecular Signaling Laboratory, Monash University Department of Immunology, Central Clinical School, Faculty of Medicine, Commercial Road, Prahran, VIC 3018, Australia

Hudson Institute of Medical Research, 27-31 Wright Street, Clayton, VIC 3168, Australia

Monash University Department of Molecular and Translational Science, Faculty of Medicine, 27-31 Wright Street, Clayton, VIC 3168, Australia
e-mail: jun-ping.liu@monash.edu

D. Xu (ed.), *Regulation of Inflammatory Signaling in Health and Disease*, Advances in Experimental Medicine and Biology 1024,
DOI 10.1007/978-981-10-5987-2_10

Abbreviations

A-SASP	alternative senescence-associated secretory phenotype
DAMPs	damage-associated molecular patterns
DDR	DNA damage response
SALI	senescence-associated low-grade inflammation
TPE	telomere position effect

10.1 Introduction

Telomeres are comprised of tens of thousands of repetitive DNA sequences (TTAGGG) at the end of chromosomes to prevent deterioration or fusion by losing some sequences during cell division. Telomerase is a ribonucleoprotein complex that operates to maintain telomeres (chromosomal ends), counteracting cell division-associated telomere shortening in the stem cell compartment and cancer [1, 2]. In the absence of telomerase in most differentiated cell types, consecutive shortening of telomeres in a chromosome replication-dependent fashion occurs, and the cell cycle-dependent, unidirectional catabolism of telomeres constitutes a mechanism with critically damaged telomeres terminating chromosome replication and initiating cellular senescence [3]. Loss-of-function gene mutations on telomerase subunits occur in a number of diseases [4, 5], including dyskeratosis congenita [6, 7], aplastic anemia [8, 9], liver cirrhosis [10, 11], and idiopathic pulmonary fibrosis (IPF) [12–15].

Experimental studies on animals have demonstrated certain causal roles of telomerase deficiency compromised in tissue homeostasis and dysfunctional organs [16–19]. Knockout of telomerase RNA component (TERC) or telomerase reverse transcriptase (TERT) causes the phenotypes to resemble dyskeratosis congenita [20, 21] and bone marrow failure [17, 21, 22]. In the lungs of TERC-deficient mice, elevated apoptosis of alveolar stem cells (alveolar epithelial type II cells, AECII) [23] hindered tissue growth in partial pneumonectomies [24]. The mechanisms of telomerase deficiency-associated diseases have not been fully elucidated, including the signaling of telomerase and telomeres in different conformations. The potentials of telomerase as antigens and in putative responses to vaccinations have been demonstrated under various conditions including immune resistance, evasion, and tolerance in animals [25].

Recently, deficiency in TERC or TERT causes a remarkable elevation in various proinflammatory cytokines, including IL-1, IL-6, CXCL15, IL-10, and TNF-α, and monocyte chemotactic protein 1 (chemokine ligand2, CCL2), decrease in TGF-β1 and TGFβR1 receptor in the lungs, and spillover of IL-6 and CXCL15 into bronchoalveolar lavage fluids [26]. Intriguingly, mice with TERC deficiency and short telomeres do not show obvious pulmonary fibrosis [27], whereas mice that developed pulmonary fibrosis from bleomycin insult require telomerase activity [28–30], mirroring the finding that the majority of IPF lung samples showed increased telomerase activity [30]. In the following sections, we will discuss the phenotypes of telomerase deficiency and telomere dysfunction-associated cellular responses, especially pulmonary alveolar stem cell senescence-associated low-grade inflammation (SALI) [26].

10.2 Telomerase Deficiency, Telomere Shortening, and Lung Epithelial Aging

Evidence indicates that telomerase operates in normal AECII [23, 31, 32] and is stimulated by silica inhalation and bleomycin instillation causing pulmonary fibrosis in rodents, suggesting that telomerase activity is involved in pulmonary fibrotic lesion [33, 34]. This scenario is in a sharp contrast to the relationship between telomerase gene mutations and IPF in humans [12–15]. Since genetic disruption of *TERC* or *TERT* leads to loss of telomerase activity and shortening of telomeres in mouse AECII cells [23, 35], we directly tested the hypothesis that the damaging effect of telomere ablations contributes to IPF in mice in vivo. Interestingly, disruption of *TERC* or *TERT* initiates telomere DDR with decreased telomere length; increased TIFs; increased p15, p16, and p21 and inductions of AECII replicative senescence characteristic of reduced total numbers of AECII; and increased populations of AECII positive for HP1γ and β-Gal staining [26]. Previous studies showed that telomerase deficiency renders pulmonary tissues susceptible to damage caused by cigarette smoke, potentially underpinning emphysema with or without pulmonary fibrosis [27, 36, 37]. It is thus feasible that telomerase is required for AECII alveolar stem cell renewal and repair of damaged differentiated cells in pulmonary interstitium [38, 39]. However, IPF myofibroblast markers were not evidenced in the telomerase-deficient mouse lungs, suggesting that telomerase deficiency and telomere shortening are responsible for pulmonary senescence but not alone sufficient in causing pulmonary fibrotic lesion [26]. It is possible that under the telomere dysfunctional pressure, additional signals are required for senescence-associated myofibroblast transdifferentiation in pulmonary fibrotic lesion.

Recent studies indicate that regulation of AECII stem cells by extracellular signaling predominantly dictates AECII response to injury signals and that only about 1% of mature AECII cells divide intermittently, with about a 40-day doubling time, supporting an overall renewal rate of 7% of alveoli per year [39]. Injuries to the AEC1 epithelial cell population, and together with EGF receptor activation, induce AECII broad stem cell functions [39]. Whereas activation of the Ras-ERK pathway stimulates proliferation [39] and inhibits differentiation [40] of AECII, inhibition of the Ras-ERK pathway is essential for TGF-β1-induced AECII differentiation [40]. Interestingly, previous studies by us and others showed that whereas mitogens upregulate telomerase activity by Ets transcriptional activation of the *TERT* gene [41, 42], TGF-β family members downregulate telomerase activity by Smad3 tran-

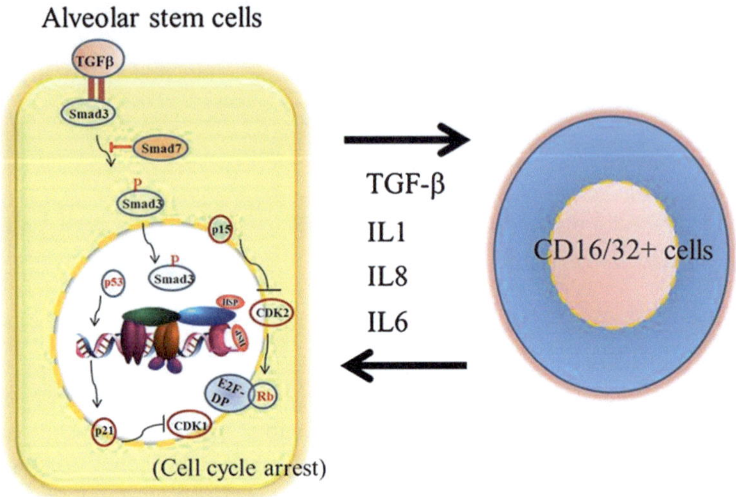

Fig. 10.1 Schematic of alveolar stem cell aging triggered by TGF-β intracellular signaling and p53- and Rb-mediated cell cycle arrest in response to cytokine-mediated cellular interaction with innate immune cells

scriptional repression of *TERT* in epithelial cell lines [43–45]. These data together suggest that the EGF-Ets and TGF-β-Smad3 pathways regulate the *TERT* gene and proliferative potential of AECII in a reciprocally opposing manner. Thus, TGF-β signaling induction of telomerase inhibition, cell proliferative suppression, and myofibroblast transdifferentiation could be a key event in initiating fibrotic lesion.

TGF-β is a key player in antagonizing AECII proliferation but stimulating AECII differentiation [46–53]. To date, it is still unclear which transcriptional targets downstream of TGF-β signaling mediate AECII differentiation and whether TGF-β-induced Smad3-mediated telomerase downregulation plays a part in suppressing AECII proliferation and allowing AECII to undergo differentiation (Fig. 10.1). In the telomerase-deficient mice where we found no evidence of significant fibrotic lesion, TGF-β and its receptor are downregulated [26]. It is possible that activation of the TGF-β signaling pathway may boost α-SMA and Col1α1 increases in the telomerase-deficient lungs. Previous studies showed increased expression of α-SMA in AECII as a marker of AECII undergoing the gene expression involved in myofibroblasts by EMT [40, 49, 54]. TGF-β stimulates α-SMA gene expression through Smad3 interaction with β-catenin on the α-SMA gene promoter in AECII [55]. In addition, TGF-β activates the transient receptor potential vanilloid 4 (TRPV4) channels and actin polymerization, resulting in the formation and nuclear translocation of the myocardin-related transcription factor (MRTF-A)/serum response factor complex and the subsequent stimulation of the α-SMA gene transcription [56]. Since anomalous expression of α-SMA and Col1α1 in AECII is among the features of AECII transdifferentiation to myofibroblasts and suggestive of a fibrotic lesion of pulmonary fibrosis [49, 54], further investigations are required to characterize the regulatory mechanisms underlying the limited temporospatial processes and scales of AECII transdifferentiation triggered by telomere dysfunction.

10.3 Telomere Dysfunction and Inflammation

Aging tissues and age-related diseases are intimately associated with low levels of chronic inflammation [57, 58]. In this regard, inflammation has been demonstrated to contribute to aging in mouse skin [59] and in mouse model of DNA damage-driven progeria [60–62], suggesting synergistic interactions between DNA damage responses (DDRs) and inflammatory signals. Telomere dysfunction-induced DDR is a major cause of cellular senescence [63]. Severe telomere dysfunction is induced by telomere shortening in late-generation telomerase (*terc*$^{-/-}$) knockout mice, where it compromises the function of tissue-specific stem and progenitor cells, limits tissue regenerative capacity, and accelerates aging [64]. Since factors secreted by senescent cells participate in mediating chronic inflammation [65–67], it is anticipated that telomere shortening that causes replicative senescence is associated with SASP directly or indirectly. It is noteworthy that chronic low-grade inflammation enhances telomere dysfunction by increasing ROS-mediated DNA damage and thus accelerates accumulation of senescent cells [68]. A recent study also found that GATA4 is a key regulator of crucial senescent phenotypes and connects DDR to senescence and inflammation through IL1A and TRAF3IP2 activation of NF-κB [69].

However, telomere dysfunction in AECII triggered an inflammatory response with upregulations of cytokine signaling pathways known to provoke inflammation in the lung [26, 70]. A marked increase in proinflammatory cytokines occurs to the telomerase-deficient lung tissues, resulting in a significant spillover into the BAL fluids [26]. The concentrations of IL-6, CXCL15, and TNF-α in the pulmonary parenchyma are increased several folds with spillovers into the BAL fluids significantly in telomerase-deficient mice [26]. These findings of telomere dysfunction-caused SASP profiles of altered cytokines and growth factors in the mouse pulmonary tissues are consistent with persistent telomere DDR in both human and mouse studies [71–73]. Furthermore, the findings of markedly increased IL-1α and IL-1β are consistent with their regulatory roles in increased IL-6 and IL-8 [73], and increased CCL2 in the microenvironment of telomere-induced AECII senescence is consistent with a fundamental role in the recruitment of NK cells [74]. Among the differentially upregulated genes are Il17c, which encode interleukin 17c, and MIF encoding macrophage inhibitory factor, consistent with their known roles in epithelial-derived innate immune responses in the lung [70]. Thus, consistent with recruitment of inflammatory cells in the innate immune pathways, the proinflammatory cytokines constitute a signature of chronic low-grade inflammation incurred by telomerase deficiency and telomere damages [26].

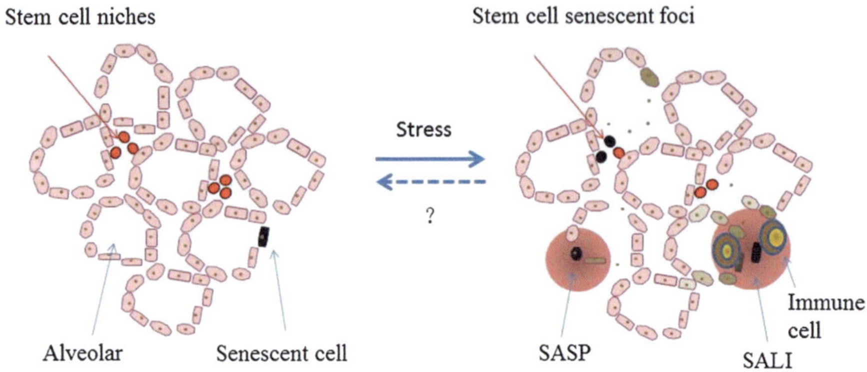

Fig. 10.2 Graphic representation of SALI foci in pulmonary tissue. SALI involves both inflammatory cytokines and cells in the foci of senescent stem cells, by contrast to SASP that defines secreted molecules from the senescent cell. SALI functions to mediate senescence spreading with the fundamental effectors of innate immune cells that interact with senescent cells and generate a progressive radial pattern of cytokine gradients

10.4 Molecular Mechanisms of Telomere-Associated Inflammation

Although senescent AECII cells show no significant increase in proinflammatory cytokine gene expressions, an alternative pathway that leads to SASP cannot be ruled out as an intermediate event in the sterile SALI [26]. Consistently, a large heterogeneous population of mononuclear inflammatory cells is positive for CD45 and/or CD16/32 in the TERC-deficient lungs [26]. It is possible that telomere attrition is a primary damage involved in the low-grade inflammation, a novel phenotype we preferred to term telomere-induced, stem cell senescence-associated, low-grade inflammation (tSALI) [26]. Since telomere DDR triggers the permanent cell cycle arrest of the stem cells lacking telomerase, SALI may generally serve as a potential mechanism to spread cellular senescence (Fig. 10.2). In line with this hypothesis, recent studies have shown that chronic inflammation suffices to induce telomere dysfunction and accelerate aging in mice [68]. Thus, as a primary means of prompting AECII senescence and undermining AECI repair, SALI represents a cellular mechanism of the focal circuit that mediates the spreading of cellular replicative senescence from the original senescent foci in pulmonary aging (Fig. 10.2).

In mediating senescence transmission, SALI assumes a fundamental feature of innate immunity which requires recruitment and activation of inflammatory cells. The participation of inflammatory cells in SALI may be central to the process of senescence progression (Fig. 10.2) and play an obliging role in the full development of aging-related pathologies by provoking cellular transition from senescence to transdifferentiation, immortalization, and transformation. Therefore, by bridging and mediating the development of tissue pathological changes, SALI corresponds to a critical window of intervention in aging-related disease.

Although it remains unknown how telomere-induced alternative SASP (tSASP or A-SASP) and subsequent tSALI are evolved during senescence progression, it is possible that at least four pathways are involved through which A-SASP progresses to tSALI. First, telomere shortening unchecks the telomere position effect (TPE) on the transcriptions of the genes encoding inflammatory cytokines and pathogenic factors [75]. Second, telomere shortening renders epigenetic alterations of heterochromatin formation resulting in activation of a specific group of gene expressions [76]. Third, telomere DNA fragments shedding off chromosomes into the *milieu externe* serve as damage-associated molecular patterns (DAMPs) that perpetuate the tSALI response, by analogy to mitochondrial and mammalian DNA molecules acting as DAMPs [77, 78]. Fourth, compromised NF-κB signaling results in depressed regulatory factors in controlling innate immune response (see below). It has previously been indicated that telomere shortening is associated with telomere DNA and protein accumulations that are involved in complex molecular signaling [3].

Premature telomere erosion in peripheral blood mononuclear cells is a common characteristic of autoimmune syndromes [79], additional to aging-associated diseases including chronic obstructive pulmonary diseases (COPD), neurodegeneration, obesity, and vascular diseases. Telomere loss has been shown to increase the susceptibility to autoimmune disease as telomerase overexpression in T cells serves as a promising therapy for the treatment of autoimmune disease [80]. Several studies have shown that hTERT overexpression in T cells extends their replicative lifespan while maintaining normal cell function [81]. Controversially, two studies have demonstrated chromosomal abnormalities in hTERT-transduced T cells [82, 83], and long-term culture of hTERT-transduced T cells results in accumulation of the cyclin-dependent kinase inhibitors p21 and p16 Ink4a, molecules that mediate cell cycle arrest and replicative senescence [84].

In addition, chronic inflammation aggravates telomere dysfunction and cell senescence, decreases regenerative potential in multiple tissues, and accelerates aging of mice. Anti-inflammatory or antioxidant treatment, specifically COX-2 inhibition, rescued telomere dysfunction, cell senescence, and tissue regenerative potential, indicating that chronic inflammation may accelerate aging at least partially in a cell-autonomous manner via COX-2-dependent hyperproduction of ROS [68]. Persistent inflammation aggravates telomere dysfunction by increasing oxidative stress at least partially through COX-2 activation [68]. This then accelerates accumulation of senescent cells, which intensifies proinflammatory and prooxidant signaling by the SASP response and by induction of mitochondrial dysfunction [85, 86], spreading DNA damage and senescence toward bystander cells [87]. As the cardinal transcriptional regulator of inflammation-related genes including proinflammatory interleukins, chemokines, cytokines, adhesion molecules, and others, NF-κB is not only activated by proinflammatory, stress, and cell senescence signals [88] but also regulated by telomerase. Mice lacking functional telomerase are defective in mounting an acquired immune response following an LPS challenge [89]. Binding directly to the NF-κB p65 subunit, telomerase regulates NF-κB p65 recruitment to a subset of NF-κB promoters such as those of IL-6 and TNF-α [89]. Furthermore, NF-κB transcriptionally upregulates telomerase activity, suggesting a feed-forward regulation between NF-κB and telomerase [89]. How loss of telomerase in AECII would compromise NF-κB-dependent gene expressions and which NF-κB downstream gene(s) might be involved in mediating tSALI require further investigations. Consistent with telomerase involvement in proinflammatory response, a recent report shows that circulating peripheral blood mononuclear cells in patients with metabolic syndromes produce enhanced levels of TNF-α and IL-6 and have high levels of telomerase activity [90].

10.5 Perspectives

Prevention of SALI from occurring would be of significant importance in prohibiting tissue premature aging, including the case of IPF. A major interceptive step to intervene senescence spreading by SALI would be to arrest the trigger of such self-antigen as telomeric DNA, thereby disallowing the generation of cytokine gradients and immune cell chemotactic selection. It is foreseeable that the mechanisms underlying A-SASP will be uncovered to provide the key molecular interfaces for targeting. Investigations on the mechanisms by which tSALI takes place with the involvement of tSASP will include characterization of the molecules that bind with telomeric DNA and telomere-associated DAMPs and the molecules that are involved in innate immune response unchecked by compromised NF-κB signaling in the absence of telomerase as an upregulator. New discoveries of the mechanisms mediating telomere molecular processing and recognition will also involve the cellular receptors and presenters for telomere-triggered innate immune responses. Antagonizing these sites will provide novel molecular targets for effectively designing prophylactic and therapeutic modalities.

References

1. Blackburn EH (2005) Telomeres and telomerase: their mechanisms of action and the effects of altering their functions. FEBS Lett 579:859–862
2. Nandakumar J, Cech TR (2013) Finding the end: recruitment of telomerase to telomeres. Nat Rev Mol Cell Biol 14:69–82. doi:nrm3505 [pii]10.1038/nrm3505
3. Liu JP (1999) Studies of the molecular mechanisms in the regulation of telomerase activity. FASEB J 13:2091–2104
4. Armanios M (2013) Telomeres and age-related disease: how telomere biology informs clinical paradigms. J Clin Invest 123:996–1002. doi:10.1172/JCI66370
5. Kong CM, Lee XW, Wang X (2013) Telomere shortening in human diseases. FEBS J 280:3180–3193. doi:10.1111/febs.12326
6. Mitchell JR, Wood E, Collins K (1999) A telomerase component is defective in the human disease dyskeratosis congenita. Nature 402:551–555
7. Vulliamy T et al (2001) The RNA component of telomerase is mutated in autosomal dominant dyskeratosis congenita. Nature 413:432–435

8. Vulliamy T, Marrone A, Dokal I, Mason PJ (2002) Association between aplastic anaemia and mutations in telomerase RNA. Lancet 359:2168–2170. doi:10.1016/S0140-6736(02)09087-6
9. Yamaguchi H et al (2005) Mutations in TERT, the gene for telomerase reverse transcriptase, in aplastic anemia. N Engl J Med 352:1413–1424
10. Hartmann D et al (2011) Telomerase gene mutations are associated with cirrhosis formation. Hepatology 53:1608–1617. doi:10.1002/hep.24217
11. Calado RT et al (2011) Constitutional telomerase mutations are genetic risk factors for cirrhosis. Hepatology 53:1600–1607. doi:10.1002/hep.24173
12. Armanios MY et al (2007) Telomerase mutations in families with idiopathic pulmonary fibrosis. N Engl J Med 356:1317–1326. doi:10.1056/NEJMoa066157
13. Tsakiri KD et al (2007) Adult-onset pulmonary fibrosis caused by mutations in telomerase. Proc Natl Acad Sci U S A 104:7552–7557. doi:0701009104 [pii], 10.1073/pnas.0701009104
14. Cronkhite JT et al (2008) Telomere shortening in familial and sporadic pulmonary fibrosis. Am J Respir Crit Care Med 178:729–737. doi:10.1164/rccm.200804-550OC
15. Tsang AR, Wyatt HD, Ting NS, Beattie TL (2012) hTERT mutations associated with idiopathic pulmonary fibrosis affect telomerase activity, telomere length, and cell growth by distinct mechanisms. Aging Cell 11:482–490. doi:10.1111/j.1474-9726.2012.00810.x
16. Blasco MA et al (1997) Telomere shortening and tumor formation by mouse cells lacking telomerase RNA [see comments]. Cell 91:25–34
17. Rudolph KL et al (1999) Longevity, stress response, and cancer in aging telomerase-deficient mice. Cell 96:701–712
18. Wong KK et al (2003) Telomere dysfunction and Atm deficiency compromises organ homeostasis and accelerates ageing. Nature 421:643–648
19. Chang S (2005) Modeling aging and cancer in the telomerase knockout mouse. Mutat Res 576:39–53. doi:10.1016/j.mrfmmm.2004.08.020
20. He H et al (2009) Pot1b deletion and telomerase haploinsufficiency in mice initiate an ATR-dependent DNA damage response and elicit phenotypes resembling dyskeratosis congenita. Mol Cell Biol 29:229–240. doi:MCB.01400-08 [pii], 10.1128/MCB.01400-08
21. Strong MA et al (2011) Phenotypes in mTERT(+)/(−) and mTERT(−)/(−) mice are due to short telomeres, not telomere-independent functions of telomerase reverse transcriptase. Mol Cell Biol 31:2369–2379. doi:10.1128/MCB.05312-11
22. Ju Z et al (2007) Telomere dysfunction induces environmental alterations limiting hematopoietic stem cell function and engraftment. Nat Med 13:742–747
23. Lee J et al (2009) Lung alveolar integrity is compromised by telomere shortening in telomerase-null mice. Am J Physiol Lung Cell Mol Physiol 296:L57–L70. doi:10.1152/ajplung.90411.2008
24. Jackson SR et al (2011) Partial pneumonectomy of telomerase null mice carrying shortened telomeres initiates cell growth arrest resulting in a limited compensatory growth response. Am J Physiol Lung Cell Mol Physiol 300:L898–L909. doi:10.1152/ajplung.00409.2010
25. Liu JP, Chen W, Schwarer AP, Li H (2009) Telomerase in cancer immunotherapy. Biochim Biophys Acta. doi:S0304-419X(09)00059-6 [pii], 10.1016/j.bbcan.2009.09.001
26. Chen R et al (2015) Telomerase deficiency causes alveolar stem cell senescence-associated low-grade inflammation in lungs. J Biol Chem 290:30813–30829. doi:10.1074/jbc.M115.681619
27. Alder JK et al (2011) Telomere length is a determinant of emphysema susceptibility. Am J Respir Crit Care Med 184:904–912. doi:10.1164/rccm.201103-0520OC
28. Liu T et al (2007) Telomerase activity is required for bleomycin-induced pulmonary fibrosis in mice. J Clin Invest 117:3800–3809. doi:10.1172/JCI32369
29. Nozaki Y, Liu T, Hatano K, Gharaee-Kermani M, Phan SH (2000) Induction of telomerase activity in fibroblasts from bleomycin-injured lungs. Am J Respir Cell Mol Biol 23:460–465. doi:10.1165/ajrcmb.23.4.3958
30. Liu T et al (2013) Telomerase and telomere length in pulmonary fibrosis. Am J Respir Cell Mol Biol 49:260–268. doi:10.1165/rcmb.2012-0514OC

31. Driscoll B, Buckley S, Bui KC, Anderson KD, Warburton D (2000) Telomerase in alveolar epithelial development and repair. Am J Physiol Lung Cell Mol Physiol 279:L1191–L1198
32. Reddy R et al (2004) Isolation of a putative progenitor subpopulation of alveolar epithelial type 2 cells. Am J Physiol Lung Cell Mol Physiol 286:L658–L667. doi:10.1152/ajplung.00159.2003
33. Kim JK et al (2000) Activation of telomerase by silica in rat lung. Toxicol Lett 111:263–270
34. Fridlender ZG et al (2007) Telomerase activity in bleomycin-induced epithelial cell apoptosis and lung fibrosis. Eur Respir J 30:205–213. doi:10.1183/09031936.00009407
35. Degryse AL et al (2012) Telomerase deficiency does not alter bleomycin-induced fibrosis in mice. Exp Lung Res 38:124–134. doi:10.3109/01902148.2012.658148
36. Nunes H et al (2014) Is telomeropathy the explanation for combined pulmonary fibrosis and emphysema syndrome?: report of a family with TERT mutation. Am J Respir Crit Care Med 189:753–754. doi:10.1164/rccm.201309-1724LE
37. Stanley SE et al (2015) Telomerase mutations in smokers with severe emphysema. J Clin Invest 125:563–570. doi:10.1172/JCI78554
38. Barkauskas CE et al (2013) Type 2 alveolar cells are stem cells in adult lung. J Clin Invest 123:3025–3036. doi:10.1172/JCI68782
39. Desai TJ, Brownfield DG, Krasnow MA (2014) Alveolar progenitor and stem cells in lung development, renewal and cancer. Nature 507:190–194. doi:10.1038/nature12930
40. Watanabe-Takano H, Takano K, Hatano M, Tokuhisa T, Endo T (2015) DA-Raf-mediated suppression of the Ras-ERK pathway is essential for TGF-beta1-induced epithelial-mesenchymal transition in alveolar epithelial type 2 cells. PLoS One 10:e0127888. doi:10.1371/journal.pone.0127888
41. Maida Y et al (2002) Direct activation of telomerase by EGF through Ets-mediated transactivation of TERT via MAP kinase signaling pathway. Oncogene 21:4071–4079
42. Xu D, Dwyer J, Li H, Duan W, Liu JP (2008) Ets2 maintains hTERT gene expression and breast cancer cell proliferation by interacting with c-Myc. J Biol Chem 283:23567–23580., doi:M800790200 [pii]10.1074/jbc.M800790200
43. Cassar L, Li H, Jiang FX, Liu JP (2010) TGF-beta induces telomerase-dependent pancreatic tumor cell cycle arrest. Mol Cell Endocrinol 320:97–105., doi:S0303-7207(10)00060-2 [pii]10.1016/j.mce.2010.02.002
44. Li H, Xu D, Li J, Berndt MC, Liu JP (2006) Transforming growth factor beta suppresses human telomerase reverse transcriptase (hTERT) by Smad3 interactions with c-Myc and the hTERT gene. J Biol Chem 281:25588–25600
45. Yang H, Kyo S, Takatura M, Sun L (2001) Autocrine transforming growth factor beta suppresses telomerase activity and transcription of human telomerase reverse transcriptase in human cancer cells. Cell Growth Differ 12:119–127
46. Zhang F, Nielsen LD, Lucas JJ, Mason RJ (2004) Transforming growth factor-beta antagonizes alveolar type II cell proliferation induced by keratinocyte growth factor. Am J Respir Cell Mol Biol 31:679–686. doi:10.1165/rcmb.2004-0182OC
47. Kasai H, Allen JT, Mason RM, Kamimura T, Zhang Z (2005) TGF-beta1 induces human alveolar epithelial to mesenchymal cell transition (EMT). Respir Res 6:56. doi:10.1186/1465-9921-6-56
48. Bhaskaran M et al (2007) Trans-differentiation of alveolar epithelial type II cells to type I cells involves autocrine signaling by transforming growth factor beta 1 through the Smad pathway. J Biol Chem 282:3968–3976. doi:10.1074/jbc.M609060200
49. Alipio ZA et al (2011) Epithelial to mesenchymal transition (EMT) induced by bleomycin or TFG(b1)/EGF in murine induced pluripotent stem cell-derived alveolar type II-like cells. Differentiation 82:89–98. doi:10.1016/j.diff.2011.05.001
50. Zhao L, Yee M, O'Reilly MA (2013) Transdifferentiation of alveolar epithelial type II to type I cells is controlled by opposing TGF-beta and BMP signaling. Am J Physiol Lung Cell Mol Physiol 305:L409–L418. doi:10.1152/ajplung.00032.2013
51. Warburton D, Shi W, Xu B (2013) TGF-beta-Smad3 signaling in emphysema and pulmonary fibrosis: an epigenetic aberration of normal development? Am J Physiol Lung Cell Mol Physiol 304:L83–L85. doi:10.1152/ajplung.00258.2012

52. Song X et al (2013) All-transretinoic acid ameliorates bleomycin-induced lung fibrosis by downregulating the TGF-beta1/Smad3 signaling pathway in rats. Lab Invest 93:1219–1231., doi:10.1038/labinvest.2013.108
53. Li M et al (2011) Epithelium-specific deletion of TGF-beta receptor type II protects mice from bleomycin-induced pulmonary fibrosis. J Clin Invest 121:277–287. doi:10.1172/JCI42090
54. Willis BC et al (2005) Induction of epithelial-mesenchymal transition in alveolar epithelial cells by transforming growth factor-beta1: potential role in idiopathic pulmonary fibrosis. Am J Pathol 166:1321–1332
55. Zhou B et al (2012) Interactions between beta-catenin and transforming growth factor-beta signaling pathways mediate epithelial-mesenchymal transition and are dependent on the transcriptional co-activator cAMP-response element-binding protein (CREB)-binding protein (CBP). J Biol Chem 287:7026–7038. doi:10.1074/jbc.M111.276311
56. Rahaman SO et al (2014) TRPV4 mediates myofibroblast differentiation and pulmonary fibrosis in mice. J Clin Invest 124:5225–5238. doi:10.1172/JCI75331
57. Chung HY et al (2009) Molecular inflammation: underpinnings of aging and age-related diseases. Ageing Res Rev 8:18–30. doi:10.1016/j.arr.2008.07.002
58. Franceschi C et al (2007) Inflammaging and anti-inflammaging: a systemic perspective on aging and longevity emerged from studies in humans. Mech Ageing Dev 128:92–105. doi:10.1016/j.mad.2006.11.016
59. Adler AS et al (2007) Motif module map reveals enforcement of aging by continual NF-kappaB activity. Genes Dev 21:3244–3257. doi:10.1101/gad.1588507
60. Kawahara TL et al (2009) SIRT6 links histone H3 lysine 9 deacetylation to NF-kappaB-dependent gene expression and organismal life span. Cell 136:62–74. doi:10.1016/j.cell.2008.10.052
61. Osorio FG et al (2012) Nuclear lamina defects cause ATM-dependent NF-kappaB activation and link accelerated aging to a systemic inflammatory response. Genes Dev 26:2311–2324. doi:10.1101/gad.197954.112
62. Tilstra JS et al (2012) NF-kappaB inhibition delays DNA damage-induced senescence and aging in mice. J Clin Invest 122:2601–2612. doi:10.1172/JCI45785
63. d'Adda di Fagagna F et al (2003) A DNA damage checkpoint response in telomere-initiated senescence. Nature 426:194–198., doi:10.1038/nature02118nature02118 [pii]
64. Choudhury AR et al (2007) Cdkn1a deletion improves stem cell function and lifespan of mice with dysfunctional telomeres without accelerating cancer formation. Nat Genet 39:99–105. doi:10.1038/ng1937
65. Campisi J (2013) Aging, cellular senescence, and cancer. Annu Rev Physiol 75:685–705. doi:10.1146/annurev-physiol-030212-183653
66. Coppe JP, Desprez PY, Krtolica A, Campisi J (2010) The senescence-associated secretory phenotype: the dark side of tumor suppression. Annu Rev Pathol 5:99–118. doi:10.1146/annurev-pathol-121808-102144
67. Tchkonia T, Zhu Y, van Deursen J, Campisi J, Kirkland JL (2013) Cellular senescence and the senescent secretory phenotype: therapeutic opportunities. J Clin Invest 123:966–972. doi:10.1172/JCI64098
68. Jurk D et al (2014) Chronic inflammation induces telomere dysfunction and accelerates ageing in mice. Nat Commun 2:4172. doi:10.1038/ncomms5172
69. Kang C et al (2015) The DNA damage response induces inflammation and senescence by inhibiting autophagy of GATA4. Science 349:aaa5612. doi:10.1126/science.aaa5612
70. Alder JK et al (2015) Telomere dysfunction causes alveolar stem cell failure. Proc Natl Acad Sci USA 112:5099–5104. doi:10.1073/pnas.1504780112
71. Kojima H, Kunimoto H, Inoue T, Nakajima K (2012) The STAT3-IGFBP5 axis is critical for IL-6/gp130-induced premature senescence in human fibroblasts. Cell Cycle 11:730–739. doi:10.4161/cc.11.4.19172

72. Savale L et al (2009) Shortened telomeres in circulating leukocytes of patients with chronic obstructive pulmonary disease. Am J Respir Crit Care Med 179:566–571. doi:10.1164/rccm.200809-1398OC
73. Rodier F et al (2009) Persistent DNA damage signalling triggers senescence-associated inflammatory cytokine secretion. Nat Cell Biol 11:973–979., doi:ncb1909 [pii], 10.1038/ncb1909
74. Iannello A, Raulet DH (2014) Immunosurveillance of senescent cancer cells by natural killer cells. Oncoimmunology 3:e27616. doi:10.4161/onci.27616
75. Robin JD et al (2014) Telomere position effect: regulation of gene expression with progressive telomere shortening over long distances. Genes Dev 28:2464–2476. doi:10.1101/gad.251041.114
76. Arnoult N, Van Beneden A, Decottignies A (2012) Telomere length regulates TERRA levels through increased trimethylation of telomeric H3K9 and HP1alpha. Nat Struct Mol Biol 19:948–956. doi:10.1038/nsmb.2364
77. Kaczorowski DJ et al (2012) Mammalian DNA is an endogenous danger signal that stimulates local synthesis and release of complement factor B. Mol Med 18:851–860. doi:10.2119/molmed.2012.00011
78. Zhang Q et al (2010) Circulating mitochondrial DAMPs cause inflammatory responses to injury. Nature 464:104–107. doi:10.1038/nature08780
79. van der Harst P et al (2007) Telomere length of circulating leukocytes is decreased in patients with chronic heart failure. J Am Coll Cardiol 49:1459–1464. doi:10.1016/j.jacc.2007.01.027
80. Weng NP (2008) Telomere and adaptive immunity. Mech Ageing Dev 129:60–66., doi:S0047-6374(07)00181-9 [pii], 10.1016/j.mad.2007.11.005
81. Rufer N et al (2001) Transfer of the human telomerase reverse transcriptase (TERT) gene into T lymphocytes results in extension of replicative potential. Blood 98:597–603
82. Schreurs MW et al (2005) Genomic stability and functional activity may be lost in telomerase-transduced human CD8+ T lymphocytes. Blood 106:2663–2670. doi:10.1182/blood-2004-09-3742
83. Roth A et al (2003) Telomerase levels control the lifespan of human T lymphocytes. Blood 102:849–857
84. Menzel O et al (2006) Mechanisms regulating the proliferative potential of human CD8+ T lymphocytes overexpressing telomerase. J Immunol 177:3657–3668
85. Kuilman T et al (2008) Oncogene-induced senescence relayed by an interleukin-dependent inflammatory network. Cell 133:1019–1031. doi:10.1016/j.cell.2008.03.039
86. Freund A, Orjalo AV, Desprez PY, Campisi J (2010) Inflammatory networks during cellular senescence: causes and consequences. Trends Mol Med 16:238–246. doi:10.1016/j.molmed.2010.03.003
87. Passos JF et al (2007) Mitochondrial dysfunction accounts for the stochastic heterogeneity in telomere-dependent senescence. PLoS Biol 5:e110. doi:10.1371/journal.pbio.0050110
88. Salminen A et al (2008) Activation of innate immunity system during aging: NF-kB signaling is the molecular culprit of inflammaging. Ageing Res Rev 7:83–105. doi:10.1016/j.arr.2007.09.002
89. Ghosh A et al (2012) Telomerase directly regulates NF-kappa B-dependent transcription. Nat Cell Biol 14:1270–1281., doi:ncb2621 [pii]10.1038/ncb2621
90. Rentoukas E et al (2012) Connection between telomerase activity in PBMC and markers of inflammation and endothelial dysfunction in patients with metabolic syndrome. PLoS One 7:e35739. doi:10.1371/journal.pone.0035739

Chapter 11
The Development and Diversity of ILCs, NK Cells and Their Relevance in Health and Diseases

Yuxia Zhang and Bing Huang

Abstract Next to T and B cells, natural killer (NK) cells are the third largest lymphocyte population. They are recently re-categorized as innate lymphocytes (ILCs), which also include ILC1, ILC2, ILC3, and the lymphoid tissue inducer (LTi) cells. Both NK cells and ILC1 cells are designated as group 1 ILCs because they secrete interferon-γ (IFN-γ) and tumor necrosis factor (TNF). However, in contrast to ILC1 and all other ILCs, NK cells possess potent cytolytic functions that resemble cytotoxic T lymphocytes (CTL). In addition, NK cells express, in a stochastic manner, an array of germ line-encoded activating and inhibitory receptors that recognize the polymorphic regions of major histocompatibility class I (MHC-I) molecules and self-proteins. Recognition of self renders NK cell tolerance to self-healthy tissues, but fail to recognize self ('missing-self') leads to activation to neoplastic transformation and infections of certain viruses. In this chapter, we will summarize the development of NK cells in the context of ILCs, describe the diversity of phenotype and function in blood and tissues, and discuss their involvement in health and diseases in humans.

Keywords NK cells • Development • NK receptors • Human disease

Abbreviations

CHILP	Common helper ILC precursor
CLP	Common lymphoid progenitor
CTL	Cytotoxic T lymphocytes
EILP	Earliest ILC progenitors

Y. Zhang (✉) • B. Huang
Guangzhou Institute of Paediatrics, Guangzhou Women and Children's Medical Centre, Guangzhou Medical University, Guangzhou, Guangdong 510623, China
e-mail: yuxia.zhang@gwcmc.org

D. Xu (ed.), *Regulation of Inflammatory Signaling in Health and Disease*,
Advances in Experimental Medicine and Biology 1024,
DOI 10.1007/978-981-10-5987-2_11

Eomes	Eomesodermin
Ets1	ETS proto-oncogene 1
Gata3	GATA-binding protein 3
ID2	Inhibitor of DNA binding 2
IFN-γ	Interferon gamma
ILC	Innate lymphocyte
ILCP	ILC progenitors
JAK1/3	Janus kinase 1/3
LTi	Lymphoid tissue inducer cells
MCMV	Murine cytomegalovirus
Mef	Myeloid elf-1-like factor
MHC-I	Major histocompatibility class I
mTOR	Mechanistic target of rapamycin
NFIL3	Nuclear factor interleukin 3
NK	Natural killer
PD-1	Programmed cell death-1
PDK1	3′-Phosphoinositide-dependent kinase 1
PLZF	Promyelocytic leukemia zinc finger
S1P1	Sphingosine-1-phosphate receptor 1
T-bet	T-cell-specific T-box transcription factor
TCF-1	T-cell factor 1
TNF	Tumor necrosis factor
TOX	Thymocyte selection-associated high-mobility group box
TRAIL	TNF-related apoptosis ligand
Zeb2	Zinc finger E-box-binding homeobox 2

11.1 NK Cells Are a Group of Innate Lymphocytes that Secrete Adaptive Immune Cytokines

The innate immune system is constituted with granulocytes, monocytes, macrophages, and dendritic cells that secrete inflammatory cytokines, as well as innate lymphocytes that secrete adaptive cytokines such as IFN-γ, interleukin (IL)-4, and IL-17. NK cells are the prototypic ILCs, and they were first described in 1975 as being able to naturally kill mouse leukemia cells [1]. Since 2008, the concept of ILCs [2] has been expanded and now includes the related subsets of NK, ILC1, ILC2, ILC3, and the lymphoid tissue inducer (LTi) cells [3]. ILCs are characterized as having lymphoid morphology but lack rearranged antigen-specific receptors and myeloid and dendritic cell phenotypical markers. ILCs develop initially from progenitors in the fetal liver [4, 5] and, later, in the adult bone marrow [6–8]. They subsequently seed mucosal tissues, where they continue to proliferate and become tissue-resident cells and maintain tissue homeostasis. ILCs and T cells share similar transcription factors that govern their differentiation and produce similar key cytokines [2, 9]. Thus, in analogy to T cells, ILCs are subdivided into cytotoxic (NK) and all other "helper"-like subsets that resembles IFN-γ/Th1-, interleukin 4 (IL-4)/Th2-, and IL-17/Th17-secreting $CD4^+$ T helper cells [10].

11.2 ILCs Are Generated from Progenitors Downstream of the Common Lymphoid Progenitor

All ILCs initially derive from the common lymphoid progenitor (CLP). The transition from CLP to ILC-specific transcriptional program is accompanied with differential expression of over 400 genes [4, 5, 11], with temporal requirements for Nfil3 (nuclear factor interleukin 3, also known as E4bp4), TCF-1(T-cell factor 1, encoded by *Tcf7*), and ID2 (inhibitor of DNA binding 2). Nfil3 expression is essential for the development of ILC progenitors prior to their commitment, and it is induced by mesenchymal-derived IL-7 [12–14]. NFIL3 also directly activates ID2 [14, 15]. TCF-1 represses genes critical for stem cell (*Hhex* and *Lmo2*) and pro-B cell (*Spib, Irf8, Ly6d*) function [11], and its loss affects the differentiation of both NK and other ILC subsets [16–18]. ID2 induces a major regulatory shift with broad repression of progenitor cell transcription factor genes and upregulation of critical regulators including Tox (thymocyte selection-associated high-mobility group box) and Gata3 (GATA-binding protein 3) [11]. Thus, immedicably downstream of the CLP, the earliest ILC progenitor (EILP) is $TCF\text{-}1^+$ [17], which further becomes $ID2^{hi}$ common helper ILC precursor (CHILP) when NK cell potential is lost [6, 14, 19, 20]. After acquisition of promyelocytic leukemia zinc finger protein (PLZF, encoded by *Zbtb16*), ILC progenitor (ILCP) loses the capacity to differentiate into LTi cells [5, 6]. Programmed cell death-1 (PD-1) is co-expressed with PLZF and can be used as a cell surface marker to identify ILCP [11] (Fig. 11.1).

11.3 NK Cells Develop Through Immature and Mature Stages

In the adult mouse bone marrow, pre-NK cell progenitor (pre-NKP) downstream of CLP (Lin^-Flt3^+ $CD27^+CD244^+$ $CD127^+CD122^-Ly6D^-$) has a $Lin^-Flt3^-CD27^+CD244^+CD127^+CD122^-$ surface phenotype, which further develop into rNKP (recently re-defined NK progenitor) that expresses CD122 [21, 22]. CD122 couples with the common γ-chain (CD132) and forms the IL-2/IL-15 receptor, allowing NK cells to respond to IL-15 and activate JAK1/3 and STAT5 [23–25]. IL-15 also activates 3′-phosphoinositide-dependent kinase 1 (PDK1)-mTOR and regulates Nfil3 and CD122 expression [26]. rNKP develops through an immature NK cell (iNK) stage to become mature NK (mNK) cells. iNK expresses NK1.1 but does not express CD49b (antigen to DX5). The expression of Ly49 receptors on the developing iNK cells is critical for NK cell education and maturation and for the detection of

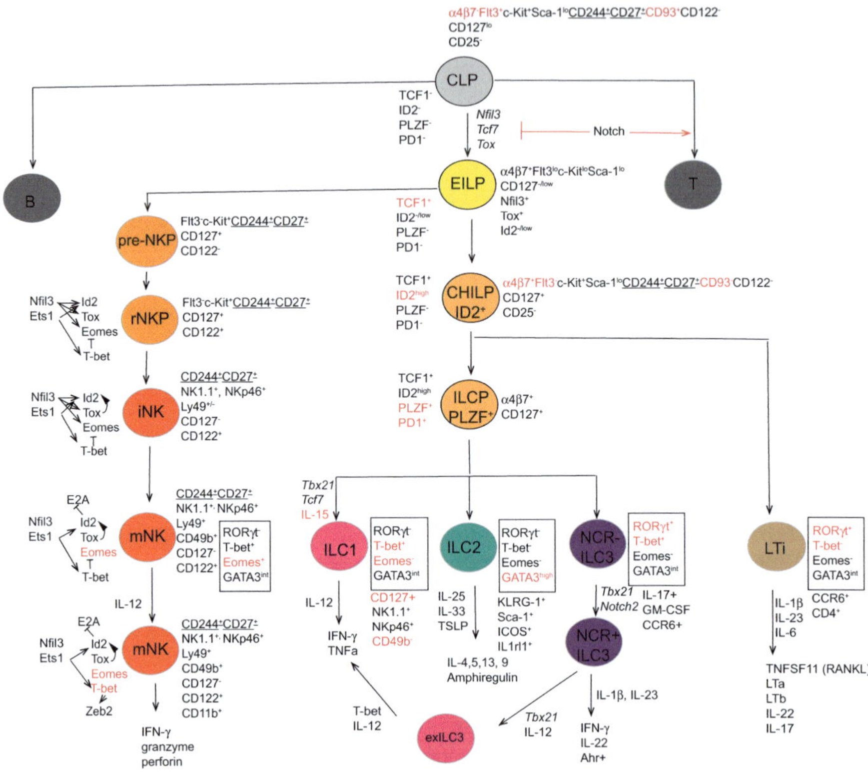

Fig. 11.1 NK and helper ILCs development in mice
T, B, NK and all other helper ILCs develop from common lymphoid progenitor (CLP). NK and ILCs development accompany with sequential differential acquirement of hundreds of transcription factors: Nfil3 and Tcf1 are required for the development and commitment of early ILC progenitors (EILP). Expression of ID2 leads to the commitment of common helper ILC precursor (CHILP), which is not able to further develop into NK cells. When PLZF is expressed, ILC progenitor (ILCP) is formed and its LTi potential is lost. Downstream of EILP, pre-NK progenitor (pre-NKP) develops into re-defined NK progenitor (rNKP) that expresses CD122, which couples with CD132 to form the IL-2/IL-15 receptor, allowing NK cells to respond to IL-15. rNKP then develops through an immature stage (iNK) to become mature NK cells (mNK). Nfil3, Tcf1, Ets1, Id2, Eomes, T-bet and Zeb2 governs NK cell development from EILP to mNK cells

invading pathogens, such as murine cytomegalovirus (MCMV) [27, 28]. The most iNK-cell-proximal mNK cells are $CD27^+CD11b^-$, produce IFN-γ and TNF-α when activated, but are not yet fully cytotoxic effector cells. Cytotoxic capacity improves with NK cell maturation by type I interferons (IFN-α or IFN-β) or proinflammatory cytokines IL-2, IL-15, IL-12, and IL-18, which upregulate CD11b through T-bet and zinc finger E-box-binding homeobox 2 (Zeb2) [29, 30]. Of note, iNK cells in the bone marrow differentiate through four stages sequentially as $CD27^-CD11b^-$, $CD27^+CD11b^-$, $CD27^+CD11b^+$, and $CD27^-CD11b^+$ [31, 32]. Besides CD11b and Dx5, mature NK cells also highly express KLRG1, CD62L, and CD43 [32].

Apart from Tcf1 and Nfil3 [8, 13–15, 17, 33] required for EILP commitment, Ets1, Id2, Eomes, and T-bet are transcription factors essential for NK cell development. Ets1, required for early NK cell lineage commitment, induces *Id2*, *Tbx21*, and *Il2rb* (CD122) expression [34–36]. Id2 suppresses E protein target genes (e.g., *Socs3*, *Tcf7*, *Cxcr5*), and the suppression of Socs3 promotes NK cell response to IL-15 [37, 38]. IL-15 is crucial for NK cell survival through the induction of the anti-apoptotic protein Bcl-2 [39, 40]. Eomesodermin (Eomes) and T-bet are members of the T-box family of transcription factors and are required by iNK and mNK cells [41]. However, tissue-resident NK cells may exhibit different developmental reliance on T-bet and Eomes [42].

NK cell maturation and function are regulated by an additional group of transcription factors. These include the Ets family protein myeloid elf-1-like factor (Mef, also known as ELF4) [43] and PU.1 (encoded by *Spi1*) [44], which respectively regulate perforin expression and NK cell proliferation in response to IL-2 and IL-12. PR domain zinc finger protein 1 (Blimp1, encoded by *Prdm1*), induced by IL-15 in a T-bet-dependent manner during early NK cell development, promotes granzyme B expression but inhibits NK cell maturation and proliferation to low concentrations of IL-15 [45]. Tox regulates mNK development partially through the induction of Id2 [46]. The Ikaros family member Aiolos (encoded by *Ikzf3*) promotes IFN-γ expression; however, its absence enhances the ability of NK cells to control tumor cells [47]. Kruppel-like factor 2 (Klf2) restricts iNK cell proliferation but is required for migration of NK cells toward IL-15-rich microenvironment [48]. IFN regulatory factor 2 (Irf2) is required for NK cell maturation in the periphery and survival in bone marrow. At homeostatic state, Gata3 is required for bone marrow NK cell maturation from $CD27^{+}CD11b^{-}$ stage and for bone marrow egress, liver migration, and IFN-γ expression. In the face of infection, Gata3-deficient NK cells demonstrated inferior control of *Listeria monocytogenes* burden in the liver [49]. However, Gata3-deficient NK cells exhibited superior activity toward MCMV due to increased CD25 expression [50]. Discrepancies regarding forkhead box protein O1 (Foxo1) exist in the literature. In one report, Foxo1 was shown to be required for iNK cell survival by inducing autophagy that removes damaged mitochondria and intracellular reactive oxygen species (ROS) [51]. In another report, however, Foxo1 inhibited late-stage NK cell maturation and function by downregulating *Tbx21* expression [52].

11.4 Tissue-Resident NK Cells Acquire Unique Phenotype and May Have Distinct Developmental Pathways

Tissue-resident NK (trNK) cells often express CD69, CD103 (αE integrin), and CD49a (α1 integrin), which are involved in retaining NK cells in the tissues. CD69 inhibits type I interferon-induced expression of sphingosine-1-phosphate receptor 1 (S1P1). S1P1 and S1P5 on NK cells binds to sphingosine-1-phosphate (S1P), which forms a gradient with the highest concentration in peripheral blood and, thereby, promotes egress of lymphocytes from tissues into the blood [53, 54]. CD103 forms a heterodimer with β7 integrin and binds to E-cadherin on epithelial cells [55]. CD49a forms a heterodimer with β1 integrin and binds to collagen [56]. The expression of both CD103 and CD49a is regulated by transforming growth factor-β (TGF-β1) [57]. Development of trNK cells may be different from conventional blood NK cells. $CD49a^+$ $DX5^-$ $Trail^+$ trNK cells in the mouse liver express higher amount of TNF-α and GM-CSF than blood and spleen conventional NK cells, and they develop in a T-bet-dependent manner in the absence of Nfil3 [41, 42]. $CD49a^+DX5^-$ NK cells that resemble liver trNK cells are also observed in the mouse uterus and skin [42]. In contrast, salivary glands [58] and uterine NK cells [59–61] develop require Eomes in the absence of Nfil3. In addition, a population of $CD127^+$ NK cells develop in Gata3- and IL-7-dependent manner independently from T-cell precursors in the mouse thymus, and thymic trNK cells demonstrate reduced granzyme B but increased IFN-γ, GM-CSF, and TNF expression [62, 63].

11.5 NK Cell Diversity and Activity Are Regulated by Variegated Surface Receptors

The activities of NK cells are regulated by various germ line-encoded activating or inhibitory receptors (Table 11.1), many of which are expressed in stochastic patterns, resulting in many subsets of functionally distinct NK cells [64–66]. The families of NK receptors that recognize MHC class I include the murine Ly49 receptors, the primate killer cell immunoglobulin-like receptors (KIRs), and the CD94-NKG2 receptors in both rodents and primates [65]. Inhibitory receptors in humans and rodents normally contain one or more intracellular immunoreceptor tyrosine-based inhibitory motifs (ITIM) that can activate downstream SHP-1, SHP-2, and SHIP phosphatase [67, 68]. Many of the activating receptors lack intracellular signaling motifs and transduce signals via the association with immunoreceptor tyrosine-based activating motif (ITAM)-containing adapters DAP12, FcεRγ, and CD3ζ, which recruit and activate Syk or ZAP70 tyrosine kinases [69]. NKG2D ligands are self-proteins related to MHC class I molecules. They are generally absent on the cell surface of healthy cells but are frequently upregulated upon cellular stress [70]. NKG2D recruits DAP10 and mediates signaling through the activation of PI3K [71, 72] and ERK [73]. Human KIRs contain either two (KIR2D) or three (KIR3D) extracellular immunoglobulin (Ig)-like domains. They are designated as KIR2DL or KIR3DL, respectively, if they possess a long cytoplasmic domain containing ITIM motif. KIR2DS and KIR3DS have short cytoplasmic domains lacking ITIM but associate through a charged residue in their transmembrane regions with DAP12 or FceRIγ, respectively. KIR2D receptors typically recognize human HLA-C alleles,

Table 11.1 The activating and inhibiting NK receptors on mouse and human NK cells

Activating NK receptors				Inhibiting NK receptors			
Gene	Common name	Ligand	Species	Gene	Common name	Ligand	Species
Klra4	Ly49D	H-2D^d	NOD Hamster	*Klra1,3,7,9*	Ly49A, C, G2, I	H-2D^d H2-M3	NOD B6 BALB
Klra8	Ly49H	MCMV-m157	NOD B6	*Klrb1b, d*	NKR-P1B, D	Ocil (Clr-b)	mouse
Klra16	Ly49P	MCMV	129	*Lilrb4*	Gp49b1	αvβ3 integrin	Mouse
Klrb1c	NK1.1 NKR-P1C		Mouse	*Pilra*	PILRa	O-glycosylated CD99	Mouse
Klrb1f	NKR-P1F	Clrg	Mouse	*SIGLEC-E*		Sialic acid	Mouse
Pilrb1	PILRβ	o-glycosylation CD99	Mouse	*CD244*	2B4	CD48	Mouse
ITGAL	LFA-1 CD11a	ICAM-1, 2, 3	Mouse Human	*KLRG1*	Mafa	E-, N-, R-cadherins	Mouse Human
KLRD1-KLRC2,3	CD94-NKG2C, E	Mouse Qa-1^b; Human HLA-E	Mouse Human	*KLRD1-KLRC1*	CD94-NKG2A	Mouse Qa-1^b Human HLA-E	Mouse Human
KLRK1	NKG2D	Mouse Rae-1, H60, MULT1; Human MICA, MICB, ULBP1-6	Mouse Human	*LAIR1*	LAIR-1 CD305	Collagen XVII	Mouse Human
NCR1	NKp46	Hemagglutinin	Mouse Human	*KLRB1*	NKR-P1A, CD161	LLT1 (CLEC2D)	Human
FCGR3	CD16	IgG	Mouse Human	*LILRB1*	ILT2 LIR1 CD85j	HLA-class I	Human

(continued)

Table 11.1 (continued)

Activating NK receptors				Inhibiting NK receptors			
Gene	Common name	Ligand	Species	Gene	Common name	Ligand	Species
CD226	DNAM-1	CD112 CD155	Mouse Human	*KIR2DL1-3,5*	CD158	HLA-C	Human
CD2	LFA-2	CD58	Human	*KIR3DL1,2*	CD158	HLA-Bw4 Some HLA-A	Human
CD244	2B4	CD48	Human	*CEACAM1*	CD66a	CD66	Human
CD48		2B4	Human	*SIGLEC7*	CDw328	Ganglioside GD3	Human
KLRF2	NKp65	KACL	Human	*SIGLEC9*		Sialic acid	Human
KLRF1	NKp80	AICL	Human	*SIGLEC10*		CD52	Human
NCR2	NKp44		Human	*TIGIT*		CD155/PVR CD112/Nectin-2/ CD112	Human
NKC3	Nkp30	B7-H6	Human	*CD96*		CD155	
KIR2DS	CD158	HLA-class I	Human				
KIR3DS	CD158	HLA-class I	Human				
SLAMF7	CRACC	CRACC	Human				
SLAMF6	NTB-A Ly108 CD352	NTB-A					

whereas KIR3D receptors recognize HLA-B or some HLA-A alleles [74, 75]. The NKG2 family contains one inhibitory NKG2A and two activating members NKG2C and NKG2E. The CD94-NKG2 receptors recognize nonclassical MHC-I that is HLA-E in humans and its ortholog Qa-1 in mice [76–78]. A subset of human NK cells express KIR-related inhibitory receptor, LILRB1, which recognizes a shared epitope in all human MHC class I proteins [79].

NK cells also express activating and inhibitory receptors that recognize non-MHC ligands [80]. For example, murine CD244 (2B4) recognizes CD48, an interaction essential for the IL-2-driven expansion and activation of NK cells [81]; human NKR-P1A (CD161) recognizes the lectin-like transcript-1 (LLT1, encoded by *Clec2d*), which is expressed on activated dendritic cells and B cells and inhibits NK cell cytotoxicity and IFN-γ expression [82, 83]; killer cell lectin-like receptor G1 (KLRG1) recognizes cadherins and mediates 'missing-self' education [84]; Gp49B1 recognizes αvβ3 integrin and inhibits IFN-γ expression [85, 86]. The activating DNAX accessory molecule-1 (DNAM-1, also known as CD226) [87–89] and the inhibiting T-cell immunoreceptor with Ig and ITIM domains (TIGIT) [90, 91] receptors both recognize poliovirus receptor (PVR, also known as CD155) and poliovirus receptor-related 2 (PVRL2, also known as nectin-2 and CD112), which are frequently expressed on transformed or stressed cells.

During NK cell development, the expression of self-MHC class I-reactive inhibitory receptors 'licenses' NK cells. Under physiological conditions, licensed NK cells engage through the Ly49 and KIR inhibitory receptors with MHC class I and prevent NK cells from attacking self, and this self-tolerance is mediated through the recruitment of SHP-1, SHP-2, and SHIP phosphatase [67, 68]. Interestingly, licensed NK cells are more potent in their cytotoxicity toward MHC class I-deficient target cells and secrete more IFN-γ and TNF-a under noninflammatory conditions [92, 93]. During infection, however, inhibitory receptor engagement impairs the ability of licensed NK cells to control cytomegalovirus (CMV) infection [93]. The absence of inhibitory receptors on NK cells can have a beneficiary effect in human leukemia patients receiving irradiation therapy followed by bone marrow transplantation. The absence or mismatch of donor NK inhibitory KIR receptors with recipient MHC-I was associated with better leukemic cell clearance and graft acceptance [94].

Activating receptors have the ability to recognize 'altered-self', which is often induced on malignant or stressed cells [95], and trigger NK cells to kill their targets. NK cells mediate target-cell killing by a number of mechanisms, including (1) the secretion of cytokines, (2) exocytosis of cytoplasmic granules containing perforin and enzyme, (3) FAS ligand and TNF-related apoptosis ligand (TRAIL)-mediated induction of apoptosis, and (4) CD16 cross-linking and antibody-dependent cell-mediated cytotoxicity (ADCC) [94]. However, when NK cells are chronically exposed to endogenous, as well as foreign ligands recognized by their activating receptors, they are tolerated through either receptor downregulation or hyporesponsiveness [65]. NK cell tolerance mediated by activating receptors is reversible and can be broken in the presence of inflammatory cytokines or infection. For instance, in C57BL/6 mice receiving MCMV infection, initially both licensed and

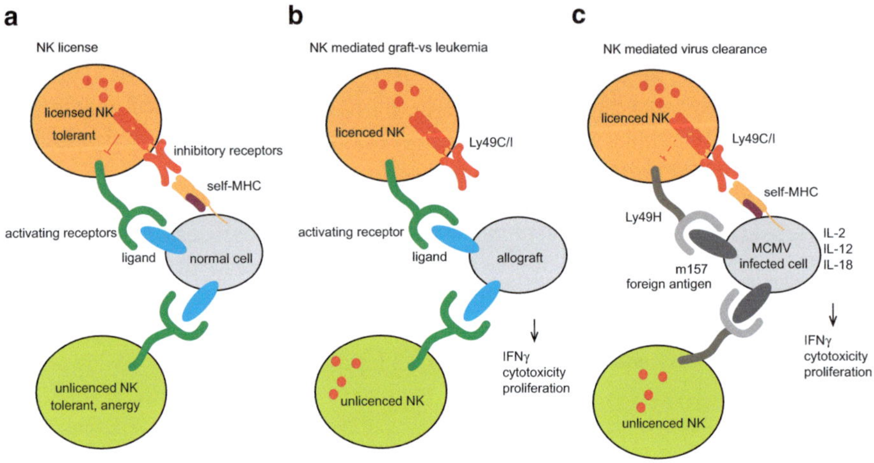

Fig. 11.2 NK cell license, activation and inhibition
(**a**) NK cell license occurs with the expression of self-MHC class I-reactive inhibitory receptors, Ly49 in mice and KIR in man. This prevents NK cells from attacking self. In the absence of inhibitory receptors, chronic exposure of activating receptors with their ligands can also render NK cell hyporesponsive. (**b**) NK cell activation takes place under instances of human leukemia patients receiving irradiation therapy followed by bone marrow transplantation. The absence or mismatch of donor inhibitory NK receptors with recipient MHC-class I promotes leukemic cell clearance by both licensed and unlicensed NK cells. (**c**) During viral infection, inhibitory receptors on licensed NK cells inhibit their proliferation burst, and under these circumstances, unlicensed NK cells are the main mediators of viral clearance

unlicensed NK cells expressed CD69 and upregulated IFN-γ and granzyme B at similar level, but, subsequently unlicensed NK cells predominated in numbers and were the main mediators of viral clearance. The engagement of the activating Ly49H receptor with MCMV-encoded glycoprotein m157 on infected cells promoted unlicensed NK cells to undergo a proliferative burst, but the inhibitory receptors on licensed NK cells restrained the proliferation through SHP-1 phosphatase signaling [68, 93] (Fig. 11.2).

11.6 NK Cells Participate in Tissue Remodeling in Humans and Undergo Clonal-Like Expansion During Viral Infection

A mouse analog of human NK progenitor has been defined as $Lin^{-}CD34^{+}CD38^{+}CD123^{-}CD45RA^{+}CD7^{+}CD10^{+}CD127^{-}$, which selectively gives rise to NK cells *in vitro* and *in vivo* [96]. Circulating human NK cells are a diverse population. In any given individual, the diversity is generated by the developmentally distinct NK cell subsets, KIR gene content, polymorphisms, and copy number variations [64], with differentiation and reprogramming in response to tissue-specific environment and infections [97]. Transcriptional, telomere length, and transfer of human NK cells into NOD/SCID/$\gamma c^{-/-}$ mice have demonstrated that circulating NK cells in human blood display sequential $CD56^{bright}$ $CD62L^{+}$, $CD56^{dim}CD62L^{+}$ $CD94^{high}$, and $CD56^{dim}CD62L^{-}$ $CD94^{low}$ developing stages [98, 99]. $CD56^{bright}$ $CD62L^{+}$ NK cells are mostly KIR^{-} $NKG2A^{+}CD27^{dim}$ $CD57^{-}CD16^{+/-}$ but express CD127 and CD117 (also known as KIT and SCFR), which are also hallmarks of non-NK ILCs [2, 100]. Upon stimulation with combinations of IL-12, IL-15, and IL-18, $CD56^{bright}$ $CD62L^{+}$ and $CD56^{dim}CD62L^{+}$ NK cells strongly proliferate and produce significantly greater amount of IFN-γ than $CD56^{dim}CD62L^{-}$ NK cells. However, engagement of the activating receptors evokes more prominent chemokine (MIP-1α, MIP-1β and RANTES) and cytokine (IFN-γ) expression and NK cell cytotoxicity in $CD56^{dim}CD62L^{+}$ and $CD56^{dim}CD62L^{-}$ cells. [98, 101, 102]. $CD56^{dim}$ NK cells can further develop with the sequential loss of NKG2A and the acquisition of KIRs and CD57 [103]. $CD56^{dim}CD57^{+}$ NK cells have increased cytotoxic capacity than $CD56^{dim}CD57^{-}$ NK when they are activated through CD16 [104].

In parallel to mice, human tissue-resident NK cells also express CD69, CD103, and CD49a, and they may derive directly from progenitors that reside within the tissues [97]. NK cells are found at high frequencies in the endometrium of human uterus and decidua in the first trimester of pregnancy. Throughout the second half of the menstrual cycle, progesterone from the ovaries acts on uterine stromal cells, which in turn secrete IL-15 and support uterine NK cell proliferation [105]. During pregnancy, a key role for $CD56b^{right}$ uNK cells is to promote trophoblast invasion and maternal spiral artery remodeling, which is mediated through the production of IL-8, interferon-inducible protein-10 (IP10), and an array of angiogenic factors including vascular endothelial growth factor A (VEGF-A), VEGF-C, and angiopoietins [99, 106]. Critically, fetal trophoblasts, which come into direct contact with maternal blood and tissues during pregnancy, are exempt from uNK-mediated cell killing. Uterine $CD56^{bright}$ $CD49a^{+}CD103^{+}CD9^{+}$ NK cells express perforin, granzymes A and B, and the activating receptors NKp30, NKp44, NKp46, NKG2D but are unable to form mature activating synapses and thus are not cytotoxic [107, 108]. Furthermore, the high expression of inhibitory KIRs (KIR2DL1, KIR2DL2, KIR2DL3), the CD94-NKG2A receptor complex, and the LILRB1 inhibit NK cell activation through the recognition of HLA-C, HLA-E, and HLA-G expressed on the extravillous trophoblasts, respectively [107, 109]. Interestingly, primary villous trophoblasts do not express HLA, and extravillous trophoblasts are devoid of HLA-A and HLA-B.

In liver sinusoids, NK cells represent up to 30–40% of all hepatic lymphocytes [110], and $CD56^{bright}$ and $CD56^{dim}$ cells are present in equal proportions [111, 112]. Hepatic $CD56^{bright}$ NK cells express CD69 and are tissue resident [113, 114]. Liver resident macrophages (Kupffer cells) interact with NK cells to keep immune tolerance to nonpathogenic antigens from food and LPS from gut commensal bacteria, but remain alert to infections by pathogens and viruses. In recognition of bacterial cell wall products via TLR2/4 -MyD88, Kupffer cells secrete IL-10 and blunt NK cell activation [115]. However, when DNA or RNA viruses activate the TLR3-TRIF-IRF-3 [115] or TLR8 pathways [116], Kupffer cells elicit potent IFN-γ and TNF expression in $CD56^{bright}$ trNK cells. Intrahepatic NK cells also mediate target-cell killing through the expression of TRAIL, whose expression is correlated with the control of hepatitis C virus (HCV) infection [117]. But during HBV infection, TRAIL also causes liver damage and can eliminate antigen-specific T cells [118, 119].

Clonal-like expansion and memory formation of NK cells have been observed in humans with cytomegalovirus (HCMV) [120–123], chikungunya virus (CHIKV) [124] and hantavirus [125] infections. Clonal-expanded cells are characterized by the expression of NKG2C, CD57, and activating KIRs (KIR2DS4, KIR2DS2, KIR3DS1), a general lack of the expression of inhibitory NKG2A and KIR3DL1 receptors (in individuals expressing its HLA-Bw4 ligand), and the decreased expression of CD161 (also known as KLRB1), NKp30, NKp46, and CD7. A subset of clonal-expanded NK cells can further acquire adaptive phenotypes that resembles more with cytotoxic $CD8^+$ T lymphocytes than conventional NK cells. The intronic region of *ZBTB16* in adaptive NK cells is hypermethylated, which is correlated with the decreased expression of PLZF and its target genes encoding $Fc\varepsilon R\gamma$, SYK, and EAT-2. Adaptive PLZF-deficient NK cells are distinct from clonal-expanded NK cells expressing CD57, NKG2C and PLZF, and produce less IFN-γ upon cytokine stimulation with IL-12 and IL-18 [126].

11.7 Conclusion

NK cells are a heterogeneous population of innate lymphocytes that develop from the common lymphoid progenitors. Tissue-resident NK cells may have different developmental origins and are phenotypically distinct from their blood counterparts. NK cells employ both inhibiting and activating receptors for 'missing-self' education, activation, and terminal differentiation. In humans, NK cells are critical for the implantation of the embryos and for the control of neoplastic transformation and viral infections, but they may also induce collateral damages to the tissues. Despite lacking rearranged antigen-specific receptors, NK cells can acquire adaptive T-cell features by clonal-like expansion and alteration in their DNA methylation profiles during viral infections.

Acknowledgments Y.Z. is supported by the Guangzhou Women and Children's Medical Center Start-up Fund (5001-3001032).

References

1. Kiessling R, Klein E, Wigzell H (1975) "Natural" killer cells in the mouse. I. Cytotoxic cells with specificity for mouse Moloney leukemia cells. Specificity and distribution according to genotype. Eur J Immunol 5:112–117

2. Spits H, Artis D, Colonna M, Diefenbach A, Di Santo JP, Eberl G, Koyasu S, Locksley RM, Mckenzie AN, Mebius RE, Powrie F, Vivier E (2013) Innate lymphoid cells—a proposal for uniform nomenclature. Nat Rev Immunol 13:145–149
3. Mebius RE, Rennert P, Weissman IL (1997) Developing lymph nodes collect CD4+CD3-LTbeta+ cells that can differentiate to APC, NK cells, and follicular cells but not T or B cells. Immunity 7:493–504
4. Chea S, Schmutz S, Berthault C, Perchet T, Petit M, Burlen-Defranoux O, Goldrath AW, Rodewald HR, Cumano A, Golub R (2016) Single-cell gene expression analyses reveal heterogeneous responsiveness of fetal innate lymphoid progenitors to notch signaling. Cell Rep 14:1500–1516
5. Ishizuka IE, Chea S, Gudjonson H, Constantinides MG, Dinner AR, Bendelac A, Golub R (2016) Single-cell analysis defines the divergence between the innate lymphoid cell lineage and lymphoid tissue-inducer cell lineage. Nat Immunol 17:269–276
6. Constantinides MG, Mcdonald BD, Verhoef PA, Bendelac A (2014) A committed precursor to innate lymphoid cells. Nature 508:397–401
7. Klose CS, Flach M, Mohle L, Rogell L, Hoyler T, Ebert K, Fabiunke C, Pfeifer D, Sexl V, Fonseca-Pereira D, Domingues RG, Veiga-Fernandes H, Arnold SJ, Busslinger M, Dunay IR, Tanriver Y, Diefenbach A (2014) Differentiation of type 1 ILCs from a common progenitor to all helper-like innate lymphoid cell lineages. Cell 157:340–356
8. Yu X, Wang Y, Deng M, Li Y, Ruhn KA, Zhang CC, Hooper LV (2014) The basic leucine zipper transcription factor NFIL3 directs the development of a common innate lymphoid cell precursor. Elife 3
9. Vivier E, Van De Pavert SA, Cooper MD, Belz GT (2016) The evolution of innate lymphoid cells. Nat Immunol 17:790–794
10. Diefenbach A, Colonna M, Romagnani C (2017) The ILC world revisited. Immunity 46:327–332
11. Seillet C, Mielke LA, Amann-Zalcenstein DB, Su S, Gao J, Almeida FF, Shi W, Ritchie ME, Naik SH, Huntington ND, Carotta S, Belz GT (2016) Deciphering the innate lymphoid cell transcriptional program. Cell Rep 17:436–447
12. Geiger TL, Abt MC, Gasteiger G, Firth MA, O'connor MH, Geary CD, O'sullivan TE, Van Den Brink MR, Pamer EG, Hanash AM, Sun JC (2014) Nfil3 is crucial for development of innate lymphoid cells and host protection against intestinal pathogens. J Exp Med 211:1723–1731
13. Seillet C, Rankin LC, Groom JR, Mielke LA, Tellier J, Chopin M, Huntington ND, Belz GT, Carotta S (2014b) Nfil3 is required for the development of all innate lymphoid cell subsets. J Exp Med 211:1733–1740
14. Xu W, Domingues RG, Fonseca-Pereira D, Ferreira M, Ribeiro H, Lopez-Lastra S, Motomura Y, Moreira-Santos L, Bihl F, Braud V, Kee B, Brady H, Coles MC, Vosshenrich C, Kubo M, Di Santo JP, Veiga-Fernandes H (2015) NFIL3 orchestrates the emergence of common helper innate lymphoid cell precursors. Cell Rep 10:2043–2054
15. Male V, Nisoli I, Kostrzewski T, Allan DS, Carlyle JR, Lord GM, Wack A, Brady HJ (2014) The transcription factor E4bp4/Nfil3 controls commitment to the NK lineage and directly regulates Eomes and Id2 expression. J Exp Med 211:635–642
16. Mielke LA, Groom JR, Rankin LC, Seillet C, Masson F, Putoczki T, Belz GT (2013) TCF-1 controls ILC2 and NKp46+RORgammat+ innate lymphocyte differentiation and protection in intestinal inflammation. J Immunol 191:4383–4391
17. Yang Q, Li F, Harly C, Xing S, Ye L, Xia X, Wang H, Wang X, Yu S, Zhou X, Cam M, Xue HH, Bhandoola A (2015b) TCF-1 upregulation identifies early innate lymphoid progenitors in the bone marrow. Nat Immunol 16:1044–1050
18. Yang Q, Monticelli LA, Saenz SA, Chi AW, Sonnenberg GF, Tang J, De Obaldia ME, Bailis W, Bryson JL, Toscano K, Huang J, Haczku A, Pear WS, Artis D, Bhandoola A (2013) T cell factor 1 is required for group 2 innate lymphoid cell generation. Immunity 38:694–704

19. Klose CS, Kiss EA, Schwierzeck V, Ebert K, Hoyler T, D'hargues Y, Goppert N, Croxford AL, Waisman A, Tanriver Y, Diefenbach A (2013) A T-bet gradient controls the fate and function of CCR6-RORgammat+ innate lymphoid cells. Nature 494:261–265
20. Seehus CR, Aliahmad P, De La Torre B, Iliev ID, Spurka L, Funari VA, Kaye J (2015) The development of innate lymphoid cells requires TOX-dependent generation of a common innate lymphoid cell progenitor. Nat Immunol 16:599–608
21. Carotta S, Pang SH, Nutt SL, Belz GT (2011) Identification of the earliest NK-cell precursor in the mouse BM. Blood 117:5449–5452
22. Fathman JW, Bhattacharya D, Inlay MA, Seita J, Karsunky H, Weissman IL (2011) Identification of the earliest natural killer cell-committed progenitor in murine bone marrow. Blood 118:5439–5447
23. Eckelhart E, Warsch W, Zebedin E, Simma O, Stoiber D, Kolbe T, Rulicke T, Mueller M, Casanova E, Sexl V (2011) A novel Ncr1-Cre mouse reveals the essential role of STAT5 for NK-cell survival and development. Blood 117:1565–1573
24. Nandagopal N, Ali AK, Komal AK, Lee SH (2014) The critical role of IL-15-PI3K-mTOR pathway in natural killer cell effector functions. Front Immunol 5:187
25. Park SY, Saijo K, Takahashi T, Osawa M, Arase H, Hirayama N, Miyake K, Nakauchi H, Shirasawa T, Saito T (1995) Developmental defects of lymphoid cells in Jak3 kinase-deficient mice. Immunity 3:771–782
26. Yang M, Li D, Chang Z, Yang Z, Tian Z, Dong Z (2015a) PDK1 orchestrates early NK cell development through induction of E4BP4 expression and maintenance of IL-15 responsiveness. J Exp Med 212:253–265
27. Babic M, Pyzik M, Zafirova B, Mitrovic M, Butorac V, Lanier LL, Krmpotic A, Vidal SM, Jonjic S (2010) Cytomegalovirus immunoevasin reveals the physiological role of "missing self" recognition in natural killer cell dependent virus control in vivo. J Exp Med 207:2663–2673
28. Lanier LL (2008) Evolutionary struggles between NK cells and viruses. Nat Rev Immunol 8:259–268
29. Kim S, Iizuka K, Kang HS, Dokun A, French AR, Greco S, Yokoyama WM (2002) In vivo developmental stages in murine natural killer cell maturation. Nat Immunol 3:523–528
30. Van Helden MJ, Goossens S, Daussy C, Mathieu AL, Faure F, Marcais A, Vandamme N, Farla N, Mayol K, Viel S, Degouve S, Debien E, Seuntjens E, Conidi A, Chaix J, Mangeot P, De Bernard S, Buffat L, Haigh JJ, Huylebroeck D, Lambrecht BN, Berx G, Walzer T (2015) Terminal NK cell maturation is controlled by concerted actions of T-bet and Zeb2 and is essential for melanoma rejection. J Exp Med 212:2015–2025
31. Chiossone L, Chaix J, Fuseri N, Roth C, Vivier E, Walzer T (2009) Maturation of mouse NK cells is a 4-stage developmental program. Blood 113:5488–5496
32. Hayakawa Y, Smyth MJ (2006) CD27 dissects mature NK cells into two subsets with distinct responsiveness and migratory capacity. J Immunol 176:1517–1524
33. Seillet C, Huntington ND, Gangatirkar P, Axelsson E, Minnich M, Brady HJ, Busslinger M, Smyth MJ, Belz GT, Carotta S (2014a) Differential requirement for Nfil3 during NK cell development. J Immunol 192:2667–2676
34. Barton K, Muthusamy N, Fischer C, Ting CN, Walunas TL, Lanier LL, Leiden JM (1998) The Ets-1 transcription factor is required for the development of natural killer cells in mice. Immunity 9:555–563
35. Lee KN, Kang HS, Jeon JH, Kim EM, Yoon SR, Song H, Lyu CY, Piao ZH, Kim SU, Han YH, Song SS, Lee YH, Song KS, Kim YM, Yu DY, Choi I (2005) VDUP1 is required for the development of natural killer cells. Immunity 22:195–208
36. Ramirez K, Chandler KJ, Spaulding C, Zandi S, Sigvardsson M, Graves BJ, Kee BL (2012) Gene deregulation and chronic activation in natural killer cells deficient in the transcription factor ETS1. Immunity 36:921–932

37. Boos MD, Yokota Y, Eberl G, Kee BL (2007) Mature natural killer cell and lymphoid tissue-inducing cell development requires Id2-mediated suppression of E protein activity. J Exp Med 204:1119–1130
38. Delconte RB, Shi W, Sathe P, Ushiki T, Seillet C, Minnich M, Kolesnik TB, Rankin LC, Mielke LA, Zhang JG, Busslinger M, Smyth MJ, Hutchinson DS, Nutt SL, Nicholson SE, Alexander WS, Corcoran LM, Vivier E, Belz GT, Carotta S, Huntington ND (2016) The helix-loop-helix protein ID2 governs NK cell fate by tuning their sensitivity to interleukin-15. Immunity 44:103–115
39. Huntington ND, Puthalakath H, Gunn P, Naik E, Michalak EM, Smyth MJ, Tabarias H, Degli-Esposti MA, Dewson G, Willis SN, Motoyama N, Huang DC, Nutt SL, Tarlinton DM, Strasser A (2007) Interleukin 15-mediated survival of natural killer cells is determined by interactions among Bim, Noxa and Mcl-1. Nat Immunol 8:856–863
40. Sathe P, Delconte RB, Souza-Fonseca-Guimaraes F, Seillet C, Chopin M, Vandenberg CJ, Rankin LC, Mielke LA, Vikstrom I, Kolesnik TB, Nicholson SE, Vivier E, Smyth MJ, Nutt SL, Glaser SP, Strasser A, Belz GT, Carotta S, Huntington ND (2014) Innate immunodeficiency following genetic ablation of Mcl1 in natural killer cells. Nat Commun 5:4539
41. Gordon SM, Chaix J, Rupp LJ, Wu J, Madera S, Sun JC, Lindsten T, Reiner SL (2012) The transcription factors T-bet and Eomes control key checkpoints of natural killer cell maturation. Immunity 36:55–67
42. Sojka DK, Plougastel-Douglas B, Yang L, Pak-Wittel MA, Artyomov MN, Ivanova Y, Zhong C, Chase JM, Rothman PB, Yu J, Riley JK, Zhu J, Tian Z, Yokoyama WM (2014) Tissue-resident natural killer (NK) cells are cell lineages distinct from thymic and conventional splenic NK cells. Elife 3:e01659
43. Lacorazza HD, Miyazaki Y, Di Cristofano A, Deblasio A, Hedvat C, Zhang J, Cordon-Cardo C, Mao S, Pandolfi PP, Nimer SD (2002) The ETS protein MEF plays a critical role in perforin gene expression and the development of natural killer and NK-T cells. Immunity 17:437–449
44. Colucci F, Samson SI, Dekoter RP, Lantz O, Singh H, Di Santo JP (2001) Differential requirement for the transcription factor PU.1 in the generation of natural killer cells versus B and T cells. Blood 97:2625–2632
45. Kallies A, Carotta S, Huntington ND, Bernard NJ, Tarlinton DM, Smyth MJ, Nutt SL (2011) A role for Blimp1 in the transcriptional network controlling natural killer cell maturation. Blood 117:1869–1879
46. Aliahmad P, De La Torre B, Kaye J (2010) Shared dependence on the DNA-binding factor TOX for the development of lymphoid tissue-inducer cell and NK cell lineages. Nat Immunol 11:945–952
47. Holmes ML, Huntington ND, Thong RP, Brady J, Hayakawa Y, Andoniou CE, Fleming P, Shi W, Smyth GK, Degli-Esposti MA, Belz GT, Kallies A, Carotta S, Smyth MJ, Nutt SL (2014) Peripheral natural killer cell maturation depends on the transcription factor Aiolos. EMBO J 33:2721–2734
48. Rabacal W, Pabbisetty SK, Hoek KL, Cendron D, Guo Y, Maseda D, Sebzda E (2016) Transcription factor KLF2 regulates homeostatic NK cell proliferation and survival. Proc Natl Acad Sci USA 113:5370–5375
49. Samson SI, Richard O, Tavian M, Ranson T, Vosshenrich CA, Colucci F, Buer J, Grosveld F, Godin I, Di Santo JP (2003) GATA-3 promotes maturation, IFN-gamma production, and liver-specific homing of NK cells. Immunity 19:701–711
50. Ali AK, Oh JS, Vivier E, Busslinger M, Lee SH (2016) NK cell-specific Gata3 ablation identifies the maturation program required for bone marrow exit and control of proliferation. J Immunol 196:1753–1767
51. Wang S, Xia P, Huang G, Zhu P, Liu J, Ye B, Du Y, Fan Z (2016) FoxO1-mediated autophagy is required for NK cell development and innate immunity. Nat Commun 7:11023
52. Deng Y, Kerdiles Y, Chu J, Yuan S, Wang Y, Chen X, Mao H, Zhang L, Zhang J, Hughes T, Deng Y, Zhang Q, Wang F, Zou X, Liu CG, Freud AG, Li X, Caligiuri MA, Vivier E, Yu

J (2015) Transcription factor Foxo1 is a negative regulator of natural killer cell maturation and function. Immunity 42:457–470

53. Shiow LR, Rosen DB, Brdickova N, Xu Y, An J, Lanier LL, Cyster JG, Matloubian M (2006) CD69 acts downstream of interferon-alpha/beta to inhibit S1P1 and lymphocyte egress from lymphoid organs. Nature 440:540–544
54. Walzer T, Chiossone L, Chaix J, Calver A, Carozzo C, Garrigue-Antar L, Jacques Y, Baratin M, Tomasello E, Vivier E (2007) Natural killer cell trafficking in vivo requires a dedicated sphingosine 1-phosphate receptor. Nat Immunol 8:1337–1344
55. Cepek KL, Shaw SK, Parker CM, Russell GJ, Morrow JS, Rimm DL, Brenner MB (1994) Adhesion between epithelial cells and T lymphocytes mediated by E-cadherin and the alpha E beta 7 integrin. Nature 372:190–193
56. Kramer RH, Marks N (1989) Identification of integrin collagen receptors on human melanoma cells. J Biol Chem 264:4684–4688
57. Cerdeira AS, Rajakumar A, Royle CM, Lo A, Husain Z, Thadhani RI, Sukhatme VP, Karumanchi SA, Kopcow HD (2013) Conversion of peripheral blood NK cells to a decidual NK-like phenotype by a cocktail of defined factors. J Immunol 190:3939–3948
58. Cortez VS, Fuchs A, Cella M, Gilfillan S, Colonna M (2014) Cutting edge: salivary gland NK cells develop independently of Nfil3 in steady-state. J Immunol 192:4487–4491
59. Boulenouar S, Doisne JM, Sferruzzi-Perri A, Gaynor LM, Kieckbusch J, Balmas E, Yung HW, Javadzadeh S, Volmer L, Hawkes DA, Phillips K, Brady HJ, Fowden AL, Burton GJ, Moffett A, Colucci F (2016) The residual innate lymphoid cells in NFIL3-deficient mice support suboptimal maternal adaptations to pregnancy. Front Immunol 7:43
60. Montaldo E, Vacca P, Chiossone L, Croxatto D, Loiacono F, Martini S, Ferrero S, Walzer T, Moretta L, Mingari MC (2015) Unique Eomes(+) NK cell subsets are present in uterus and decidua during early pregnancy. Front Immunol 6:646
61. Tayade C, Fang Y, Black GP, Paffaro VA Jr, Erlebacher A, Croy BA (2005) Differential transcription of Eomes and T-bet during maturation of mouse uterine natural killer cells. J Leukoc Biol 78:1347–1355
62. Ribeiro VS, Hasan M, Wilson A, Boucontet L, Pereira P, Lesjean-Pottier S, Satoh-Takayama N, Di Santo JP, Vosshenrich CA (2010) Cutting edge: thymic NK cells develop independently from T cell precursors. J Immunol 185:4993–4997
63. Vosshenrich CA, Garcia-Ojeda ME, Samson-Villeger SI, Pasqualetto V, Enault L, Richard-Le Goff O, Corcuff E, Guy-Grand D, Rocha B, Cumano A, Rogge L, Ezine S, Di Santo JP (2006) A thymic pathway of mouse natural killer cell development characterized by expression of GATA-3 and CD127. Nat Immunol 7:1217–1224
64. Manser AR, Weinhold S, Uhrberg M (2015) Human KIR repertoires: shaped by genetic diversity and evolution. Immunol Rev 267:178–196
65. Orr MT, Lanier LL (2010) Natural killer cell education and tolerance. Cell 142:847–856
66. Rahim MM, Makrigiannis AP (2015) Ly49 receptors: evolution, genetic diversity, and impact on immunity. Immunol Rev 267:137–147
67. Binstadt BA, Brumbaugh KM, Dick CJ, Scharenberg AM, Williams BL, Colonna M, Lanier LL, Kinet JP, Abraham RT, Leibson PJ (1996) Sequential involvement of Lck and SHP-1 with MHC-recognizing receptors on NK cells inhibits FcR-initiated tyrosine kinase activation. Immunity 5:629–638
68. Long EO (2008) Negative signaling by inhibitory receptors: the NK cell paradigm. Immunol Rev 224:70–84
69. Vivier E, Nunes JA, Vely F (2004) Natural killer cell signaling pathways. Science 306:1517–1519
70. Chan CJ, Smyth MJ, Martinet L (2014) Molecular mechanisms of natural killer cell activation in response to cellular stress. Cell Death Differ 21:5–14
71. Billadeau DD, Upshaw JL, Schoon RA, Dick CJ, Leibson PJ (2003) NKG2D-DAP10 triggers human NK cell-mediated killing via a Syk-independent regulatory pathway. Nat Immunol 4:557–564

72. Zompi S, Hamerman JA, Ogasawara K, Schweighoffer E, Tybulewicz VL, Di Santo JP, Lanier LL, Colucci F (2003) NKG2D triggers cytotoxicity in mouse NK cells lacking DAP12 or Syk family kinases. Nat Immunol 4:565–572
73. Quatrini L, Molfetta R, Zitti B, Peruzzi G, Fionda C, Capuano C, Galandrini R, Cippitelli M, Santoni A, Paolini R (2015) Ubiquitin-dependent endocytosis of NKG2D-DAP10 receptor complexes activates signaling and functions in human NK cells. Sci Signal 8:ra108
74. Campbell KS, Purdy AK (2011) Structure/function of human killer cell immunoglobulin-like receptors: lessons from polymorphisms, evolution, crystal structures and mutations. Immunology 132:315–325
75. Parham P (2005) MHC class I molecules and KIRs in human history, health and survival. Nat Rev Immunol 5:201–214
76. Fang M, Orr MT, Spee P, Egebjerg T, Lanier LL, Sigal LJ (2011) CD94 is essential for NK cell-mediated resistance to a lethal viral disease. Immunity 34:579–589
77. Orbelyan GA, Tang F, Sally B, Solus J, Meresse B, Ciszewski C, Grenier JC, Barreiro LB, Lanier LL, Jabri B (2014) Human NKG2E is expressed and forms an intracytoplasmic complex with CD94 and DAP12. J Immunol 193:610–616
78. Wada H, Matsumoto N, Maenaka K, Suzuki K, Yamamoto K (2004) The inhibitory NK cell receptor CD94/NKG2A and the activating receptor CD94/NKG2C bind the top of HLA-E through mostly shared but partly distinct sets of HLA-E residues. Eur J Immunol 34:81–90
79. Davidson CL, Li NL, Burshtyn DN (2010) LILRB1 polymorphism and surface phenotypes of natural killer cells. Hum Immunol 71:942–949
80. He Y, Tian Z (2017) NK cell education via nonclassical MHC and non-MHC ligands. Cell Mol Immunol 14:321–330
81. Lee KM, Forman JP, Mcnerney ME, Stepp S, Kuppireddi S, Guzior D, Latchman YE, Sayegh MH, Yagita H, Park CK, Oh SB, Wulfing C, Schatzle J, Mathew PA, Sharpe AH, Kumar V (2006) Requirement of homotypic NK-cell interactions through 2B4(CD244)/CD48 in the generation of NK effector functions. Blood 107:3181–3188
82. Aldemir H, Prod'homme V, Dumaurier MJ, Retiere C, Poupon G, Cazareth J, Bihl F, Braud VM (2005) Cutting edge: lectin-like transcript 1 is a ligand for the CD161 receptor. J Immunol 175:7791–7795
83. Rosen DB, Cao W, Avery DT, Tangye SG, Liu YJ, Houchins JP, Lanier LL (2008) Functional consequences of interactions between human NKR-P1A and its ligand LLT1 expressed on activated dendritic cells and B cells. J Immunol 180:6508–6517
84. Li Y, Hofmann M, Wang Q, Teng L, Chlewicki LK, Pircher H, Mariuzza RA (2009) Structure of natural killer cell receptor KLRG1 bound to E-cadherin reveals basis for MHC-independent missing self recognition. Immunity 31:35–46
85. Castells MC, Klickstein LB, Hassani K, Cumplido JA, Lacouture ME, Austen KF, Katz HR (2001) gp49B1-alpha(v)beta3 interaction inhibits antigen-induced mast cell activation. Nat Immunol 2:436–442
86. Gu X, Laouar A, Wan J, Daheshia M, Lieberman J, Yokoyama WM, Katz HR, Manjunath N (2003) The gp49B1 inhibitory receptor regulates the IFN-gamma responses of T cells and NK cells. J Immunol 170:4095–4101
87. Bottino C, Castriconi R, Pende D, Rivera P, Nanni M, Carnemolla B, Cantoni C, Grassi J, Marcenaro S, Reymond N, Vitale M, Moretta L, Lopez M, Moretta A (2003) Identification of PVR (CD155) and Nectin-2 (CD112) as cell surface ligands for the human DNAM-1 (CD226) activating molecule. J Exp Med 198:557–567
88. Chan CJ, Andrews DM, Mclaughlin NM, Yagita H, Gilfillan S, Colonna M, Smyth MJ (2010) DNAM-1/CD155 interactions promote cytokine and NK cell-mediated suppression of poorly immunogenic melanoma metastases. J Immunol 184:902–911
89. Pende D, Spaggiari GM, Marcenaro S, Martini S, Rivera P, Capobianco A, Falco M, Lanino E, Pierri I, Zambello R, Bacigalupo A, Mingari MC, Moretta A, Moretta L (2005) Analysis of the receptor-ligand interactions in the natural killer-mediated lysis of freshly isolated myeloid

or lymphoblastic leukemias: evidence for the involvement of the poliovirus receptor (CD155) and Nectin-2 (CD112). Blood 105:2066–2073
90. He Y, Peng H, Sun R, Wei H, Ljunggren HG, Yokoyama WM, Tian Z (2017) Contribution of inhibitory receptor TIGIT to NK cell education. J Autoimmun 81:1–12
91. Stanietsky N, Simic H, Arapovic J, Toporik A, Levy O, Novik A, Levine Z, Beiman M, Dassa L, Achdout H, Stern-Ginossar N, Tsukerman P, Jonjic S, Mandelboim O (2009) The interaction of TIGIT with PVR and PVRL2 inhibits human NK cell cytotoxicity. Proc Natl Acad Sci USA 106:17858–17863
92. Kim S, Poursine-Laurent J, Truscott SM, Lybarger L, Song YJ, Yang L, French AR, Sunwoo JB, Lemieux S, Hansen TH, Yokoyama WM (2005) Licensing of natural killer cells by host major histocompatibility complex class I molecules. Nature 436:709–713
93. Orr MT, Murphy WJ, Lanier LL (2010) 'Unlicensed' natural killer cells dominate the response to cytomegalovirus infection. Nat Immunol 11:321–327
94. Yokoyama WM, Altfeld M, Hsu KC (2010) Natural killer cells: tolerance to self and innate immunity to viral infection and malignancy. Biol Blood Marrow Transplant 16:S97–S105
95. Waldhauer I, Steinle A (2008) NK cells and cancer immunosurveillance. Oncogene 27:5932–5943
96. Renoux VM, Zriwil A, Peitzsch C, Michaelsson J, Friberg D, Soneji S, Sitnicka E (2015) Identification of a human natural killer cell lineage-restricted progenitor in fetal and adult tissues. Immunity 43:394–407
97. Bjorkstrom NK, Ljunggren HG, Michaelsson J (2016) Emerging insights into natural killer cells in human peripheral tissues. Nat Rev Immunol 16:310–320
98. Juelke K, Killig M, Luetke-Eversloh M, Parente E, Gruen J, Morandi B, Ferlazzo G, Thiel A, Schmitt-Knosalla I, Romagnani C (2010) CD62L expression identifies a unique subset of polyfunctional CD56dim NK cells. Blood 116:1299–1307
99. Yu J, Mao HC, Wei M, Hughes T, Zhang J, Park IK, Liu S, Mcclory S, Marcucci G, Trotta R, Caligiuri MA (2010) CD94 surface density identifies a functional intermediary between the CD56bright and CD56dim human NK-cell subsets. Blood 115:274–281
100. Carson WE, Fehniger TA, Caligiuri MA (1997) CD56bright natural killer cell subsets: characterization of distinct functional responses to interleukin-2 and the c-kit ligand. Eur J Immunol 27:354–360
101. Cooper MA, Fehniger TA, Turner SC, Chen KS, Ghaheri BA, Ghayur T, Carson WE, Caligiuri MA (2001) Human natural killer cells: a unique innate immunoregulatory role for the CD56(bright) subset. Blood 97:3146–3151
102. Fauriat C, Long EO, Ljunggren HG, Bryceson YT (2010) Regulation of human NK-cell cytokine and chemokine production by target cell recognition. Blood 115:2167–2176
103. Bjorkstrom NK, Riese P, Heuts F, Andersson S, Fauriat C, Ivarsson MA, Bjorklund AT, Flodstrom-Tullberg M, Michaelsson J, Rottenberg ME, Guzman CA, Ljunggren HG, Malmberg KJ (2010) Expression patterns of NKG2A, KIR, and CD57 define a process of CD56dim NK-cell differentiation uncoupled from NK-cell education. Blood 116:3853–3864
104. Lopez-Verges S, Milush JM, Pandey S, York VA, Arakawa-Hoyt J, Pircher H, Norris PJ, Nixon DF, Lanier LL (2010) CD57 defines a functionally distinct population of mature NK cells in the human CD56dimCD16+ NK-cell subset. Blood 116:3865–3874
105. Wilkens J, Male V, Ghazal P, Forster T, Gibson DA, Williams AR, Brito-Mutunayagam SL, Craigon M, Lourenco P, Cameron IT, Chwalisz K, Moffett A, Critchley HO (2013) Uterine NK cells regulate endometrial bleeding in women and are suppressed by the progesterone receptor modulator asoprisnil. J Immunol 191:2226–2235
106. Hanna J, Goldman-Wohl D, Hamani Y, Avraham I, Greenfield C, Natanson-Yaron S, Prus D, Cohen-Daniel L, Arnon TI, Manaster I, Gazit R, Yutkin V, Benharroch D, Porgador A, Keshet E, Yagel S, Mandelboim O (2006) Decidual NK cells regulate key developmental processes at the human fetal-maternal interface. Nat Med 12:1065–1074

107. Koopman LA, Kopcow HD, Rybalov B, Boyson JE, Orange JS, Schatz F, Masch R, Lockwood CJ, Schachter AD, Park PJ, Strominger JL (2003) Human decidual natural killer cells are a unique NK cell subset with immunomodulatory potential. J Exp Med 198:1201–1212
108. Kopcow HD, Allan DS, Chen X, Rybalov B, Andzelm MM, Ge B, Strominger JL (2005) Human decidual NK cells form immature activating synapses and are not cytotoxic. Proc Natl Acad Sci USA 102:15563–15568
109. Sharkey AM, Xiong S, Kennedy PR, Gardner L, Farrell LE, Chazara O, Ivarsson MA, Hiby SE, Colucci F, Moffett A (2015) Tissue-specific education of decidual NK cells. J Immunol 195:3026–3032
110. Racanelli V, Rehermann B (2006) The liver as an immunological organ. Hepatology 43:S54–S62
111. Burt BM, Plitas G, Zhao Z, Bamboat ZM, Nguyen HM, Dupont B, Dematteo RP (2009) The lytic potential of human liver NK cells is restricted by their limited expression of inhibitory killer Ig-like receptors. J Immunol 183:1789–1796
112. Marquardt N, Beziat V, Nystrom S, Hengst J, Ivarsson MA, Kekalainen E, Johansson H, Mjosberg J, Westgren M, Lankisch TO, Wedemeyer H, Ellis EC, Ljunggren HG, Michaelsson J, Bjorkstrom NK (2015) Cutting edge: identification and characterization of human intrahepatic CD49a+ NK cells. J Immunol 194:2467–2471
113. Heydtmann M, Lalor PF, Eksteen JA, Hubscher SG, Briskin M, Adams DH (2005) CXC chemokine ligand 16 promotes integrin-mediated adhesion of liver-infiltrating lymphocytes to cholangiocytes and hepatocytes within the inflamed human liver. J Immunol 174:1055–1062
114. Hudspeth K, Donadon M, Cimino M, Pontarini E, Tentorio P, Preti M, Hong M, Bertoletti A, Bicciato S, Invernizzi P, Lugli E, Torzilli G, Gershwin ME, Mavilio D (2016) Human liver-resident CD56(bright)/CD16(neg) NK cells are retained within hepatic sinusoids via the engagement of CCR5 and CXCR6 pathways. J Autoimmun 66:40–50
115. Tu Z, Bozorgzadeh A, Pierce RH, Kurtis J, Crispe IN, Orloff MS (2008) TLR-dependent cross talk between human Kupffer cells and NK cells. J Exp Med 205:233–244
116. Jo J, Tan AT, Ussher JE, Sandalova E, Tang XZ, Tan-Garcia A, To N, Hong M, Chia A, Gill US, Kennedy PT, Tan KC, Lee KH, De Libero G, Gehring AJ, Willberg CB, Klenerman P, Bertoletti A (2014) Toll-like receptor 8 agonist and bacteria trigger potent activation of innate immune cells in human liver. PLoS Pathog 10:e1004210
117. Stegmann KA, Bjorkstrom NK, Veber H, Ciesek S, Riese P, Wiegand J, Hadem J, Suneetha PV, Jaroszewicz J, Wang C, Schlaphoff V, Fytili P, Cornberg M, Manns MP, Geffers R, Pietschmann T, Guzman CA, Ljunggren HG, Wedemeyer H (2010) Interferon-alpha-induced TRAIL on natural killer cells is associated with control of hepatitis C virus infection. Gastroenterology 138:1885–1897
118. Dunn C, Brunetto M, Reynolds G, Christophides T, Kennedy PT, Lampertico P, Das A, Lopes AR, Borrow P, Williams K, Humphreys E, Afford S, Adams DH, Bertoletti A, Maini MK (2007) Cytokines induced during chronic hepatitis B virus infection promote a pathway for NK cell-mediated liver damage. J Exp Med 204:667–680
119. Peppa D, Gill US, Reynolds G, Easom NJ, Pallett LJ, Schurich A, Micco L, Nebbia G, Singh HD, Adams DH, Kennedy PT, Maini MK (2013) Up-regulation of a death receptor renders antiviral T cells susceptible to NK cell-mediated deletion. J Exp Med 210:99–114
120. Beziat V, Liu LL, Malmberg JA, Ivarsson MA, Sohlberg E, Bjorklund AT, Retiere C, Sverremark-Ekstrom E, Traherne J, Ljungman P, Schaffer M, Price DA, Trowsdale J, Michaelsson J, Ljunggren HG, Malmberg KJ (2013) NK cell responses to cytomegalovirus infection lead to stable imprints in the human KIR repertoire and involve activating KIRs. Blood 121:2678–2688
121. Guma M, Angulo A, Vilches C, Gomez-Lozano N, Malats N, Lopez-Botet M (2004) Imprint of human cytomegalovirus infection on the NK cell receptor repertoire. Blood 104:3664–3671
122. Hendricks DW, Balfour HH Jr, Dunmire SK, Schmeling DO, Hogquist KA, Lanier LL (2014) Cutting edge: NKG2C(hi)CD57+ NK cells respond specifically to acute infection with cytomegalovirus and not Epstein-Barr virus. J Immunol 192:4492–4496

123. Lopez-Verges S, Milush JM, Schwartz BS, Pando MJ, Jarjoura J, York VA, Houchins JP, Miller S, Kang SM, Norris PJ, Nixon DF, Lanier LL (2011) Expansion of a unique CD57(+) NKG2Chi natural killer cell subset during acute human cytomegalovirus infection. Proc Natl Acad Sci USA 108:14725–14732
124. Petitdemange C, Becquart P, Wauquier N, Beziat V, Debre P, Leroy EM, Vieillard V (2011) Unconventional repertoire profile is imprinted during acute chikungunya infection for natural killer cells polarization toward cytotoxicity. PLoS Pathog 7:e1002268
125. Schlums H, Cichocki F, Tesi B, Theorell J, Beziat V, Holmes TD, Han H, Chiang SC, Foley B, Mattsson K, Larsson S, Schaffer M, Malmberg KJ, Ljunggren HG, Miller JS, Bryceson YT (2015) Cytomegalovirus infection drives adaptive epigenetic diversification of NK cells with altered signaling and effector function. Immunity 42:443–456 doi:10.1371/journal.ppat.1002268
126. Bjorkstrom NK, Lindgren T, Stoltz M, Fauriat C, Braun M, Evander M, Michaelsson J, Malmberg KJ, Klingstrom J, Ahlm C, Ljunggren HG (2011) Rapid expansion and long-term persistence of elevated NK cell numbers in humans infected with hantavirus. J Exp Med 208:13–21

Erratum to: cGAS-STING Activation in the Tumor Microenvironment and Its Role in Cancer Immunity

Geneviève Pépin and Michael P. Gantier

Erratum to: Chapter 8 in: D. Xu (ed.), *Regulation of Inflammatory Signaling in Health and Disease*, Advances in Experimental Medicine and Biology 1024, https://doi.org/10.1007/978-981-10-5987-2_8

The original version of this chapter was revised to include the following references:

40. Baird JR, Friedman D, Cottam B, Dubensky TW Jr, Kanne DB, Bambina S, Bahjat K, Crittenden MR Gough MJ (2016) Radiotherapy combined with novel STING-targeting oligonucleotides results in regression of established tumors. Cancer Res 76(1):50–61. https://doi.org/10.1158/0008-5472.CAN-14-3619
41. Li T, Cheng H, Yuan H, Xu Q, Shu C, Zhang Y, Xu P, Tan J, Rui Y, Li P, Tan X (2016) Antitumor activity of cGAMP via stimulation of cGAS-cGAMP-STING-IRF3 mediated innate immune response. Sci Rep 6:19049. https://doi.org/10.1038/srep19049

The updated online version of this chapter can be found at https://doi.org/10.1007/978-981-10-5987-2_8

D. Xu (ed.), *Regulation of Inflammatory Signaling in Health and Disease*, Advances in Experimental Medicine and Biology 1024, DOI 10.1007/978-981-10-5987-2_12

MIX
Papier aus verantwortungsvollen Quellen
Paper from responsible sources
FSC® C105338

If you have any concerns about our products,
you can contact us on
ProductSafety@springernature.com

In case Publisher is established outside the EU,
the EU authorized representative is:
Springer Nature Customer Service Center GmbH
Europaplatz 3, 69115 Heidelberg, Germany

Printed by Libri Plureos GmbH
in Hamburg, Germany